Family Geographies

The Spaciality of Families and Family Life

Family Geographies

The Spaciality of Families and Family Life

Edited by Bonnie C. Hallman

OXFORD
UNIVERSITY PRESS
8 Sampson Mews, Suite 204, Don Mills, Ontario M3C 0H5
www.oupcanada.com

Oxford University Press is a department of the University of Oxford.
It furthers the University's objective of excellence in research, scholarship,
and education by publishing worldwide in

Oxford New York

Auckland Cape Town Dar es Salaam Hong Kong Karachi
Kuala Lumpur Madrid Melbourne Mexico City Nairobi
New Delhi Shanghai Taipei Toronto

With offices in

Argentina Austria Brazil Chile Czech Republic France Greece
Guatemala Hungary Italy Japan Poland Portugal Singapore
South Korea Switzerland Thailand Turkey Ukraine Vietnam

Oxford is a trade mark of Oxford University Press
in the UK and in certain other countries

Published in Canada
by Oxford University Press

First Published 2010

Library and Archives Canada Cataloguing in Publication

Family geographies : the spatiality of families and family life / Bonnie C. Hallman, editor.

Includes bibliographical references and index.
ISBN 978-0-19-543168-1

1. Families. 2. Human geography. 3. Spatial behavior.
I. Hallman, Bonnie C. (Bonnie Catharine), 1956–

HQ503.F3177 2010 306.85 C2010-900544-9

Cover image: Juice Images Photography/Veer

Oxford University Press is committed to our environment. This book is printed on
Forest Stewardship Council certified paper, harvested from a responsibly managed forest.

Printed and bound in Canada.

1 2 3 4 – 13 12 11 10

Contents

Detailed Contents

Acknowledgements

While it may sound a tad trite, it is true that working on this book has been a labour of love. There are many people to thank for their support, encouragement, and assistance through the years of genesis for this book. First and foremost I must thank each and every one of the contributors, who were committed, diligent, and enthusiastic in their work on their individual chapters and in their embracing of the vision of the book as a whole. I commend all of you, and look forward to working with as many of you as possible in the future.

I must also thank Kate Skene and Kathryn West at Oxford University Press. How lucky I am to have 'Kate and Katie' as my Oxford team! Thank you seems hardly enough to express to both of you my gratefulness for your enthusiasm and your understanding and patience, especially at those dark moments when I might have thrown up my hands in despair.

Thanks are also due to my University of Manitoba supporters, Dr. William Norton and Dr. Mary Benbow. First, my sincere thanks are due to Dr. Norton for his encouragement to approach Oxford with the book idea, his ongoing intellectual support and, not to be underestimated, his support during his time as acting head of our department in securing me course release time to work on the book. How can I thank you, Mary? I am indebted to you as my colleague, co-author, and friend—this book would never have seemed possible without your suggestion that there might be something for a feminist social/cultural geographer to discover at the zoo!

Lastly, but far from least, I must thank those who are the landmarks in my own personal family geography. I am most grateful for the love and care of my husband, Barry, who has weathered with me more than I can write here, and has unfailingly supported me in whatever projects I take on, in whatever places they may take us. Thanks are also due to my mother Bev, and to my sister Sara and brother-in-law James, who have shown me unequivocally that mere geographic distance is not enough to weaken true affection; the real 'family-place' is in the heart. Finally, to my son Andrew: I did not truly understand 'family' until you came into my life some 18 years ago. Wherever you are, that is my family geography.

Introduction

Placing Families, Placing This Book

Bonnie C. Hallman

Why Focus on the Geographies of Families?

> Why wouldn't the concept of the everydayness reveal the extraordinary in the ordinary? (Lefebvre 1991: 7)

In his acclaimed work, *Critique of Everyday Life*, Lefebvre stresses that who we are as human beings is not merely or only the outcome of our economic roles and responsibilities, but perhaps more importantly our identity is determined by our regular, mundane private social and family lives. Our day-to-day activities, responsibilities, and opinions regarding family life are grounded in the spaces and places in which family geographies are enacted (Aitken 1998). Thus it is imperative that we pay attention to how individual people act, and interact, with others in their everyday lives and worlds. As Holloway and Hubbard (2001) note, '. . . it is in the trivialities of the everyday that the essence of human existence can be discerned and accordingly, that there is nothing trivial about it at all' (35). The activities of the everyday, though seemingly mundane or banal, are all inherently geographical (Dyck 2005), and rarely occur in isolation. Who are we most likely, and most often, interacting with in the everyday activities and transactions of our lives but those closest to us socially, emotionally, and often, either currently or at some time in our pasts, spatially; but those persons associated with us through kinship relations that we identify as members of our 'family'. Together we actively create the spaces and places of family life as resources for the activities and responsibilities of that life together (Aitken 1998).

Behind what Holloway and Hubbard (2001) refer to as our 'being-in-the-world' (7), our experiences and interactions with other people and with the places we inhabit, are complex sets of relationships between people, and between people and places. These relationships are at the heart of much of what we study in contemporary human geography and are understood to be always changing and evolving as individuals and societies respond to social, economic, cultural, and political forces at work across geographic scales from the local to the global.

A geography that focuses on families and family life clearly recognizes that individuals are not unencumbered in their experience of the everyday, or in their activities in the spaces and places of daily life. Rather, they are mediated or negotiated through, within, and because of their family relationships to their children, spouses/partners, siblings, parents, cousins, and others, whether kin through biology or legality or not, to whom they feel attachment and commitment. In this broad view of family, at the forefront is the recognition that we 'reproduce and reconstruct our family values and politics with every new event and challenge' (Aitken 1998: 7). Definitions of family therefore must be fluid and remain broad rather than narrow. The spaces and places of family life are (re)produced and (re)constructed daily, affected by the social characteristics of race/ethnicity, class, gender, and age. It is central to investigations of family geographies to explore and interrogate how these social parameters influence and shape ways of 'knowing and doing' within particular family forms and geographical contexts (Aitken 1998; Hallman and Benbow 2007). This knowledge in turn can, and arguably should, have significant effects on social behaviour and social policy (Duncan and Smith 2002).

A focus at the scale of the family can also be understood to enlarge upon '. . . the links between everyday activity and larger regional, national and international/global processes' (Dyck 2005: 236). It is here that global processes are made 'real', at the local space of the family and in the roles and relationships of and between its members, including those proximate and those separated by distance. In fact, as noted by Massey (1995) there is increasing stretching of the social relations of families across space which changes the spatiality of home and family and disturbs and alters our understanding of these once taken-for-granted concepts. Building on this idea, Dyck states that,

> . . . the everyday activities of family life are not simply a local or private matter—they are effects of/effected by this stretching of social, political and economic relations over space, across interlocking scales from the body to the global (2005: 234).

At the same time, this multi-scaled focus on the geography of families affords us the ability to link topics generally separate in the geographical and larger social science literature (e.g., caregiving analyses, migration studies, rural and urban historical geographies, feminist geographies, etc.). In so doing we are able to more fully contextualize previously hidden women's (and men's) lives. In particular, women in families emerge, in a geography that focuses on the everyday and on family, as 'creative actors rather than passive recipients of spatial arrangements' (Dyck 2005: 236). We can begin to see the family as a social organization that incorporates a set of social relationships and strong cultural symbolism, but is anything but a static social institution. Instead, as is apparent in several chapters in this volume, families must and do react to economic changes and imperatives, and their concomitant social and cultural ramifications. Women, especially in their roles as mothers, daughters, and wives, appear to bear much of the burden of adaptation to change driven by global forces of economic and social restructuring. It is they and their families who, for example, (im)migrate for better economic and social opportunities for themselves and their children, adjust to changing work patterns and demands in a restructuring economy, or shoulder the increasing burdens of end-of-life care in a healthcare system that increasingly emphasizes family-based caregiving in a resource-scarce environment.

Given the scholarly context as outlined here, this volume seeks to contribute to a family geography that focuses on the 'web of family and community relations' (Aitken 1998: 17) and does not look for generalizations or universalizing themes in family life. Rather, the call here is to illuminate the complex, volatile, and dynamic geographies of family life and how the examination of the interactions between family members and the spaces and places of everyday life can inform our thinking about the interconnections between social, economic, and political relations across geographic scales. Following on Soja (1985), a more nuanced understanding of the spatiality of family life, of the processes involved in making the spatial, social and the social, spatial, is the project that the research presented in this book is intended to initiate.

How This Book Came to Be

> People practice family—artfully, creatively, with intention—in local settings, in more diverse groups than may always be acknowledged. And their intentions and craft are formed with material and discursive contexts that shape and channel those efforts in ways that have not been fully recognized and explored by scholars. (DeVault 2000: 501)

The quote above reflects both the diversity of families and ways of being a 'family', both in contemporary terms and in terms of looking at family forms and functions as they have changed historically. It also evokes the very 'groundedness', the embeddedness of family life in the mundane and ordinary places of everyday life—in family homes, shopping malls, local parks, schools, and the neighbourhood streets, playgrounds and sports fields of our communities. It is in these places that our lives as members of families are played out, are 'enacted'. The focus here is not only on how these spaces and places of daily life may shape, or are shaped by the practice of family life, but as the quote says, the focus must also include the contexts, material and discursive, that shape the activities of family life and the intentions and meanings that animate those activities. In this second focus, the direction of this volume has been particularly inspired by the following words of Edward Relph (1985):

> ... landscape and space are part of any immediate encounter with the world, and so long as I can see I cannot help but see them no matter what my purpose. This is not so with places, for they are constructed in our memories and affections through repeated encounters and complex associations (26).

It is these complex associations, rooted in culture and accepted social norms and practices, in gender roles and relations, in ideologies of motherhood, fatherhood, and childhood, (Hardy and Weidmer 2005) and often accepted and acted upon uncritically, even unconsciously, which must also be examined in order to understand the nature of the places of family life—of family geographies. The chapters gathered here in this volume—despite a diversity of specific topics that ranges from, for example, an exploration of the decision-making of families to migrate internationally, to an analysis of rural farm women's spaces of economic and social interaction in the 1930s as shaped by their family roles, and the influence of travel and sports culture on families with competitive youth athletes—all make considerable contributions to this second, and arguably, more significant focus. All go far beyond a discussion of the spaces of family life to explore how and why, and to what effect, the meanings of and attachments to place, and to the members of our families (however we may define their inclusion), influence our connections to, and shaping of, the places of family life.

Admittedly, the concept behind this edited volume stems at least in part from my own career as a human geographer with a long-standing interest in the ways that families, at various points in the life course, perceive and use the elements of their place context/community to meet their commitments and responsibilities to one another. My own work has shown that even when space/distance and community resources may present as many, or more, barriers as they do opportunities, families actively shape, and are shaped by, these geographies of daily life. Starting with doctoral research in the late 1990s on the spatiality of family eldercare (e.g., Hallman and Joseph 1999; Joseph and Hallman 1998), and more recently examining family leisure locations (zoos) as sites of importance in the maintenance and construction of family life (e.g., Chapter 1 in this volume, Hallman 2007; Hallman and Benbow 2007), my research and contributions to the literature exposed me to an emergent research area focusing on the geographies of families.

It seemed, in early 2007 when the idea of a book first took form, that there were several of us in the Canadian geographic community and beyond who were doing work that was interconnected through its focus on some aspect of the spaces and places of family life. We seemed to be scattered across various sub-disciplines: geographies of health and of caregiving, urban and economic geographies, feminist and gender geographies, migrations studies; the list could go on. Unsure whether this was a mere impression, or something bigger that needed attention, a special session was organized on Family Geographies for the Canadian Association of Geographers (CAG) Annual Meeting at the University of Saskatchewan for May 2007, to 'test the waters' as it were to see if there were others who felt their work would also be best presented in such a forum, and who also wanted to engage with scholars who were interested in exploring similar themes. As it happened, enthusiasm was so great, a second session was called for, and attendance was outstanding. The participants in those sessions, the vast majority of whom have gone on to become contributors to this book, agreed that this focus on the spaces and places of families and family life is an emerging area of scholarship, not only within geography but across disciplines with which they engage in their own work and which they draw upon, i.e., sociology, gerontology, migration studies, urban studies, family studies, social history, etc. As a group, it was felt that now is the time to formulate a stronger identity and indicate some directions for research within an area of scholarship called 'family geographies', and that an edited volume was an important and logical step in defining this intellectual terrain. On behalf of my contributors and myself, I trust that we have done what we set out to do in this collection of chapters centred on the spatiality of families and family life.

Thus, based on the very positive reaction and encouragement of the session participants and the audience for the CAG 2007 sessions on Family Geographies, the collection of chapters brought together here evolved, along with contributions from others that responded to a later call for contributors, to become the volume you now hold. The collection has had from the start as its unifying theme a central interest in the role of geography in the experience, practice, and ongoing social construction or 'becoming' of families and everyday family life. In their own way, each of the chapters uses the geographical concepts of space, distance, place and/or community context to understand various aspects of how families engage in and with the world in which they are situated, to meet the needs and obligations of their various members, to maintain identity as 'family' across distances and in different/new settings. Thus, the focus is unequivocally on the 'everyday', and on the ways that geographical context shapes, and is shaped by, the processes and activities of family life. Throughout, it is recognized that the term 'family' is a fluid and dynamic concept, open to variable and changeable definitions.

Structure and Organization of the Book

> What is often overlooked . . . is that families have not simply changed over time, they also vary over space at any one time . . . In short, family formations have geography, as well as a history. (Duncan and Smith 2002: 471)

What follows are eleven chapters, each of which introduces and represents possible areas for further development for family geography focused research, in terms of conceptual/theoretical advancements, methodological innovations, and in the translation to, and influence of such research into policy at all levels of administration and governance. The chapters have been organized, as close as possible, to mimic the passage of families through the family life course. This reflects an understanding that the spatial nature of family life changes at different points in the life course, which is one key element in the 'constellation of issues that define the diversity of family geographies' (Aitken 1998: 6). The opening chapters focus on family spaces/places and everyday life in the early years of family formation and childrearing, and in the building of family ties and cohesion between children and their parents. For example, Chapters 1, 2, and 3 have as their focus environments and places in which families, particularly children and parents, engage in the practice of everyday life, whether it is the social space of the zoo, the sports field, or the trip to school. As such, these chapters share a focus on places of family life as 'territories of meaning' that are interpreted and experienced intersubjectively (Norton 2006). In Chapter 1, Hallman and Benbow examine zoos as cultural landscapes steeped in family ideology, and which function as important locations for the practice of 'quality' family life through family-oriented leisure. They investigate the capturing of this understanding of, and use of, the zoo environment through interviews with parents, facilitated by a review of family photographs taken during a zoo visit. A similar focus on the places of family life is taken in a different direction in Chapter 2, where Williams and Crumplin examine the influence of sports culture on the practice of family life for families with daughters in competitive sport. Family geographies and relationships are impacted because the sport culture demands that the activities of the youth athletes (practices, games, tournaments) occur in specific places at different times or in several places at the same time. This chapter specifically focuses on the impact that travel has on families and is particularly informed by an analysis of the gendered spaces, roles, and norms of sports culture and the tensions between being a 'good parent' and/or a 'good sport parent'. The influence of cultural factors on family life continues to be examined in Cloutier's analysis of children's journeys to school in Chapter 3. She examines the influence of the living environment of parents upon their knowledge, beliefs, and perceptions of risk concerning their children's safety, and thus the likelihood of their children being able to engage in the physically and socially healthy activity of walking to school. Thus the geographical experience and understandings of the adult members of the family unit shape the experiences and behaviours of its younger members.

The following two chapters, while still focusing on the influences of space and place on family life, have as their focus family relations and family life as they intersect with the demands of other social institutions, specifically spaces of work and of school. Chapter 4, by Andrey and Johnson, specifically looks at the tensions and redefinitions associated with the combining of telework done from home, and the home space of the family. They investigate, uniquely to this area of research, how teleworking households engage in the renegotiation of family practices and the re-sculpting of their family homes, thus altering family geographies in important, though often subtle, ways. In Chapter 5 a distinctive historical perspective is taken by Summerby-Murray. This chapter explores the connection between changes in women's mobility and the dynamics of family geographies, and does this through the analysis of the diaries of two farm women living in 1920s and 1930s New Brunswick, Canada. These women provided important examples of the renegotiation and reconstruction of the farm family and its spatial dynamics in a time of socio-economic adjustment.

The next four chapters have as a common theme the geographies of family networks, movement, and migration. In various ways they examine aspects of the drivers, and effects, of migration, whether interprovincial (within Canada) or transnational, and call for a greater recognition of the role of family considerations in both of these areas of migration research. In Chapter 6, Marshall outlines the strains of adjustment to a new environment, and specifically a new and difficult social setting, on the family unit that migrates in search of employment, in a case study of families migrating from Newfoundland to New Brunswick. Importantly, the research reported upon in this chapter focuses on the process of successful (or unsuccessful) adjustment of the family, especially children, in the new community, as paramount in the decision to return in the case of seasonal migration, and to remain or not in the case of more permanent employment migration. In Chapters 7 and 8, family factors in the decision to migrate, and in the assessment of the overall success of the migration experience, are assessed for transnational migrants from India and from South Africa, respectively, to urban Canada. In Chapter 7, Samuel presents research grounded in the lived experience of first generation female immigrants from Kerala, India who arrived in Canada in the 1970s. She examines cultural retention and transmission among this group of migrant women, focusing on norms of family life, and how the success of the transmission of these traditional norms colours their assessments of their decision to migrate and their current satisfaction with that decision. Family factors in migration are further explored in Chapter 8, where Huot and Dodson discover the power of family considerations in skilled migration decision-making. Initial interest in the motives for skilled migrants leaving South Africa and for choosing Canada as a destination led to an unexpected conclusion. The chapter elucidates the dominant themes that emerged in relation to the role and influence of children in their parents' migration process, from the initial decision to move to their integration into Canadian society. A different and important perspective on the intersections between migration and family is presented by Walton-Roberts in Chapter 9. Here, she focuses on the nature and impacts of social contacts within families and across spaces and places, centred on a reflexive and personal examination of the interconnectedness of her research on transnational migrant families and her own family relations and family geography. She accomplishes this not only from her perspective as an academic researcher, but also as a wife, daughter, and mother. In all four of these chapters, the family relations impacted by space, place, and distance cut across all generations, and thus extend the family geography focus to not only the spaces and places of the nuclear family, but to the extended family of grandparents, aunts, uncles, and more distant relations.

The final two chapters, Chapters 10 and 11, examine very different aspects of the geographies of later-life families, and at the same time return to the organizing themes of places of family life, and the relationships and activities of families. In Chapter 10, Lucas and Sanders use imagery from retirement home promotional literature to critically examine the emplacement through marketing of particular ideologies of what it is to age successfully, and what the family life of those who have aged successfully ought to look like in the retirement home landscape. They are concerned with how such imagery becomes the means by which specific ideas about aging and family are made to seem natural and inevitable, though they clearly are not. The promotional materials of retirement communities are implicated in the construction, reproduction, and emplacement of the active retiree lifestyle and the promotion of successful aging, but they are also implicated in shaping, deconstructing, and reconstructing specific notions of the nature of family in later life.

In the final chapter, Crooks and Williams explore the reasons why family geographies matter in the informal provision of care to elderly family members, specifically those at end-of-life. They effectively argue that the ways in which experiences are lived out through time and in space are central considerations in the geographies of family caregiving for the elderly, and therefore of family geographies more generally. They do this by examining the experiences of those providing informal care for a dying family member. The issues raised are explored with particular emphasis for their implications for Canadian social policy.

Thus, with these last two chapters, the discussion of the spatiality of families and family life once again returns to how families 'read' understandings of family formation, family relationships, and family interactions, into specific places and environments. However, this discussion now extends to families as interactions change with age and infirmity, and as they embark on the final stages of life together as a social and emotional unit. At the same time, these last two chapters remind us of the embeddedness of families and family life in the mundane and ordinary places of daily life.

Absences and Opportunities

The purpose of this volume is to identify key themes and to 'stake out' an intellectual terrain and central themes within the emerging research area of family geographies. This is not to say that the research presented here is an exhaustive catalogue of all research that might be considered as 'family geography', nor of all methods or theories that might be employed to conduct or inform geographic study of families and family life. Rather, the goal is for the research organized together in this volume not only or merely to help influence and define central themes in existing research on the geographies of families and family life, but far more importantly to inspire new lines of inquiry and new questions that derive from, build on, refute, and extend into yet unseen areas that further not only geographic, but interdisciplinary social science research. This broader reach is intended in part because this collection brings together research from multiple human geography sub-disciplines (e.g., children's geographies, transportation geography, migration research, historical geography, cultural geography, geographies of aging, etc.) with strong thematic and conceptual ties to other academic fields such as sociology, family studies, history, rural studies, feminist and gender studies, urban studies, and social gerontology, under the mantle of 'family geographies'.

As noted above, in a volume of eleven chapters, not all of the numerous possible aspects that could be developed in a survey of family geographies, methodologies for interrogating the nature of the spaces and places of family life, or the complexity of contemporary families as they interact within and across geographical space, can be encapsulated. There are noteworthy absences that must be acknowledged. These absences are no doubt a product of the processes by which contributors were identified and came into the project, as outlined above. However, these same absences can be interpreted as fruitful avenues for new research and further development of the concepts, theories, and topics encompassed within and useful in the further exploration and better comprehension of the geographies of families and family life. The application of the concept of 'family geographies' will thereby be broadened as more scholars contribute to, challenge, and/or reconfigure family geographies as an analytical category.

Perhaps the most glaring absence in the current volume is a lack of representation from researchers focusing on gay and lesbian family geographies. This is regrettable and hopefully shines a strong light on a much needed area of expansion for research in family geographies. An example of one direction this research might take can be found in a very recent paper by Myrdahl (2009), who calls for a critical engagement with (hetero) normativity in social spaces. In particular, she examines the active construction of a 'family-friendly' image by the Women's National Basketball Association (WNBA) and the Minnesota Lynx team specifically, through a 'ubiquitous presence and focus on children' and (heterosexual) families (Myrdahl 2009: 292) despite the known dominance of lesbians in the WNBA fan base. She argues that this creates a 'game space' that largely erases lesbian fans in order to win team support from a smaller 'mainstream and traditional' family audience, and that this is a decidedly political process of spatial production that openly favours one set of social relations and one family form over any other.

Other areas absent that are particularly relevant to the Canadian context, include aboriginal family geographies and the family geographies of visible minorities and new immigrants. While the later two groups are represented to a certain degree in Chapters 7 and 8, and to a lesser degree in Chapter 9 of this volume, examination of the spatiality of family life for these groups needs to move beyond a central focus on immigration studies and the ongoing effects of transnational identities. A possible area for future research might, for example, examine the spaces and places important to the construction of ethnic/racial identities of families in the second generation and beyond, after initial arrival in North America. Other areas that should be developed include more topics that focus on the geographies of families with disabled members and later-life families, moving beyond the current representation of geographies of caregiving/eldercare and the social construction through media of images of the aging family. One area which could prove fruitful lies in research that examines the concept and creation of 'age-friendliness' in communities (Hallman, Menec, Keefe, and Gallagher 2008), and particularly the place of social interaction and family networks in making communities of all sizes safe, inviting, and supportive places for aging citizens. Lastly, it is apparent that the focus in the current volume is on the North American, and particularly, the Canadian context which is, at least in part, a reflection of the connections and linkages between the researchers who have contributed to the book. There is a real need, however, to reach out to geographers and researchers in cognate areas whose studies extend beyond the North American/Canadian social setting. Perhaps some comparative work looking at the spatiality of families and family life in other social, cultural, political, and economic contexts is one way forward in this regard.

This is a far from expansive list of absences/opportunities to take the terrain that is outlined here for examining geographies of families and family life and extend the boundaries of this focus for analysis. It is hoped by all the contributors to this volume that our work can and will inspire others, students and researchers alike, to build on what we have presented here; that this may be something of a 'launch pad' for further contributions to, contestations of, and reworking of the family geographies concept.

Bonnie C. Hallman
Editor
15 October 2009

References

Aitken, S. 1998. *Family Fantasies and Community Space.* New Brunswick NJ: Rutgers University Press.

Duncan, S., and D. Smith. 2002. 'Family geographies and gender cultures'. *Social Policy & Society* 1,1: 21–34.

Dyck, I. 2005. 'Feminist geography, the everyday, and local-global relations: hidden spaces of place making'. *The Canadian Geographer* 49, 3: 233–43.

Hallman, B.C. 2007. 'A family-friendly place: family leisure, identity and wellbeing—the zoo as therapeutic landscape'. Pp. 133–45 in A. Williams, ed. *Geographies of Health—Therapeutic Landscapes.* Aldershot: Ashgate Publishing Limited.

Hallman, B.C., and S.M.P. Benbow. 2007. 'Family leisure, family photography and zoos: exploring the emotional geographies of families'. *Social and Cultural Geography* 8, 6: 871–88.

Hallman, B.C., and A.E. Joseph. 1999. 'Getting there: mapping the gendered geography of care giving to elderly relatives'. *Canadian Journal on Aging* 18, 4: 397–414.

Hallman, B., V. Menec, J. Keefe, and E. Gallagher. 2008. 'Making small towns age-friendly: what seniors say needs attention in the built environment'. *Plan Canada* 48, 3: 18–21.

Hardy, S., and C. Weidmer, eds. 2005. *Motherhood and Space: Configurations of the Maternal through Politics, Home, and the Body.* New York: Palgrave Macmillan.

Holloway, L., and P. Hubbard. 2001. *People and Place: The Extraordinary Geographies of Everyday Life.* London: Prentice Hall.

Joseph, A.E., and B.C. Hallman. 1998. 'Over the hill and far away: distance as a barrier to the provision of assistance to elderly relatives'. *Social Science & Medicine* 46, 6: 631–39.

Lefebvre, H. 1991. *Critique of Everday Life.* London: Verso.

Massey, D. 1995. 'The conceptualization of place', in D. Massey and P. Jess, eds. Pp. 45–85, *A Place in the World? Places, Culture and Globalization.* Oxford: The Open University Press.

Myrdahl, T.K. Muller. 2009. '"Family-friendly" without the double entendre: a spatial analysis of normative game spaces and lesbian fans'. *Journal of Lesbian Studies* 13, 3: 291–305.

Norton, W. 2006. *Cultural Geography: Environments, Landscapes, Identities, Inequalities,* 2nd ed. Toronto: Oxford University Press.

Soja, E. 1995. 'The spatiality of social life: towards a transformative re-theorization', in D. Gregory and J. Urry, eds. Pp. 90–127. *Social Relations and Spatial Structures.* New York: Palgrave Macmillan.

Chapter One

'Seeing if you can catch the one picture that just makes it': Placing Family Life Through Family Zoo Photography

Bonnie C. Hallman and S. Mary P. Benbow

Introduction: Family Geographies, Family Life, and Zoos as Family Places

Zoos, both historically and in contemporary times, are family spaces that place family members in close proximity with each other and with animals. They are one of the few places where parents can take their children to both educate and entertain them. In fact, they are places where parents are targeted and encouraged by the zoo community to bring their children for these very reasons (Turley 2001). Marketers and advertisers increasingly recognize that zoos across North America offer access to a 'family-friendly audience of 143 million people who visit zoos each year' (Hampp 2006), and are understood by them to be educational and enriching environments for children with a 'green halo' message of care for the environment and for other living things (Hampp 2006; Morey & Associates 2004).

However, like all cultural institutions, zoos are both reflective and constitutive of social norms. The research discussed in this chapter reflects our larger exploration of zoos as **cultural landscape**s (Hallman and Benbow 2006). We see contemporary North American zoos as cultural landscapes with clearly delimited geographies, that have been organized and planned to evoke certain experiences that portray particular ways of seeing and understanding the natural world, and of conceiving of human relationships with nature and with the beings that inhabit it with us (Norton 2006; Wolch 2002).

The zoo environment also reflects other values and meanings applied or 'read' into it by, and reflected back to, the larger (Western) society in which it is situated. We have specifically explored and sought to improve our understanding of zoos, similar to other cultural institutions in urban centres, as culturally laden places with a specific and unique role in the practice of family life and family-oriented recreation. We are also interested in the place of zoos in the creation of healthy family identity through the relationship between constructions of self/group identity and **place identity** (Hallman 2007; Hallman and Benbow 2007). Specifically, we understand the zoo as an enduring urban cultural landscape that is a context for enacting family relationships, marking quality time and social interaction amongst family members, and fostering positive emotional connections between family members. In this way, the local zoo can be an important location in the geographies of families as they engage in and develop family connections between individuals.

In our work we have been careful to define the 'family' we are focused upon as any single adult or pair of cohabitating adults with children living in their household for which they are the primary or shared caregiver(s), including post-divorce 'blended' families. In some cases this family unit may include additional extended members and **fictive kin** (adults and children) when they are engaged in activities in the zoo environment. Such extended family are not the primary focus of our research however, at least not at this time. As we expand this work, our focus will broaden to more actively include later-life families, minority families, and families of same-sex partners. Through this, we hope to contribute to an expansion of the areas of focus within family geographies research, the project to which this volume is aimed.

Therefore, here we seek to better understand the significance of zoos in the geographies of families raising young children. We do this by combining analysis of family photography of zoo visits with interviews carried out with parents who take their children to the local zoo in our community (Assiniboine Park Zoo, Winnipeg, Manitoba). Family photographs, particularly those taken of the family at leisure and of children at play, have a significant purpose in capturing the 'high points' of family functioning. We have observed that the zoo environment is 'full of stages for practicing family and then capturing and memorializing this spatial behaviour through **family photography**' (Hallman and Benbow 2007: 872). We take a step further than in earlier related analyses by not only examining family zoo photographs as 'texts' to be 'read' and analyzed in and of themselves (Hallman and Benbow 2007), but also by interviewing the primary family photographer (most often, as it happened, the mother) in order to develop an understanding of the intentions and motivations of the image producer. At the same time, this allowed us to discuss with the family photographer the values and meanings attached to the images produced, to the zoo trip experience, and to the zoo itself as the site of that production (the zoo as 'family-place'). We agree with Gillespie (2006) that the interaction between family photographer and subject (in our work, photograph subjects are most often the image producer's own children, as well as the animals and sights at the zoo) is a dynamic site of identity construction, of the family unit and of the zoo as a 'family-place'. Thus we seek with this study to contribute to family geographies by illuminating the use of, and meanings attached to, the mundane and everyday locales of family life. In this way we can not only better plan for and protect such places, but we can also gain a better understanding of contemporary family life and parenting through an examination of the use of urban landscapes like zoos that make up the geographies of families.

The chapter is organized into several sections. The first two sections outline concepts that shape our research, focusing first on notions of parenting, especially motherhood, which can be seen as culturally embedded into the zoo landscape. In the next section, we contend that zoos are important, yet largely unexamined, locations for enacting or performing 'quality' family life, and quality mothering/fathering through family-centred recreation. We describe the importance of places for leisure in contemporary parenting, and the role of zoos as family-leisure destinations, which leads into a brief discussion of why zoos are considered '**family-friendly**' places. This is followed by a section that discusses family photograph analysis, focusing on the capturing and memorializing of (spatial) behaviour (such as the family trip to the zoo) through family/domestic photography. A more in depth discussion of this topic can be found in our previously published research paper (Hallman and Benbow 2007), where we analyze a large collection of family zoo photographs.

We then outline the study results from interviews with a group of parents of young children about their family zoo photographs, the meanings and values they attach to the zoo visit, and to the zoo space itself within their family geographies.

Here we report on the first round of interviews with families who responded to our call for participants, who visited our local zoo, and shared their experiences and their zoo family photographs with us. We conclude the chapter with a section that discusses what we have learned about the meaning of the zoo landscape in the geographies of families with young children. At this point we also make some recommendations for further investigations of the meanings, values, and uses made of zoos, as well as similar cultural institutions and leisure destinations in the geographies of families and of family life. In particular, we call for research across all stages of the family life course, and with a greater diversity of family types.

Constructions of Mothers/ Motherhood and Parenting in the Zoo Space

Since the 1970s, studies influenced by environmental feminism (Griffin 1978; Ortner 1974) have asserted that women, because of their biology, have been thought to be closer to nature. This is the view shared by some populist notions of ecofeminism, which emphasize the strength of matriarchy, particularly in a spiritual sense (Domosh and Seager 2001). More recent research by geographers and other social scientists has sought to challenge this idea of 'natural motherhood' by developing the notion that human–environment relations, and within this, human–animal relations, are explicitly *constructed* as gendered (Nesmith and Radcliffe 1993; Sabloff 2001). Such work coincides with the emergence in the mid-1990s of a geographic analysis of human–animal relations, centred on concepts such as space, place, and landscape. This new 'animal geography' is only beginning to show how the spaces and places in which animal–human interactions occur (i.e., zoos, marine parks, etc.), constitute those relations, and to demonstrate and critique how practices that generate animal representations (i.e., family zoo visits) reveal the social and cultural constructs that shape animal–human relations (Philo and Wilbert 2000), and are extrapolated to or associated with human interrelationships.

Feminist studies of the environment inform the new animal geography by showing how gender is reproduced through dominant understandings of what the natural environment (and, therefore, the animals 'placed' there) is or should be. Examining specifically constructed landscapes or sites, such as the zoo, where 'the natural' has been defined and displayed for some 150 years according to the norms and purposes of Western societies, uncovers certain assumptions about gendered identities (of nature, of animals, and of humans), and their deployment when these sites are designed, planned, and used (Rose, Kinnaird, Morris, and Nash 2002; Sabloff 2001).

Thus zoos, as constructed landscapes blending nature and culture, in their very design and purpose are steeped in assumptions, and not always reflective of the most progressive or egalitarian perspectives, about gender roles and identities (Anderson 1995; Hanson 2002; Rose et al. 2002). Specifically, we suggest that zoos are inscribed with a feminized depiction of nature as a bountiful and caring mother/Madonna, particularly in the species conservation reproduction imperative, and through the attention in zoo exhibits and marketing of animal families, especially the focus on animal mothers and 'babies'. Lastly, we see the zoo as a feminized cultural institution due to its strong identification with the entertainment and education of children (DeVault 2000; Turley 2001), traditionally central duties in the practice of mothering, but as we discuss later in this chapter, increasingly evident in modern fathering as well, at least to a certain degree.

Understanding that zoos are cultural institutions which reflect and reinforce the (admittedly, sometimes outdated, biased, and/or prejudicial) norms and perceptions of the times and societies in which they exist, we can see that the zoo 'reflects not nature itself, but a human adaptation of "nature"' (Anderson 1995: 275). The enclosures we build, the texts we provide (e.g., zoo maps, informational placards, media, and advertising), even the choices made about what species to display, cater to cultural demands and expectations about animals and human–animal relations. We suggest that this includes gender roles and relations amongst humans, depicted as being modelled in 'nature', specifically a construction of 'natural motherhood', however controversial this model may be.

Research on zoo visitor patterns has shown that the majority of family zoo visits, as in most family recreation decisions, are suggested and planned by mothers (DeVault 2000; Howard and Madrigal 1990; Turley 2001). Part of the reason for this, we suggest, is that they see reflected back to them in the display of animal mothers and mothering in the zoo, a 'naturalization' of family relations, specifically the role of the female as nurturer and an inherent quality to the parent–child bond. For some, this may coincide with and/or verify their own family organization or perhaps a family and gender role orientation that is recognized as a desired goal. For some, on the other hand, it may be that they see in the displays of animal 'families' something that they wish to contest in their everyday practice of parenting.

Just as Anderson (1995; 1998) argues that zoos initiate the public into a way of seeing the human capacity for order and control over nature, it can be argued that central to this taming and converting of nature is the control of reproduction. Depictions of animal motherhood are constructed in a manner that illustrates a patriarchal capacity to intervene and control the feminine. We can see this in the assessment of good vs. bad animal mothering in the former practice of separating animal infants, particularly primates, and raising them away from their own species in zoo nurseries, apparently all in the interest of species conservation. While this past common practice is no longer the case unless the infant is ill or the mother is unresponsive or does not feed it, when infant animals are removed there is often considerable press attention paid, including photographs of keepers holding the baby animals, often wrapped in a blanket like a human child.

Through depictions of animal motherhood and animal 'families', a traditional patriarchal model of the family with the female in the role of carer/nurturer is reified, and indicated as central to female identity. This includes gendered perceptions of roles within families, particularly the role and place of women and the social politics of reproduction. For instance, Frame (1994) writes in great detail about the death of an octopus, the usual result in giving birth in that genus, as they do not feed while brooding their eggs. Frame's paper describes a 'heroic' mother *octopus vulgaris,* as she first arrived at the National Zoo (in Washington, DC) injured, and then refused food for many weeks while brooding her eggs. Many other female animals in zoos are viewed as central motherly figures, the most notable being the matriarch elephant. Many such elephants have been observed using caring behaviours even beyond their own offspring. One elephant in the Calgary Zoo (in Alberta, Canada) became ill following the birth of her calf, leaving her unable to stand. Another female elephant would come to her assistance and hold her up when needed. Often these same animals feature as 'marquee' animals in the promotion of a specific zoo.

Ordinary 'everyday' practices inscribe gender roles and relations. Just as in other institutions and locations where family life is played out (home, church, school, the shopping mall, etc.), the zoo is part of the larger urban landscape reflective of the social and cultural practices of the larger society. Many zoos offer animals a variety of unusual foods in order to celebrate holidays, such as pumpkins at Halloween. And Mother's Day is a very special day at numerous zoos, with promotions and exhibits featuring new zoo mothers and babies, and special souvenirs and even free admission for mothers of all ages (American Association of Zoos and Aquariums www.aza.org).

Based on this admittedly rather essentialist notion of mothers, motherhood, and parenting, it would seem that zoos influence family geographies by institutionalizing in situ highly gendered and patriarchal notions of family roles and relations. While we can argue that this function is apparent, we should not negate the active construction of family roles and relations, of family life itself, engaged in by parents of young children. We now turn to a discussion of family-oriented recreation in contemporary parenting, and more specifically the place of zoos in family geographies of leisure and recreation.

The Role of Family-Oriented Leisure in Family Life

> People practice family—artfully, creatively, with intention—in local settings . . . their intentions and craft are formed within material and discursive contexts that shape and channel those efforts. (DeVault 2000: 501)

The value of families spending time together, quality leisure time, is stressed as desirous, even normative, in North American/Western culture. Innumerable parenting and lifestyle self-help books (e.g., Baicker-McKee 2003; Ellis 1999; Ehman, Hovermale, and Smith 2005; Partow and Partow 1996), lifestyle/reality television programs (e.g., *SuperNanny, Nanny 911, Shalom in the Home*), and articles in the many parenting and women's magazines to be found in supermarkets, corner stores, and bookstores, extol the numerous virtues of spending time together 'as a family' doing 'fun' things together. Advice centres around the importance of getting away from daily routines to (re)connect and (re)engage with each member of the family unit. Generally, emphasis is placed on how time spent as a family is positive for children's physical, mental, social, and even spiritual, development. Strategies are conveyed for making family outings educational for children of all ages, and fun for everyone involved, including parents, other accompanying adults, such as grandparents or other extended family members, or family friends. Particularly, and strongly, invoked is the value or importance of family leisure as a means of creating and promoting family togetherness as a cultural, sometimes even religious or spiritual, value (Daly 1996; Shaw and Dawson 2001).

This culturally embedded construction of family leisure views it as an integral practice in successful parenting, making those places that support or encourage time spent in recreational or leisure activities for families, and which may be identified as 'family places', valuable spaces in geographies of family life. Family leisure activities such as a trip to the zoo provide a context that links the practice of family-oriented leisure and the outcome of successful parenting, of being a 'good parent', and the creation and sustaining of a positive, healthy, 'successful' family life. Those recreational spaces identified as positive, safe, and educational (at least to some degree) reflect, affirm, and constitute the values of family connection, closeness, and togetherness parents are both inspired by, and are seeking to transmit or sustain for their children through quality family time spent in family-oriented activities and environments.

Research (Shaw and Dawson 2001) demonstrates the significant, in some cases nearly overwhelming, sense of purpose attached to family participation in public leisure practices by both mothers and fathers, especially those of preschool and school-aged children. Considerable determination and commitment to this aspect of family life is apparent, to the point that good parenting at times is nearly equated with the active and always ongoing planning of, and participation in, leisure activities undertaken with and for children. These activities, which can become central to the recreational lives of parents of young children, may well supersede all other leisure-time activities (Maher 2005; Samuel 1996; Such 2006). The places in which families engage in these pursuits, for a certain period of time in their lives at least, become central sites in the leisure/recreational geographies of these families.

Studies have indicated that mothers report consciously organizing family leisure activities (such as zoo trips, trips to parks and museums, historical sites etc.), in order to facilitate and build family cohesion and closeness (Shaw 1992; Shaw and Duncan 2001). Mothers largely recognize that they do so consciously, purposively, and at least in part as a preventative measure to guard against the possibility of family dysfunction and collapse (DeVault 2000; Howard and Madrigal 1990; Shaw and Duncan 2001; Turley 2001).

In the few studies that have examined the relationship between fathering and leisure (e.g., Burns, Mitchell, and Obradovitch 1989; Shaw 1992; Such 2006), it is argued that family leisure has become central to fathering as well in contemporary times, a trend that is reflective of late twentieth-century changes in the dominant image of fatherhood which '. . . now highly values active participation that facilitates child development and family functioning . . . [this] supersedes traditional [individual] leisure goals of personal enjoyment, development, and relaxation' (Such 2006: 119). Thus, good parenting, by mothers and fathers, increasingly is defined by time spent in recreational family activities that parents undertake in order to willingly fulfill what is understood as a duty to promote positive family functioning (communication, togetherness, closeness, cohesion) and well-being. Those public places/cultural institutions which reflect the values of family togetherness, that make spending time together as a family enjoyable and beneficial for the children, their parents, and any other accompanying adults, will figure prominently in the geographies of families with young and school-age children, and will be valued locations in their communities. This is at least in part because they transmit a place identity ('family–place') which reinforces a positive 'good parent' identity desired by or constructed by families through family-oriented leisure in a specific (e.g., zoo) environment.

Therefore, we understand zoos as public spaces that family members inhabit together, immersed in a discourse of (potentially patriarchal, certainly highly gendered) family life that metaphorically 'swirls' around them (DeVault 2000). An outing to the zoo is the sort of activity seen to be, by experts and many parents alike, as central to satisfactory family life. As such, the family outing, and by extension the family and the roles and relations that define it, are constituted by both its members and those who create, maintain, and market such sites for family leisure activities. In short, 'good parents' seek out specific characteristics in the places that constitute their everyday geographies of family life that can be summed up as 'family-friendly'. We now turn to a discussion of how we can think of zoos as important places in the recreational geographies of families.

Identifying Zoos as Family-friendly Places

As noted earlier, we understand that built environments, including zoos, over time are imbued with personal, social, and cultural meanings that collectively form a place identity. People, alone and in groups, use places to communicate qualities of the self/group to others and to themselves. As outlined above, constructions of 'good parenting' including rather patriarchal notions of the 'good mother' and mothering, figure significantly in readings of zoos as a 'family places'. Moral and value judgements may be transferred to specific places, giving them a personality or quality with which they are, often uncritically, identified and labelled (Hallman and Benbow 2006; Williams 1999). Individuals and groups may synthesize, even internalize these subjective meanings given to the objective reality of place (Hagerstrand 1982; Pred 1984); what Relph (1976) termed 'existential insideness'. Relph explained this as '. . . insideness . . . in which a place is experienced without deliberate and self-conscious reflection, yet is full of significance' (1976: 55).

Zoos, therefore, are public places inscribed with meanings about private family life and about the nature of good parenting, where families can enact their own 'family fantasies' (Aitken 1998). Where there is a sense of fit between the meanings reflected in the zoo space, and the self/group identity of parents and family groups, then the zoo can be considered a positive 'family place' and therefore figure prominently in the geographies of family life for a given family unit. For others, the meanings reflected in the environment may be something to be contested against, or which (re) inforce a vision of family or family life that is at odds with their experience or way of life.

Descriptions of zoos as 'good family places' are based on a place identity so culturally embedded in Western societies as to be largely unquestioned and almost automatic in its recognition. Following on Relph's notion of 'insidedness', zoos are almost universally 'read' as family-friendly without conscious or deliberate thought. A tour of the children's book section of any sizeable bookstore is a quick and thought-provoking exercise that illustrates this point: the number of picture books and early-reader titles about events at zoos, zoo animals, and visits to zoos is really quite astounding![1]

Given this identification with the place identity of zoos as 'family friendly', we argue in our research that the capturing of the zoo visit in family photographs is a behaviour that both constitutes and reflects the value of this site, and the activities in this space for building and maintaining a positive family life (Hallman and Benbow 2007). This assessment of the zoo as 'family-friendly' may even extend to a reading of the zoo as a healthy place for those families that benefit from reinforcing, or working towards, a close fit between their (perceived) identity as a 'good family', as 'good mothers' and 'good fathers', and their experiences together as a family unit (whether or not this includes extended and fictive family members, or all or only part of the nuclear family group) in the zoo environment (Hallman 2007). Therefore, where there is congruence between the place and the identity, feelings of well-being, of being a 'good parent/family' may well be engendered (Hallman and Benbow 2007).

We now turn to a discussion of the practice of family photography, as a means of recording and practicing this connection between place, activity, and interaction in the construction of positive family relations and family life in the zoo environment.

Photography and the 'Family Gaze': Seeing Geographies of Family Life

As observed by Haldrup and Larsen with respect to tourist photography, we see family photography 'as a theatre of life where people in concert perform places, scripts, and roles to and for themselves' (2003: 23). This is because we agree with Schwartz and Ryan who note in their book *Picturing Place: Photography and the Geographical Imagination*, that

'Through photographs, we see, we remember, we imagine: we "picture place"' (2003: 1).

There is an inherent subjectivity in what, where, and whom an individual family photographer decides to record in even the most banal family snapshot. Examining this can shed light on the links between vision, memory, and identity (Schwartz and Ryan 2003). This is one of the roles of photography; to transform the space on the ground (e.g., the zoo space) into a place in the mind (e.g., zoo as family place). Family photographers at the zoo use landmarks (such as ponds, fountains, and statues), sites of activity (such as playgrounds and picnic areas), and animal exhibits to set the subject of the photograph in context. The photograph then becomes a remembrance of not only the featured person(s) but also the notable events of the zoo visit. Photographic practice, including family photography 'plays a central role in constituting and sustaining both individual and collective notions of landscape and identity . . . giving such social imagery solid purchase as part of the "real"' (Schwartz and Ryan 2003: 6). Family values and meanings are narrated and 'made real' through the production, viewing, and sharing of family zoo photographs, forming a visual dialogue between public and private discourses of space (Chambers 2003).

Photographs record visual facts that are not only representations of sites and of the associations and relationships depicted in them, but are sites themselves where those visual facts are interpreted and invested with meaning. They are not simply looked at, but are read, uniquely by their creators and those who have a connection to the people pictured, steeped as they are in the values and perceptions of their particular '**family gaze**'. Depiction, picturing, and seeing are all features of the process by which people come to know the world and their place in it, as it really is for them (or at least how they hope/wish it to be). As noted by Bourdieu (1990: 19) the '. . . practice of photography functions to solemnize and immortalize the high points of family life and to reinforce the integration of the family group by reasserting the sense that it has both of itself and its unity'. Family values and meanings as well as the very view or sense that the family has of itself, its character, and interrelationships, are narrated and made 'real' through the 'gaze' of family photography (Chambers 2003; Lury 1998).

The creation of photographs in a zoo visit objectifies the gaze of the family photographer, however, it is also a form of the 'reverse gaze' (Gillespie 2006: 343) and is informed by the relationship between the photographed subject and the photographer. This relationship is also informed by the site of the image production. This is distinct from John Berger's reference to the animals' gaze that to Berger represented the marginalization of animals and humans' loss of connection with them (Berger 1980). From comments from our research participants, the animals do indeed often seem secondary if not invisible. The gaze is also not only the production of the photographer but also the zoo itself, guided by the layout and design of the features that are used as photographic backgrounds and props. We see many features common among our research participants' photographs that indicate common expectations for, and conceptions of family photographs and what ought to be featured in them. The photographs also give rise to a narrative of not only the single pictured zoo visit, but also the later reminiscing about the events of the day.

In our work, this includes the family's presence at the zoo, the act of photograph production itself, and the review of family zoo photos and reminiscing about the experience depicted, at a later date. The 'family gaze' of the family photographer at the zoo is informed by the memories and expectations of the parents who have planned, organized, and undertaken the zoo visit as a family event. The resulting images capture the significant moments of the visit, placing them spatially and emotionally close to the photographer seeking to remember the people and events captured in the images they produce. The family photographer's gaze is also selective, as they edit the visual narrative being created, using the photographic frame to include details that add to the overall story of the image, and placing people and objects in different parts of the frame to create emphasis. Items that place the family at the zoo (e.g., animals, signs, memorable statues, etc.) are included, while other visitors not part of the family group are generally not included whenever possible. This creates the impression of the family unit 'together' in the space to the exclusion of others; the gaze is on 'their' family in the zoo leisure space, engaged in activity deemed important and worthy of recording. Therefore, we could say that the family zoo photograph is a tangible representation of the value placed on various elements in the zoo environment, and on the zoo as a site within family leisure/recreational geographies.

Research Methods

> To explore the instrumentality of photographs in shaping notions of landscape and identity, we must inquire into the meanings with which they are invested by their authors, and the meanings that they generated in their audiences. (Schwartz and Ryan 2003: 18)

The larger research project on which this chapter is based draws on analysis using visual discourse methods to study family photographs, and qualitative content analysis of project participant in-depth interviews to examine the local zoo as a valued site in family geographies, and the place or value of this leisure destination in the family lives of participants. Thus, we explored 1) the meanings families ascribe to the zoo visit; 2) the intended purpose of zoo family photographs and the use/display of these images; and 3) the relationship between the production and use of these photos of the family at leisure in the zoo environment and the role of the zoo within the geographies of family life. Our specific focus was on the use made of this environment and the meanings or values attached to this place by individual families with young children.

With the assistance of the Zoological Society of Manitoba and the Assiniboine Park Zoo in Winnipeg, Manitoba, families identified through 'family memberships' with the zoological society were recruited for participation in the study. In order to preserve member confidentiality, zoological society staff initially made contact with their members via email. Interested individuals allowed their contact information to be released to the research team, at which time a telephone call was made to ensure participation in both a photograph exercise and consent to be interviewed at a later date about their family zoo photographs and their zoo visit. Recruitment was a slow process, however those who agreed to participate consented to both the photograph exercise and to the interview. The sample is not a representative one, but rather is a small-scale, in-depth analysis intended to explore the range of meanings and values associated with zoo visits and the family photographs produced in the zoo environment, amongst the parent volunteers. All six of the participating families in the first round of interviews are white, two-parent, heterosexual couples, with preschool- or elementary school-aged children (one family includes a teenage stepchild who lives in the household only on alternate weekends), of various cultural and linguistic backgrounds and from a range of 'blue collar' and 'white collar' occupations. While we cannot be certain, it is quite likely that this group does in fact represent the social and racial/ethnic mix of those who hold zoological society family memberships. We recognize that this analysis does not address the place of zoos in the family geographies of visible minorities, later-life families or gay/lesbian households; this shortcoming is an area for future, much-needed research in family geographies.

Over the summer and early fall of 2007, six families participated in the first phase of the research project, the photograph exercise. Each participating family received a packet by mail that included a 24-image disposable camera, illustrated instructions, a written description of the project, a return mailer, and a consent form to be signed and returned in the provided mailer along with the used camera for processing. Families were instructed to use our camera on their next visit to the zoo, to have the usual family photographer take the pictures, and to go about their visit and their photograph-taking as they normally would when using their own cameras. These images were then the ones reviewed and discussed with the interviewer during the subsequent interview with the designated family photographer.

For each participating family two CDs were produced from each set of images, one of which was returned to the participating families, along with a set of prints and a written reminder that a member of the research team would soon be calling to arrange for an interview. Interviews took place through the fall and early winter, approximately six weeks after the individual family's zoo visit. These interviews were all conducted by one member of the research team to maintain consistency. Following the work of Rose (2003) and McIntyre (2003), the interview discussion centred on exploring the images produced (intentions, experiences, memories evoked, etc.) and the use/display of those images subsequent to the zoo visits, and afforded these parents the chance to discuss their experiences, perceptions and understandings of the zoo trip, the photographs produced, and their motivations for and interpretations of both. Interview prompts developed ideas expressed concerning the value and importance of the zoo experience and of aspects of the zoo environment based on what was depicted in the individual family's zoo photographs and from the reactions of the interview subjects as they reviewed their images and remembered details of the specific time and place portrayed in their photographs.

Interviews ranged in duration from 45 to 90 minutes and were generally conducted in the evening in the family home. In one instance the interview was conducted over the lunch hour in the mother's office at her request. In most cases the interview was with only the primary family photographer, however in some instances the other parents inserted their own comments on the zoo visit and the photos taken as they felt appropriate, often adding details or reminiscing about the zoo visit and what they did together, or what a child saw or said during their family trip to the zoo.

The results presented here are from a 'first round' of camera submissions and interviews. After receiving some local print and television media attention, as well as revisiting our selection criteria with the zoological society, a 'second round' of camera submissions was undertaken in late 2008 and 2009. We are now in a position to begin interviewing 10 to 15 more families as this chapter is being written. The findings from our first six interviews do, however, give us significant, and consistent, information on the geographical understanding families have of the zoo space, the use they make of the zoo and the meanings and value attached to the zoo outing and the images they create to mark or capture the experience of a family trip to the zoo.

Study Findings

Each interview began, once introductions were made and the informed consent reviewed, with a thorough check of the zoo photographs of the family being interviewed. This was facilitated by having all 24 images printed out in sequence on a single poster-sized sheet that could be tacked up on a wall or spread out on a table for ease of viewing. In each case the primary family photographer was the person interviewed; in five instances this was the mother while one was a stay-at-home father. Overall, and similar to our earlier analysis of family zoo photography (Hallman and Benbow 2007) the most prominent features photographed were each family's children playing in the several playgrounds and on play structures throughout the zoo grounds. This coincided with responses to questions about the reasons these families went to the zoo.

Kids Having Fun and 'Being Kids'

Perhaps surprisingly, parents of these young children were not motivated by a desire to show their children the animals on exhibit, to have them experience nature, nor to give them an environmental education experience—something 'good parents' might be expected to do and certainly one of the main features used by zoos to market themselves to parents. One parent photographer reasoned quite eloquently, 'I mean the kids think that's why we go to the zoo. . .because the animals are there. It's a good reason for them, for us. . .it's a reason for them to go, it's family time, burns off some energy, it gives us something to do'.

The zoo's playgrounds and green space specifically were seen to be safe, interesting for the kids, and a way for them to get outside, play, and exhaust some of their energy. These sites in the zoo allowed kids to have fun and 'run like animals and nobody really notices'. They have the freedom to run around, to 'just go' and 'just to have some fun, an outing' in a safe environment where others are not bothered by rambunctious children. In addition, picturing a number of children together also reinforces the sense of family relationships. Coupled with this desire, as noted in one of the quotes below, was the identification of the zoo trip as a family time, as a time to be a family, *together*.

Thus families took lots of photographs of their children on the zoo play equipment, because using these facilities was a primary reason for going to the zoo. As repeatedly noted, letting their children play and being a family together are their motivators for going to the zoo; interacting with animals or learning about the natural world are a bonus. Some typical comments from parents were as follows:

> 'Getting them out of the house and we really enjoy going to the playground. I have taken them there just to go to the playground now, since we are members';

> 'Kids burn off their energy at the zoo. We like going in the morning to get it out of the way and have a nice relaxing afternoon after that—and for the family time';

> 'The animals are just a plus being there but you see us going to the playground and being outside and just being a family together'.

Capturing the Uniqueness of their Child(ren)

Numerous other pictures were explained to us in the interviews as being taken because they seemed to capture something unique about an individual child or group of children. This might be their attraction to a favourite animal in some cases, or the importance of seeing or being photographed with a particular feature or landmark on the zoo grounds. One mother spoke at length while fondly looking at an image of her daughter in front of the vicuna exhibit. Apparently this llama-like animal is her favourite, and the four-year-old regularly stops there for 20 minutes or more. As her mother poignantly said, '. . .she just stands there and watches them and loves them, and talks to them'.

Another mother, when asked what her 'favourite' was of their zoo family photographs, selected

Photo 1.1. She then discussed the ritual she and her husband have with their daughter, who is fond of the polar bear statue/donation boxes scattered throughout the zoo grounds.

'The one with my daughter and the bear [is my favourite]. We have a picture of every bear in the park every time we go. This is a thing with her. She likes the bears and we have to put money in the bears [polar bear donation boxes] and we have to have a picture with each bear. The fact that they are all the same she hasn't realized. That's not me planning that, that's a demand. She goes and

Photo 1.1 Girl posing with polar bear donation box at the zoo

puts the money in and if I walk off she says "no mom, you have to do a picture" so you have to do a picture or you can't leave the bear!'

Thus, those pictures identified as the ones 'most liked' are the ones that captured the unique personalities and interests of the interviewed parents' child or children, as well as their physically unique growth and change; what one mom referred to as her 'real' kids. Zoos offer a rare opportunity for children to determine what the family group does and where they go, and this reveals the distinctive and unique qualities that parents want to see and record. This can be considered as another way of saying that what is valued in the zoo visit is the opportunity for children to demonstrate their uniqueness; in this space they are free to be themselves. The zoo environment itself then is read as a place of freedom and discovery, a relaxed environment that is a healthy and positive place for children, and therefore for the entire family as they engage with each other in this family-oriented recreational landscape.

For some parents, an area of the zoo is featured because it is a good backdrop for spontaneous action that captures the essence of who their child is, that they were looking particularly well or 'cute', or the fact that they were there together with certain family members or close friends. These notions are illustrated in the following quotes:

> 'I think my favourite one is this one with her and the bear. She is a ham when she gets with these bears [polar bear shaped donation collection figures] and every picture is a different one, she poses differently with each bear and sometimes she will give the bear a kiss and so . . . it's all the colour, she has the blue on with the blue sign and she has her arm on the bear and they look like they are kind of buddies';

> 'You know I haven't gone over these recently but I like this because obviously it has N— and C— and aww does he ever look so much younger';

> 'All of them are cute, when B—'s hair was so much shorter. I remember when I got her that haircut and my husband was just devastated, he was like "you cut all her hair off". She hasn't had a haircut since and it's nice to see how short it was because I am always saying her hair never grows';

> 'Because they are having fun and enjoying themselves. I like this one, they are such good friends these two. My daughter B— and E—, my friend's daughter. And the peacocks, because they chase the peacocks';

> 'I like all of them, they are all showing my kids doing things and being themselves'.

The theme of 'uniqueness' came up regularly to explain why photographs were taken and why they might be considered special or valuable. If zoo animals were in the photo it was a bonus that 'placed' the photo, in the family-friendly zoo environment, but was not an absolute necessity. One parent summed this up in this way,

> 'I didn't take specific pictures of animals. I find that they are always the same so we've got all the animals. I have lots of pictures of the animals from years prior so I don't take them; the animals don't change. If you take a picture from 1999 or 2005 the bear looks the same. *The kids are what you want pictures of to look at*'. [Emphasis added by the authors.]

The 'gaze' of the parent photographer at the zoo is, of course, informed by his or her memories and expectations of his or her child's development. Comments such as 'I had forgotten how little she was', or 'You can see the changes in your kids' could not only be remarking on physical growth and change, but also indicate pride in this evidence of a properly functioning family, in which children grow up healthy and capable. Recording growth and change is a motivation for taking the specific photograph, perhaps adjacent to a known landmark or backdrop in the zoo 'for scale'. The image also forms the link, perhaps subconsciously, between the evidence of a happy, growing child or multiple children and the uncritical understanding of the zoo as a positive, family-friendly place; as a place that is part of the individual family's geographical world.

Thus, those pictures identified by the interviewed parent/family photographer as those that were 'good', were 'most liked', and most likely to be displayed at work, at home, or to be shared (most often electronically) with extended and absent family members or fictive kin, were those which captured the unique personalities and interests of an individual child or group of children. Those images which captured the children unfettered by the need to 'behave' or 'be quiet', as is usually the case when in public spaces, were especially favoured. Others were noted as special because they marked physical growth and change.

Pictures That 'Just Make It'

However, the types of images noted above are not the only 'good' or 'special' family zoo photographs. The other images that drew comments from parents in the interviews, as we reviewed together the images they had taken during their recent zoo visit, were those images that for them seemed to demonstrate the nature of the zoo trip as family time. Parents talked fondly and occasionally with strong emotion about how they and the child were active and happy together and having fun together. Parents talked quite clearly about the zoo trip as family time, as an outing purposively engaged in to be together, in that space and time, as a family, focused on the interests and activities of their child/ren. This is what seemed to be valued, even actively sought in the acts of going to the zoo and of taking family photographs while there. It was those moments which were captured most often in family zoo photos and elicited the most comment from parents when viewing the images again at a later time. Thus, beyond the uniqueness of each child, it was apparent that the practice of taking family photographs (in our case, specifically in the zoo space) was about capturing the specialness of one's specific family, and the 'good time' of positive relations and happiness through togetherness in family life that binds the family as a unit. One mother in particular spoke about these connections and purposes very clear.

> 'A trip to the zoo is not an event for us; we do it frequently. It's something we enjoy as a family; it's an outing. But every time we go we take pictures and yeah, it's capturing the everyday, seeing if you *can catch the one picture that just makes it*'. [Emphasis added by the author.]

Other parent photographers had similar things to say as they reviewed their zoo photographs with us, again evoking the importance of the zoo visit as family time, and how this valuation of the place and time spent there came through to make certain photos special. Significantly, many looked at their photographs and would start their comments with phrases such as 'I remember' or 'That was the time' and took obvious pleasure in sharing stories about their families and the fun they had at the zoo, doing things together, even if sometimes it is a challenge to get everyone organized and out the door for the outing.

> 'I remember having fun with them and S— it was good. It's hard to say because a lot of times there's frustration in trying to get everybody organized.'

> 'It was just that we were having a good day as a family. It was a chance for us to go out. It was a beautiful day. You know it was nice to . . . it's not often that we have time to go out and do something fun and not having to worry about having to be here or get there. It was a fun afternoon; we enjoyed ourselves.'

> 'Oh yeah, just like how great my kids are and how much fun we can have, most of the time.'

> 'Yeah, we always have fun when we go to the zoo, it's just an adventure because he is so, you kind of wish he would slow down and have a peek and stop maybe running.'

Sometimes, the parent interviewed would get a little emotional as they reviewed the zoo photographs with us, and remembered the time spent in the zoo space with the child with great fondness. One parent was particularly moved by, and recounted several anecdotes from the zoo visit, because it was special 'one on one' time with their youngest child. It was clear that this was a highly meaningful time, connecting with this child in a way not often able to be done in a busy multi-child household with two working parents. Quite poignantly, it is special to this parent that they remember this time together at the zoo, and that they have the pictures to look back on, even though the parent expects that the child will not remember the event.

> 'Just that it was time with just one kid which is nice sometimes. I mean it's nice having all of them but one-on-one time with each of them in their own particular time zone. My one-on-one time with her [youngest daughter]. . .making her smile and making her have a good time. She probably doesn't remember it but that's okay, I do.'

Empowering the Child Through Photography

As noted above, the opportunity to allow children to be active, spontaneous, and free to 'act like kids' is a key factor in parents' attraction to the zoo environment and a theme in their photographs. They 'read' the zoo space as one that is safe and therefore liberating for their children, unlike much of urban public space (see Chapter 3 in this volume, for example). Often, this results in parents letting children guide the zoo visit and decide where the family group goes, what it does, and what features to look at. It is the children who decide what animals to see, for how long, and what will be done (e.g., take another picture with another 'bear bank'). Another example of the zoo experience being child-led—not only in the choice of where to go, how long to spend in any one area, and in 'goofing around' and 'running around'—is child agency extending to the recording of the experience in photographs. For one of the interviewed families, their elementary school-aged daughter also contributes to the memorialization of the zoo visit as family time through taking her own pictures.

> 'This one, even some of the others, are pictures she has taken. She likes to try and take pictures too so the pictures that don't quite look like 'good' pictures are usually her doing the pictures.'

While not considered 'good' photographically by the parent, these images were still prized, for the interest the child showed in the 'grown-up' activity of family photography, and for the images themselves as they indicate the child's perspective and the valuation of family time, especially with the parents who are the primary subjects of her pictures.

Posed Pictures: 'Placing' Family at the Zoo

In other studies examining family photography (e.g., Chambers 2003; Rose 2003, 2004) it has been emphasized that spatial proximity is demonstrative of the emotional togetherness (real or desired) which is a central purpose of both the act of family photography and the event being depicted. This is what makes photos such important and emotionally resonant objects, and such rich sources for explicating the meanings and importance of places valued in family geographies. As in Rose's (2004) study, those pictures that showed groups of siblings, siblings with friends, and parents and children together were considered 'favourites' by many parents and were the images most likely to be shared with extended family and displayed at work and in the family home. This is particularly true for those zoo family photographs posed by the family photographer. These images were purposively created to show 'togetherness', to symbolize and reiterate the integration of the family unit (Rose 2004).

Posed pictures (see Photos 1.2 and 1.3) can be anywhere in the zoo space that is appealing to the parents, or more often, at exhibits, statues or playground areas children stop at, ask to see, or for which they demonstrate a real fondness or interest. These qualities become central to the stories told later when reviewing the photographs taken at the zoo; about what the child or children did that day, what interested them, or anything fun that made the time spent at the zoo special for that specific family. For one family, for example, a picture had to be taken on every zoo visit beside a specific large old tree, as a sort of family tradition.

Photo 1.2 Group of children posed in front of a fountain at the zoo

Photo 1.3 Girl standing in front of antelope exhibit, a favourite place in the zoo

The father had been taken to the zoo as a child, and had his picture taken by this same tree. With a significant amount of nostalgia, when interviewed the parents discussed the importance of this continuity with generations before, how much they valued being able to compare pictures, compare experiences of being together as a family at the zoo, and talk about the differences and similarities between their children and themselves as children. In this way they also demonstrated the value of the zoo visit as family time, spent being active and having fun together, 'just like when we were kids'. However, the organization of the zoo visit to mark certain traditions and the posing of photographs also provides a measure of predictability and control that is often missing from modern family life.

Concluding Comments

> By creating photographic images of events, people and places . . . domestic [family] photography traditionally serves the purpose of constructing the family as myth, by capturing a preferred version of family life. These structures set the parameters within which agents create their visual family "life stories". (Chambers 2003: 13)

Our examination of family zoo photographs and interviews with the primary family photographer who produced the images has afforded us a window onto the ways families use both the process of family photography, and the places in which those photos are taken, to come to know their world and situate themselves as families in space and time (Schwartz and Ryan 2003). Through our analysis of the photographs taken by families who participated in our research, and especially through the interviews and the review of photographs with our participant families, it is readily apparent that the zoo visit, and the photographs that are taken to record it, are practices which help shape and define the meaning of 'family' to these parents of young preschool- and school-aged children.

The camera appears to act as a kind of witness to their outings and presence in public space, engaged in positive and emotionally valued family-oriented recreational experiences and activities. Through the photographs, the family-oriented zoo trip becomes fixed, and through looking at them, sharing them, talking about them, and displaying them, the images are then transformed into (desired) memory. As Sanders (1980) states, '. . . the image produces a form of memory that takes an ideological form: the way one would like to be remembered' (as quoted in Chambers 2003: 109).

In the process of conducting the interviews, we learned much about the ways that such middle-class and working-class families with young children incorporate the zoo space into their family geographies. For all of our interview subjects, the zoo is a valued recreational space that, unlike most other public places, affords their children an unprecedented freedom and autonomy of action. The presence of animals, the lure of environmental education or information about nature and natural processes are at best secondary draws to the zoo space for this cohort of zoo-goers; an interesting outcome when parents with children under the age of 12 represent 80 to 85 per cent of zoo visitors in North America (American Association of Zoos and Aquariums www.aza.org). The zoo space becomes valued for the opportunity it gives children to be themselves, unfettered by the expectations and strictures to 'behave' which are normally attributed to being 'out' in public. This specialness is then reflected in both the production and identification of family photographs that capture this expression of what one mother called her 'real' child.

It does seem however, given the strong, uncritical identification parents made with the zoo as a 'family place' that has inherent importance to their lives as parents, that the zoo is a significant part of their family geographies also because it taps into their sense of what it is to be a good parent. As discussed earlier in this chapter, it was apparent from the interviews, especially as the parents re-examined their zoo photographs and spoke fondly about specific memories, activities, and changes they saw in their children, that the trip to the zoo, and the recording of the trip in the photographs, are clearly activities they engage in because they believe it builds and tightens the emotional bonds between themselves and their children. This is clearly, for this group of parents, the primary and overwhelming purpose to the inclusion of the zoo space in their family geographies. This is where they can be a family, together, doing things everyone enjoys, building memories and positive emotions that they can return to, even relive, through the photographs they take at that time, in that place.

Where else might families engage in activities that facilitate their goals of positive family functioning, cohesion, and togetherness? What other public, recreational, or leisure landscapes may fulfill similar needs for families, not only in the early years of family formation and childrearing, but as children grow older and become teenagers, and later still as parents age and children are 'launched' and form families of their own? Do families at different life stages, of different ethnic/racial groups, or sexual orientations, use the same public places, but differently? Or are there other locations in their communities that become valued in their family geographies as places for building and maintaining connection and relationships across the generations? It is to questions such as these that we turn in future research. We encourage other researchers, inspired by our explorations, to contribute to a growing and expanding family-geographies research focus in human geography.

Questions for Further Thought

1. What other environments allow for families to spend quality time together, and what are their characteristics? How do they compare to the zoo as 'family-friendly' places?
2. If family zoo photographs both reflect and construct the values perceived in the zoo experience, what do they say about the ways that parents 'read' the zoo landscape that might differ from other 'readings' of that place?
3. How do family photographers use the highly selective nature of the gaze to exclude some features and create their desired images?

Further Reading

Anderson, K. 1995. 'Culture and nature at the Adelaide zoo: at the frontiers of "human" geography'. *Transactions, Institute of British Geographers* New Series 20: 275–94.

Fraser, J., and J. Sickler 2009. 'Measuring the cultural impact of zoos and aquariums'. *International Zoo Yearbook.* 43: 103–12.

Larsen, J. 2005. 'Families seen sightseeing: performativity of tourist photography'. *Space and Culture.* 8, 4: 416–34.

Mullan, B., and G. Marvin. 1999. *Zoo Culture: The Book about Watching People Watch Animals.* 2nd edition. Urbana IL: University of Illinois Press.

Endnotes

1. A recent (October 2009) children's book search of the word 'zoos' on the online bookseller amazon.com generated a list of 14,896 current titles.

References

Aitken, S.C. 1998. *Family Fantasies and Community Space.* New Jersey and London: Rutgers University Press.

American Association of Zoos and Aquariums. www.aza.org

Anderson, K. 1995. 'Culture and nature at the Adelaide Zoo: At the frontiers of "human" geography'. *Transactions, Institute of British Geographers* NS 20: 275–94.

Baicker-McKee, C. 2003. *Fussbusters on the Go: Strategies and Games for Stress-free Outings, Errands, and Vacations with your Preschooler.* Atlanta: Peachtree Publishers.

Berger, J. 1980. *About Looking.* New York: Pantheon.

Bishop, R. 2000. 'Journeys to the urban exotic: embodiment and the zoo-going gaze'. *Humanities Research* 11, 1:106–24.

Bourdieu, P. 1990. *Photography: A Middle Brow Art.* Stanford CA: Stanford University Press.

Burns, A.L., G. Mitchell, and S. Obradovitch. 1989. 'Of sex roles and strollers: female and male attention to toddlers at the zoo'. *Sex Roles* 20, 5/6: 309–15.

Chambers, D. 2003. 'Family as place: family photograph albums and the domestication of public and private space'. Pp. 96–114 in J. Schwartz and J. Ryan, eds. *Picturing Place: Photography and the Geographical Imagination.* London and New York: I.B. Tauris.

Daly, K.J. 1996. *Families and Time: Keeping Pace in a Hurried Culture.* London: Sage.

DeVault, M. 2000. 'Producing family time: practices of leisure activity beyond the home'. *Qualitative Sociology* 23, 4: 485–503.

Domosh, M., and J. Seager. 2001. *Putting Women in Place: Feminist Geographers Make Sense of the World.* New York and London: Guilford Press.

Ehman, K., K. Hovermale, and T. Smith. 2005. *Homespun Memories for the Heart: More Than 200 Ideas to Make Unforgettable Memories.* Grand Rapids MI: Revell.

Ellis, G. 1999. *The Big Book of Family Fun.* Grand Rapids MI: Revell.

Frame, M. 1994. 'Octopus Swan Song'. *Sea Frontiers* 40, 5: 38–42.

Gillespie, A. 2006. 'Tourist photography and the reverse gaze'. *Ethos* 34, 3: 343–66.

Griffin, S. 1978. *Woman and Nature.* New York: Harper & Row.

Hagerstrand, T. 1982. 'Diorama, path and project'. *Tijdschrift voor Economische en Sociale Geographie* 73: 323–39.

Haldrup, M., and J. Larsen. 2003. 'The family gaze'. *Tourist Studies* 21, 1: 4–21.

Hallman, B.C. 2007. 'A family-friendly place: family leisure, identity and wellbeing—the zoo as therapeutic landscape'. Pp. 133–45 in A. Williams, ed. *Geographies of Health—Therapeutic Landscapes.* Aldershot: Ashgate Publishing Limited.

Hallman, B.C., and S.M.P. Benbow. 2006. 'Canadian human landscape examples: Naturally cultural—the zoo as cultural landscape'. *The Canadian Geographer* 50, 2: 256–64

———. 2007. 'Family leisure, family photography and zoos: exploring the emotional geographies of families'. *Social and Cultural Geography* 8, 6: 871–88.

Hampp, A. 2006. 'Animal attraction: advertising at the zoo—sponsorships climb as marketers go after families at low cost'. *Advertising Age.* 29 October. http://adage.com.

Hanson, E. 2002. *Animal Attractions: Nature on Display in American Zoos.* Princeton and Oxford: Princeton University Press.

Howard, D.R., and R. Madrigal. 1990. 'Who makes the decision, the parent or the child? The perceived influence of parents and children on the purchase of recreational services'. *Journal of Leisure Research.* 22, 3: 244–58.

Lury, C. 1998. *Prosthetic Culture: Photographs, Memory and Identity.* London and New York: Routledge.

Maher, J. 2005. 'A mother by trade: women reflecting mothering as activity not identity'. *Australian Feminist Studies.* 20, 46: 17–29.

McIntyre, A. 2003. 'Through the eyes of women: photovoice and participatory research as tools for reimagining place'. *Gender, Place and Culture: A Journal of Feminist Geography* 10, 1: 47–66.

Morey & Associates. 2004. *National Zoo and Aquarium Attitude and Usage Survey Report.* Conducted for the American Zoo and Aquarium Association. Charleston S.C.: Morey & Associates.

Nesmith, C., and S.A. Radcliffe. 1993. '(Re)mapping Mother Earth: A geographical perspective on environmental feminisms'. *Environment and Planning D: Society and Space* 11: 379–94.

Norton, W. 2006. *Cultural Geography: Environments, Landscapes, Identities, Inequalities, 2nd ed.* Toronto: Oxford University Press.

Ortner, S. 1974. 'Is female to male as nature is to culture?' Pp. 67–88 in M. Rosaldo and L. Lamphere, eds. *Woman, Culture and Society.* Stanford CT: Stanford University Press.

Partow, C., and D. Partow. 1996. *Families that Play Together Stay Together: Fun and Healthy TV-Free Ideas.* Minneapolis MN: Bethany House Publishers.

Philo, C., and C. Wilbert. 2000. 'Animal spaces, beastly places: an introduction'. Pp. 1–34 in C. Philo and C. Wilbert, eds. *Animal Spaces, Beastly Places: New Geographies of Human-Animal Relations.* London and New York: Routledge.

Pred, A. 1984. 'Place as historically contingent process: structuration and the time-geography of becoming places'. *Annals of the Association of American Geographers* 74: 279–97.

Relph, E. 1976. *Place and Placelessness.* London: Pion Limited.

Rose, G. 2003. 'Family photographs and domestic spacings: a case study'. *Transactions of the Institute of British Geographers* NS 28: 5–18.

———. 2004. ' "Everyone's cuddled up and it just looks really nice": an emotional geography of some mums and their family photos'. *Social and Cultural Geography* 5, 4: 549–64.

Rose, G., V. Kinnaird, M. Morris, and C. Nash. 2002. 'Feminist geographies of environment, nature and landscape'. In Women and Geography Study Group, ed. *Feminist Geographies: Explorations in Diversity and Difference.* New Jersey: Prentice Hall/Pearson Education Ltd.

Sabloff, A. 2001. *Reordering the Natural World: Humans and Animals in the City.* Toronto: University of Toronto Press.

Samuel, N. 1996. *Women, Leisure and the Family in Contemporary Society: A Multinational Perspective.* Wallingford UK: CAB International.

Sanders, N. 1980. *Family Snaps: Images, Ideology and the Family.* New South Wales Institute of Technology, Faculty of Humanities and Social Sciences. Australia: New South Wales Institute of Technology.

Schwartz, J., and J. Ryan. 2003. 'Introduction: photography and the geographical imagination'. Pp. 1–18 in J. Schwartz and J. Ryan, eds. *Picturing Place: Photography and the Geographical Imagination*. London and New York: I.B. Tauris.

Shaw, S.M. 1992. 'De-reifying family leisure: an examination of women's and men's everyday experiences and perceptions of family time'. *Leisure Sciences* 14: 271–86.

Shaw, S.M., and D. Dawson. 2001. 'Purposive leisure: examining parental discourses on family activities'. *Leisure Sciences* 23: 217–31.

Such, E. 2006. 'Leisure and fatherhood in dual earner families'. *Leisure Studies*. 25, 2: 185–99.

Turley, S.K. 2001. 'Children and the demand for recreational experiences: the case of zoos'. *Leisure Studies*. 20: 1–18.

Williams, A. 1999. 'Place identity and therapeutic landscapes: the case of home care workers in a medically under services area'. Pp. 71–96 in A. Williams, ed. *Therapeutic Landscapes: The Dynamic Between Place and Wellness*. New York: University Press of America.

Wolch, J. 2002. 'Anima urbis'. *Progress in Human Geography* 26, 6: 721–42

Chapter Two

The Impact of Travel on Families with Youth Competitive Athletes

Donna Williams and William Crumplin

> America has become a society of excess: super-sized everything. Youth sports are no exception: more of just about everything; practices, games, tournaments, competition, select teams, travel, media coverage, money, burnout, and injuries. The only thing youth sports have less of is kids having fun and just being allowed to be kids . . . we think that if we don't do everything possible to help them get ahead of their peers, even in preschool, they—and by extension, we—will be branded as failures, as losers, and sent, literally and figuratively to the sidelines. (de Lench 2006: xii)

Introduction

Sports, and sports for tween and teenaged youth in particular, have become important elements in western society (Coakley 2009; Coakley and Donnelly 2004; Côté and Fraser-Thomas 2007; de Lench 2006; Horn and Horn 2007). Competitive youth sports, defined here to be for persons between the ages of 11 and 18, place demands on athletes and their families alike, through a complex and oftentimes invisible web of relationships between coaches and managers, young athletes, parents, and other non-athlete family members. Competitive sports are seen to be different from recreational sports on a number of fronts. They demand significant time commitments for practice in addition to games. Competitive teams travel extensively to participate in league and tournament games. Organized leagues have been established to oversee competition and teams compete regionally, provincially, nationally, and, in some instances, internationally. Teams are ranked at various geographic scales due to their relative success in the local league and appropriate tournaments. The travel demands of participating at this level are significant and represent major investments of time and money and, we argue, have significant impacts on family life and, therefore, family geographies.

Family geographies and relationships are shaped or reconfigured as the demands between sport and non-sport activities are negotiated and accommodated. Some impacts are positive, such as the opportunity to cultivate parent–child relationships based upon a common interest/activity (Anderson et al. 2003; Coakley 2009; Côté and Fraser-Thomas 2007). Trade-offs must be made between time spent all together as a family and time required to accompany a single youth athlete to practices, games, or distant tournaments that can produce lengthy separations.

It follows that many decisions made at the family level are influenced by the requirements of the team and the **sporting spaces** it occupies. For instance, as youth athletes' space demands expand through competitive sport, family members make decisions between occupying various spaces, including arenas, fields, and courts, and the family home and vehicle. This study illustrates that, for families, these decisions are influenced by a variety of factors of which sports culture and emotions about the spaces and parental activities are important. Supporting youth competitive athletes is found to shape the geographies of such families and gendering of spaces is seen to occur.

This chapter specifically focuses on the impacts that travel have on families supporting youth in competitive sports. This is informed by concepts of **sports culture**, **emotional geography**, **gendered spaces**, '**good parents**', and '**good sport parents**' each of which is explained in following sections. The analysis is informed by participant observations and interviews of parents whose teenaged daughters were involved in competitive hockey, soccer, and volleyball. The methodology is described later, however 18 personal interviews were conducted with parents known by one of the authors. The interviews identified individual family decisions, tradeoffs, and concerns about the impact of participation in competitive sport on their families and family life. When seen as a whole, it became apparent that sports culture and emotions of place were important, if invisible, influences in many family decisions. Often opposing rhythms became clear as parents attempted to be 'good sports parents' and 'good parents', and differences emerged between mothers and fathers as mothers reported feeling conflicted between these dual roles while fathers did not.

Conceptual Framework: Sports Culture and Family Spaces of Supporting Competitive Athletes

Families make decisions trying to balance many things including finances, time, and the needs of all family members including those of the youth athlete while respecting the expectations of the sports culture of the athlete. Sports culture varies depending on the sport, club, and team, but there are commonalities. As in any culture, subtle and frequently unspoken norms exist that demand certain forms of behaviour and these can vary depending on the location or space being occupied by the athletes and/or parents at a particular point in time. Family geographies, therefore, are created and re-created with decisions that are made by the family or imposed directly or indirectly by the sports culture. The family geographies represented in supporting youth in competitive sport consist of two distinct spaces: sports spaces and home spaces. These two spaces are, however, linked by a third—that of the family vehicle. Populating these three spaces depends on various factors. Other places, including restaurants, hotels, and billets, are occupied by athletes and parents, but these can be either extensions of family space or team space and, therefore, were seen to be places and spaces outside the realm of this investigation.

Sports or activity spaces for families with youth competitive athletes are the easiest to conceive. These include the places where the practices and competitions take place, such as arenas, pools, courts, fields, and gyms, including those in distant places for tournaments and competitions. Home spaces consist not only of the home but can be extended to include the family vehicle. The home is used as a base to support many activities such as storing, maintaining, and cleaning the athlete's equipment, and feeding, sheltering, and even healing the athlete. The family vehicle is used to transport the athlete to practices, games, and/or competitions and as a place to parent. As such, the family vehicle is the site where many family and sport-related functions occur. For instance, the car is a family space for building relationships, consuming food, and monitoring children. In other instances, the vehicle is like the participation space where lectures about the sport are given, loud music is blared to 'psych up' the athlete and post-mortems on the games are held.

Sports Culture

Sports cultures vary across sports and teams and they can influence decisions made by families with competitive athletes. Parents are very important in youth decisions to participate in sports (Anderson et al. 2003; Coakley and White 1999; Côté and Fraser Thomas 2004; de Lench 2006; Horn and Horn 2007; Varpalotai 1995). The reasons for introducing children to sports and for continuing to support them in these endeavours are many and include promoting physical and motor skills development, increasing levels of fitness, learning teamwork skills, gaining self-confidence, improving responsibility and time management, appreciating goal orientation and success, and expanding friendships while honing socializing skills (Côté and Fraser-Thomas 2004; de Lench 2006; Horn and Horn 2007).

There are other broader reasons, however, that also steer parents to encourage their children to enter and remain in sports and they serve to encourage many parents to volunteer and to uphold team expectations of themselves and their children. In Western society there is much value placed upon children being involved in and succeeding at organized sports (Coakley 2009). 'In fact, many mothers and fathers feel that their moral worth as parents is associated with the visible achievements of their children—a factor that further intensifies parental commitment to youth sports' (Coakley 2009: 126). In addition, '. . . parents expect their children to receive college scholarships, professional contracts as athletes, or social acceptance and popularity in school and among peers' (Coakley 2009: 132). This means that sports have become serious activities for parents and athletes alike.

Sports in general and teams in particular are subcultures of the society in which they belong and reflect and encourage selected social processes, outcomes, and desired behaviours (Coakley and Donnelly 2004). Formal and informal sets of rules, regulations, and norms work in concert to promote the sport, maintain its integrity, and guide the participation and behaviour of those involved. Administrators, coaches, managers, and athletes are the obvious persons targeted and influenced by these sets of rules or codes of behaviour. But, in the case of youth competitive sports, another significant group of individuals includes the families of the athletes and, in particular, the athletes' parents. In most instances, parents learn what the team or club expects of them both formally and informally, and frequently act in ways to support the team in order to enhance the sporting experience of their children. At the same time, they demonstrate their own value as good parents (Coakley 2009). Parents' decisions to support their children and adopt or adapt to the sport culture are influenced by their emotions.

Emotional Geography

Emotions serve as a thread influencing the decision-making process to occupy sports, vehicle, and home spaces. Emotion of place is a recent area of study that builds on the ideas of having positive or negative feelings for a place (topophilia or topophobia [Tuan 1974]). Bondi et al. (2005) expand on Tuan's concepts and argue that emotions matter; that emotions play an important role in how people make decisions to view and use space since they serve '. . . as relational flows, fluxes or currents, in between people and places . . .' (Bondi et al. 2005: 3). People not only feel various emotions about places, but these emotions can be produced or modified through relations with other persons and/or environments (Bondi et al. 2005). Athletes and parents are no exception and their decisions to occupy different spaces are influenced by their emotions.

Parents' behaviours can be influenced by their feelings and perceptions of what other parents or coaches expect of them. Recognizing the importance of emotions as they are invoked by spaces and social expectations assists in understanding the concept of the 'good sport parent' in general, and the 'good sport mother' and 'good sport father' in particular. For instance, '. . . understanding of what constitutes "good mothering" varies significantly between different social groups in different geographical areas and that these understandings are, in part, maintained and reproduced through local social networks where deviant behaviour is unsupported and conforming behaviour is rewarded' (Duncan and Smith 2002: 21). Emotions of place, gendering of spaces, and the concepts of good sports parents influence each other but do so within the expectations and demands of sports culture.

Sports and the organizations, upon which they are constructed, occupy spaces and territories. Social organizations deal with issues of territory or space, control of that space, and the persons that occupy and use it (Marotta 2005; Rabinow 1984; Valentine 2001). But sports and their organizations can also be powerful shapers of interactions within society at large and within and between smaller societal components thereby serving as an arena for the establishment, maintenance, and enforcement of accepted cultural norms (Coakley 2009). These norms impact not only the persons directly involved in the sports but also include the family of the athletes (Coakley 2009; Coakley and White 1999; Côté and Fraser-Thomas 2007; de Lench 2006; Lenskyj 2003; Marotta 2005; Varpalota 1995) with different pressures, expectations, and impacts on various family members, particularly the parents. The norms of sports spaces can, and do, shape family decisions.

Examples of varying expectations can include, but are not limited to, parents driving to practices and competitions, attending games, providing proper meals, adhering to schedules, knowing the rules of the game, and so on. Parents define the roles they will play in the athlete's sports, vehicle, and home spaces. Making these space–time decisions results in parents choosing to occupy one, two, or more spaces and the decision can produce feelings of guilt about, for instance, making sacrifices concerning time together as a family at meal time or investing more time with one child than another. Emotions can influence decisions and actions to support, or not to support, the athlete at any given time.

Gendered Spaces

Sports reflect and are tied to dominant cultural ideologies and, therefore, tend to favour the interests of the dominant groups in a society. In Western societies, these dominant groups and the dominant ideology tend to favour male interests (Coakley 2009; Coakley and Donnelly 2004; Hardy and Wiedmer 2005; Lenskyj 2003; Marotta 2005; Valentine 2001). As sports have evolved and become more organized, they reflect society at large and have tended to adopt the gender logic of society thereby generally disadvantaging women (Coakley and Donnelly 2004; Donnelly and Harvey 1999; Lenskyj 2003; Marotta 2005). Many relationships '. . . revolve around sports, especially among men' (Coakley and Donnelly 2004: 17). In fact, '. . . sport is key for establishing a masculine identity' (Donnelly and Harvey 1999: 49).

Sports can, therefore, shape the personal lives of those involved. As Duncan and Smith contend (2002), family geographies are related to the existence of regional and local norms for family life and can impact gender roles and shape what '. . . is the "proper thing to do" as far as women and men, and in particular mothers and fathers, are concerned' (31). Significant here is that sports and the family are closely related in many cases as '. . . relationships between family members are nurtured and played out during sport activities or in conversation about these activities' (Coakley and Donnelly 2004: 20).

One way to account for the influence of sports and sport culture is through the lens of **control theory**. Some scholars argue that social facilitations, like learning and adopting sports culture, can be explained in control theory terms (Baxter et al. 1990). Control theory holds that the audience serves to focus the attention of the performer on '. . . discrepancies between what he or she is actually doing and the requirements of some currently salient standard of behaviour, producing "enhanced conformity to the standard" ' (Carver and Scheier 1981 in Baxter et al. 1990: 352). Therefore, the performer or the parents behave so as to reduce perceived discrepancies between the expected norm and their behaviour in activity, home, and/or gendered spaces. This phenomenon can help explain the desire of parents, particularly mothers, to achieve 'good mother' status in the eyes of other mothers. This is reminiscent of women constantly observing other women and themselves in cultural contexts and through the eyes of the dominant cultural group (Doucet 2006; Duncan and Smith 2002; Heenan 2005). Marsha Macotta (2005) takes this one step further when she discusses the concept of 'MotherSpace' since particular spaces seem to attract mothers *or* fathers.

While the playing field has become more level for female athletes in Canada (Lenskyj 2003; Stevens et al. 2007), there are areas in and around sports where gender does matter (Coakley 2009; Lenskyj 2003; Varpalota 1995). For example, '. . . the administration, coaching, and officiating of sport is, for the most part, in male hands, even in many sports that have high female participation as athletes' (Lenskyj 2003: 56). Coakley (2009) analyzed the 156 team American Youth Soccer Organization (AYSO) and noted the gender bias in the organization. He discovered that the commissioner and assistant commissioner were men, 21 of the 30 board members were men, and 85 per cent of the head and assistant coaches were men (39–41). '[W]hereas 86 per cent of the team managers, or "team moms" as most people referred to them, were women' (Coakley 2009: 39). Messner discovered that many women were more qualified for the board and coaching positions than were some of the men holding these positions, but the women were less likely to volunteer.

Doucet (2006), writing on years of participant observation of recreational and competitive soccer in Ottawa, noticed that '. . . the majority of volunteer coaches, assistant coaches, organizers, and managers of children's soccer are fathers' (150). Doucet also states that networking by fathers around sports '. . . provides an opportunity for fathers to connect with other fathers and link dominant masculine activities with networking and facilitating children's social and physical growth' (150). She feels that context matters for 'mothering' and it can influence gendering of spaces. As she states, '. . . the ways in which men and women conduct themselves in their domestic and community lives are simultaneously informed by and form part of their moral identities, which are conceived as the shoulds and oughts of gendered social behaviours and norms' (33).

Gender differences also exist with respect to driving, and this is important when considering the importance of the family vehicle as a space that links sports and home spaces. The literature clearly illustrates differences in driving attitudes between males and females with males being more comfortable with many aspects of driving, including long-distance driving (Baxter et al. 1990; Bergdahl 2005; Harré et al. 1996; Kostyniuk & Molnar 2008; Özkan and Lajunen 2006; Sundström 2007). Males regardless of age were significantly more comfortable driving in unfamiliar territory, at night, on weekends in denser traffic, and in bad weather (Bergdahl 2005; Kostyniuk and Molnar 2008). Kostyniuk and Molnar (2008) found that with respect to driving over 200 miles (322 km) to attend an important appointment or in an unfamiliar area, nearly 20 times as many women as men claimed they would not drive at all.

While driving children to school or soccer practice is part of acceptable behaviour for mothers (Marotta 2005), motherhood and mothering are not 'natural' in that biology alone is not responsible for these behaviours; they are socially defined and are created and re-created through cultural forces. Marotta (2005) believes '. . . that the contemporary material and discursive spaces of Western mothers reveal the links among the powers, practices, and subjects that discipline mothers to the point of tyrannizing them' (15). For instance, 'When certain spaces are identified as appropriate and proper for "good" mothers—home, spaces around children, especially school, and so on—the messages to mothers are clear: this is where you belong, this is where it is appropriate for you to be, this is how you should think about yourself' (Marotta 2005: 17). These ideals and norms are also passed on by other mothers since concepts of gender and behaviour can be relationally constituted, such that other people's gender enactment is intimately involved with constellating and changing our own (Shotwell 2008). Shotwell (2008) feels that social norms are demonstrated and learned through interpersonal micro-practices and intimately affect relationships. Sport culture is one way that the concept of good sport parenting is demonstrated and passed along to other parents.

Borrowing the concept of the '**Court of Motherhood**' from a popular novel[1], Doucet (2006) uses it to add a moral dimension to understanding how emotions fuel mothers' and fathers' feelings as to how they should act, and how they perceive that these actions will be viewed by others. Mothers in Doucet's (2006) study admitted that they felt they were being judged against social standards. They felt negatively judged if they worked outside the home while their spouses were stay-at-home dads, or if they did not plan things like birthday parties for their children. This general concern was also expressed by mothers who provided long-term healthcare to family members at home and were troubled that their homes had come to resemble hospitals more so than the homes of their friends (Yantzi and Rosenberg 2008).

The invisible standards for being a 'good mother' are effectively imposed by the invisible social 'Court of Motherhood' as if they existed in statutes and were administered by magistrates. A comprehensive list of characteristics that is universally held to define a 'good mother' does not exist. However, mothers in Doucet's (2006) study mentioned activities that they, as working mothers who lived with stay-at-home fathers, felt a need to perform, including baking for the family, knowing what the kids are doing every night, being involved in the children's schooling, caring for hurt children, and taking over domestic responsibilities when they come home from paid work (Doucet 2006). Women as 'good mothers' are in the first instance '. . . viewed and judged according to a long tradition that defines women as caregivers' (Doucet 2006: 194). Marotta (2005) states that 'good mothers' are those who operate within the society's boundaries of 'MotherSpace' and it is seen to include the home, the children's school, grocery shopping, and taxiing children to medical appointments and their extracurricular activities.

Another important duty that frequently falls to mothers is ensuring the family meals are prepared and, therefore, nutritional requirements of the family are met. A study by Roy and Petipas (2008) determined that mealtime is the cornerstone of the family. This is also supported in a 2008 study by Sarah Woodruff who found regularly having a family meal resulted in healthier children (Proudfoot 2009). While some fathers prepare family meals, the kitchen is still seen as a place for mothers (Marotta 2005).

In the context of supporting youth, just being a 'good mother' is not enough as there are additional responsibilities and expectations. This leads to the concept of being a 'good sport mother'. However, clearly and universally defining a 'good sport mother' is just as difficult as defining a 'good mother'. Nonetheless, Donnelly and Harvey (1999) found that mothers '. . . often make sacrifices in terms of time, money, or access to the family vehicle(s) to facilitate the participation of their husbands and children' (49). But being a 'good sport mother' increases the area or spaces to mother. Providing care at activity spaces and in the family vehicle are obvious 'sport' extensions to the 'good mother' concept. These extensions can prove difficult due to conflicting time demands. In addition, some expectations of being a 'good sport mother' only seem definable through observing or experiencing the behaviours and subtle judgements between mothers in activity, home, and gendered spaces.

Doucet (2006) feels a 'Court of Fatherhood' also exists and that certain activities and roles are assigned to and expected of fathers. The fathers in Doucet's study, all stay-at-home caregivers, wanted to be seen as men while being accepted as equals in mothering situations and spaces. From these fathers, she learned their perception of being a good father included the ability to earn money, being handy around the house, and talking with other guys about sports, tools, and home renovations. To dispel concerns that they might be gay or un-men, '. . . most fathers emphasize masculine qualities of their caregiving such as promoting their children's physical and outdoor activities, independence, risk taking, and the fun and playful aspects of care' (Doucet 2006: 196). The study also showed that fathers formed networks with other fathers around children's outdoor activities and sports.

There are several avenues, therefore, for gender to play a role in the support of youth competitive athletes. General attitudes about driving differ according to gender. Administrative and coaching positions in sports organizations are largely dominated by men. Sports themselves tend to support male interests and attitudes and many male relationships revolve around sports and percolate into sports culture which is also an important component of decision-making as to who occupies sports, vehicle and home spaces, and what roles are expected in those spaces.

Control theory, therefore, the hidden social courts of motherhood and fatherhood, and emotions can be used to understand decisions such as which parent drives the athlete, who stays home with other children, or who provides for the nutritional needs of the athlete and family, among other decisions. Specific reasons for decisions taken by this study's participants to occupy sports, vehicle, and home spaces are presented after describing the methodology used to understand the impact that supporting competitive athletes has on families and family geographies.

The Study

The initial intent of this research was to present the findings of some basic questions that evolved from Donna Williams' daughters' participation in competitive sport. Friends noted that her family geography was distributed over a large area because of the daughters' participation in competitive sports. The stresses and strains of travelling to various volleyball, hockey, and soccer games and tournaments were impacting the relationships in the family. This raised questions of what the broader social impacts of travel with competitive athletes have on families. This exploratory study has been developed to produce this chapter and share the findings of this work.

Donna Williams developed the initial interview method. She had experience with interview methods through surveys done in a different context (Williams 1987; Williams et al. 2003).

Sample interview guides from other work were also reviewed. The work then took on a life of its own. As gaps needed to be filled and time was running short, both focus groups (run with four to eight parents answering questions together and building on one another's answers) and e-mail questionnaires were used as described below.

The parents who participated in this research were from the three teams on which the author's two daughters participated between February and May 2007. This included an under-17 volleyball team, an under-14 soccer team, and a peewee hockey team (girls aged 11 and 12 years). This volleyball team had 12 athletes and their families participating in the sport whereas the soccer and hockey teams each had 16 athletes and their families. The parents were informed verbally of the study over the course of several games and practices for the various teams in early February 2007 and many volunteered to participate. As the study continued, several parents from other teams asked if they could participate and three were selected to increase numbers and bring different perspectives.

All parents interviewed were known to the researcher and had a level of familiarity with her and her family. This enabled a quick rapport to be built and frank discussion of the questions. The interviews were held by the side of fields, in gyms, at arenas, in cars, and in hotel rooms. A structured interview was used to elicit responses and develop discussion. The questions of importance to this chapter were mainly open questions relating to the positive and negative aspects of travelling for competitive sports as well as on time spent in the sporting activities. It is important to note that none of the athletes whose parents were interviewed, were drivers. The parents driving cars were the primary means of transportation for these athletes. One parent in this study used public transportation.

A series of follow-up questions were e-mailed to a group of mothers and fathers who wished to participate but were unable to do so within the time constraints of the interviews. These questions were used to further explore specific trends that were seen through the interviews. In total, seven e-mail responses were received from a total of 12 sent out.

Analysis took place in two major phases. Initial findings were recorded through review of all interview, focus groups, and e-mail notes. A focus group was held with a group of soccer parents to review the initial findings from the interviews, focus group, and e-mail questionnaires. The group was used to validate findings, raise questions, and discuss meaning of some observations. The second phase was a more systematic review of interviews. The more open questions were analyzed using common generalized themes that were developed with the co-author. Some of the respondents stated their points more strongly or they made a single point several times for emphasis; this was also recorded. The table that resulted from this analysis (Table 2.1) is the central piece for the results section below. The quotations from the interviews are used to support the generalized themes.

The analysis also incorporates participant observations made over time by Donna Williams. She has been involved with her daughters in competitive sports since 2004. This knowledge of the competitive sports environments of volleyball, hockey, and soccer informs the work included in this study.

The Analysis: Spaces of Family Life and Youth Competitive Sport

Families who participated in this study shared their feelings and experiences about supporting youth in competitive sport. In analyzing these responses, a third intermediary space was discovered, the family vehicle, that serves to link sports and home spaces. The impacts of sports culture on the decision-making process of who occupies these spaces is

Table 2.1 Summary of Questionnaire/Interview Results

	Mothers												Fathers					
	1	2	3	4	5	6	7	8	9	10	11	12	1	2	3	4	5	6
Better relationship with athlete	x		x				x	x		x	x	x	x	x	x	x	x	xx
Learn from in-vehicle talking				x	x					x		x	x	x	x			
Positive family activity				x								x	x				x	
Compromised family dinners	xx		xx	x					x		xx			x	x			
Nutrition a concern	xx		xx					x	xx		x	x						
Little time with other children	x		x	x	xx	xx	x	x			x	x	x	x			x	
Loss of family time	x		x		xx	xx	x	x			x		x	x				x
Less time for spouse	xx				x	x	x			x	xx				x			xx
Less time for self	x		x	x				x	xx	x	x	x			x	x		xx
Cannot attend sport events		x	x	x	x	x		x	xx			x			x			
Stress	xx	x	x		xx			x	x	x	x							
Driving problematic			x	x	x	x		x	x		x							

xx used to signify emphasis through mentioning several times or strong emotional response.

Note: A group discussion was held with 6 soccer fathers. Their responses were important to the study, but are not reflected in this table.

important and is borne out in the results below. These decision-making processes are seen to be shaped by the emotions of parents as they negotiate between sports culture expectations and their own priorities in providing a positive environment in which their children can develop social and physical skills. The emotions identified here are not associated with those around the competition itself such as the highs and lows of winning and losing. Sports culture and emotions are found to influence how mothers and fathers decide where to be and what roles to assume in sports, vehicle, and home spaces. While these decisions are ultimately made at the family level, trends emerge in this investigation that show gendering of spaces occurs.

Concepts introduced earlier including 'good sport mother', 'good sport father', and driving behaviour are used in the following analysis. Evidence is also found to support that mothers experience a higher degree of conflict between the roles of being a 'good mother' and a 'good sport mother' than do fathers with their corresponding roles. This conflict or discomfort is often reflected in the gendering of sports, vehicle, and home spaces.

Sports Culture

Aspects of sports culture and its importance in influencing parental behaviours in general were presented earlier and these are reflected in the responses of study participants. Answers to interview questions and participant observations indicate that sports culture exists and that it is an important element influencing family decisions to support youth competitive athletes and family geographies. The interview responses are summarized in Table 2.1 and illustrate the emotions and gendering that are evident in parents' decisions. Sports are seen to offer benefits to youth and many parents feel these are worth the investments of time.

Supporting youth athletes can be costly in terms of time. Table 2.2 presents the average weekly time

Table 2.2 Average Weekly Time Commitment by Sport Per Child

Sport	No. of Months in season	Total Weeks	Total Hrs/ Season [a]	Avg Hrs/ Week [b]	Net Family Hrs [c]	% Net Hrs on Sport [d]
Hockey	7	28	414	14.8	72	20.5
Volleyball	8	32	854	26.7	72	37.1
Winter Soccer	7	28	107	3.8	72	5.3

Notes:
[a] Total Hrs/Season is calculated from interview results and represents the average reported.
[b] Avg Hrs/Week can be misleading since tournament play can require 48 to 72 hours for a weekend tournament, hence the time commitment can vary dramatically week to week.
[c] Net Family Hrs is calculated by subtracting 40 working hours and 56 sleeping hours per week from the total number of hours in a week (168).
[d] % Net Hrs on Sport assumes the same parent is attending all practices, games, and tournaments.

commitment for each sport indicating that nearly 40 per cent of family time can be required for volleyball. These times were calculated from averaging data received in the interviews. This commitment is even higher in cases where one athlete competes in multiple sports or more than one child from the same family participates in a competitive sport.

Although participants did not specifically identify it as a requirement to be a good sport parent, a family had to have the ability to afford the financial costs and time to support competitive athletes. There were comments from many respondents in the current study about the high cost in dollars of participating in sports. Parents commented that tournaments served as personal or mini family vacations. Sean commented that, 'In regards to our other children, it has meant less time on their activities, however, we try to integrate them into these soccer weekends by using them as mini-vacations'. This is supported by Faith's comments, 'There are no extras, it all goes to tournaments'.

Respondents reiterated in various ways the benefits from competitive sports and most seemed to feel they were worth the investment of time and money, and many, but not everyone, also supported the changes in family interaction that resulted. Several parents mentioned that they felt their children benefitted from competitive sports by learning how to set goals, be better organized, work on a team, build self confidence, make friends, learn how to win and lose, and improve physical fitness. These reflect benefits cited in sports psychology literature and were the reasons why they chose to continue to support their children in sports and accept sports' cultural norms and the financial and time costs.

Sports and individual teams develop particular sets of expectations and acceptable and/or desirable behaviours for parents and athletes. Many of the expectations come to exist as normative behaviour and have been developed over time. In short, a culture develops that is frequently particular to a sport and individual teams, especially if the team is a competitive one. The fact that these expectations or behaviours may not be formalized in writing does not mean that they are trivial or that they can be ignored by parents or athletes. Sometimes the athlete has to suffer the consequences of an infringement even though he/she has little or no control over the infraction or team expectations.

Different parents and even parents of the same athlete can demonstrate different approaches to supporting the athlete while attempting to meet the expectations of the sport culture. It is important to recognize that some parents learn what is required of them and fulfill these responsibilities without question. Others, however, accept, endure, and/or cope with elements of competitive sports culture that as independent individuals they might not. One of the major motivators for accepting or enduring some aspects of the cultural expectations is the belief that the perceived benefits for the child to belong to the sport are great and justify the investments and adjustments to family life.

Sports, therefore, can and do shape the personal lives of those involved directly or indirectly. This was captured well by Janet, who felt that, 'The structure of our life [due to sport] becomes a family value', thus supporting Coakley and Donnelly's (2004) assertion that, '. . . relationships between family members are nurtured and played out during sport activities or in conversation about these activities' (20).

Sport Spaces

At the competitive level, these cultures are important and are likely understood by parents and athletes alike. However, there can be much self-regulating between parents through subtle or obvious comments to encourage others to follow or respect these expectations. Attendance at practices, games, and tournaments is one such expectation and this study found many examples when this was reinforced by comments from other parents. As Linda commented,

> 'I often attend [other daughter's] soccer games and when I haven't been around [at the pool for a swim meet] I can't believe how many people comment that I haven't been around. I find that I get a bit defensive and feel I have to explain . . .'

Sometimes the expectations are clear and well understood. For instance several respondents reported that their 12- and 13-year-old athletes were informed that they had to arrive an extra 15 minutes early and the coach added that if they were late for games or practices they would be benched for a shift of the following game. This created great anxiety amongst some athletes who had little or no control over the time they arrived, which was translated to the parents and added to their stress.

The strongest positive expression of feelings related to the importance of the sport in terms of the impact on the family. Specifically, the competitive sport experience was seen to strengthen the relationship between the parents and the athlete. Most parents identified this as one of the most tangible positive benefits that resulted from their investment of time, and of being together in the spaces and places of competition. Mike expressed what many of the fathers felt when he commented, 'Travelling together has given us a special bond, where we have so much in common'.

While parents saw benefits from their involvement with their children, they also expressed the importance they saw in having the opportunity to see their child mature through the social networks they developed and the resulting friendships. This led to feelings of gratification. Parents saw themselves as fulfilling their roles as good parents, in broad social terms, by preparing their child for life, and keeping them from harm. Parents also frequently noted that they themselves had formed social connections with other parents that made the entire experience enjoyable for themselves at an individual level. Anna, however, felt that the social connections with other parents were negative. She felt the connections were necessary for her child to participate fully in the sport as she did not drive. This led to her feeling she had to 'perform' for the other parents and stay on good terms to garner drives to tournaments out of town and clearly supports the value of control theory in understanding some parents' behaviours.

Good Sport Parents

A good sport parent, as indicated by information gleaned from study participants, is at the most macro level someone who is very concerned with providing for the needs of the youth athlete. In general, the concept of a good sport parent has its roots in the popular interpretation of the 'Soccer Mom', a term that was actually coined in the 1996 US presidential election to identify a specific voting demographic that was seen to consist of white, female, middle-class, middle-aged, voters who drove mini-vans (Marotta 2005). However, it has come to represent the fact that 'Soccer Moms take their children to practice and games, no matter what else might be on their schedules. They make their children's priorities their priorities . . .' (Marotta 2005: 28). Respondents echoed these ideas many times as characteristic of being a good sport parent, but other expectations of 'good sport parents' emerge too.

Analyzing responses suggested that being a 'good sport mother' or 'good sport father' included a few common and distinct requirements. These include attending games, practices, and tournaments regularly, if not frequently. Knowledge of team standings, the upcoming league and tournament schedules, and athlete statistics also served as criteria in the eyes of many respondents. Related to this was the expectation that parents deliver the athlete on time and this necessitates driving. Driving behaviours, however, were seen to differ between mothers and fathers and this is discussed in the following section.

This investigation determined that as women and men occupy the required parental spaces associated with organized sports they frequently have different emotions about occupying these spaces. This strongly suggests that there are different pressures on or expectations of mothers versus fathers and this leads to the development and use of the terms 'good sport mother' and 'good sport father'. Mothers expressed experiencing more stress and guilt about the mutually exclusive conflicts of missing sport events or losing non-sport family time than did fathers. This illustrates that tensions exist between the roles of being a 'good sport mother' and that of being a 'good mother'. The following statements exemplify the stress of four mothers:

> 'I can't leave my other daughter alone so I have no time to see [sporting daughter] play. I find we're split down the middle' (Lucy);
>
> 'I do the sports with [daughter] and [husband] does the sports with [son] . . . I hardly connect with [son]' (Janet);
>
> 'Mine and [husband's] relationship . . . lots of passing by in the hall, not a lot of family time together all four [of us]' (Faith);
>
> 'The impacts on other relationships in the family are mostly negative and need to be balanced with time with other children . . . leaves even less time for myself . . .' (Linda).

Again, comments like these were absent from the fathers in the study even though they were asked the same set of questions and given the same prompts. The absence of similar concerns or emotions about family stresses from the fathers suggests that the roles of being a 'good sport father' and a 'good father' are more similar to each other and different from those expected from and by mothers in their distinct roles.

It is important to mention that parents expressed some negative emotions related to many aspects of their activities surrounding sport and their child: fear of doing the wrong thing for their children; worry about others in the family being ignored; anger at the way their lives were lived because of sport; and concern about damaged and/or broken relationships with their spouse and other children. Evidence of this is presented in Table 2.1. Parents also articulated strong positive emotions with respect to their need to support their children in their chosen sports; this was regularly reported to overshadow any and all negative aspects of the participation in, and travel to, their sport activities. Dominique's comment is typical of how many parents summed up how they felt. She said that she, '. . . would do anything to support [her daughter] in volleyball. She loves it and it's so good for her'. For these respondents there was enough justification of the positive to negate the detrimental effects of the travel for sport on the family. These justifications, however, became unconvincing during evaluation of the interviews. In some ways, the accepted benefits seemed to take on almost myth-like status, and were not questioned by the parents. However parents, and especially mothers, found themselves in conflict over their roles as 'good mother' and as 'good sport mother', and it is this conflicted state that seems to be at the root of the emotional responses for them.

In sport spaces, this conflict was quite clear at times. For instance, some parents in the study maintained a mental attendance sheet for other parents and made subtle and not so subtle remarks to the 'delinquent' parents about missing a certain number of games in a row. Anna mentioned that she felt she could not miss more than three games in a row without jeopardizing benefits of belonging to the team. Lynn shared that parents had commented on the long length of time since they last saw her at a volleyball game. Attendance of parents at games and practices is, it seems, expected by other parents and is part of what constitutes a 'good sport parent'. While some mothers reported this stress about maintaining attendance, not one father mentioned it. This observation of peer or cultural pressure clearly reflects those identified by Marotta (2005), Doucet, (2006) and Duncan and Smith (2002). It is unclear if the absence of this concern from the fathers means that they attended more regularly or whether they did not feel stress or guilt from being away from the home space in particular. Again this hints at similarities between the roles of 'good sport father' and 'good father'.

Some respondents divided driving responsibilities with their spouses. Fathers tended to do more and made comments that seemed to indicate they felt they were fulfilling a 'good sport father' expectation. For example, Claude expressed much gratification from having the sport in common with his daughter as this strengthened his relationship with her. This comment supports Coakley and Donnelly's (2004) assertion that many men use sports as a conversation opener and many relationships with men often revolve around sports. In situations where the father did the driving this frequently resulted in the mother remaining at home to attend to the rest of the family. While this decision allowed mothers to fulfill their 'good mother' role at home it simultaneously induced stress and guilt for not attending the tournaments or games and supporting their youth athletes.

The idea of conflict in roles of 'good sports mothers' with their athletes at the fields, arenas, and gyms and their roles as 'good mothers' at home is related to the distance of travel. The observation of fewer mothers than fathers attending distant tournaments, discussed further below, suggests the existence of an underlying cultural and/or emotional force at play regarding driving long distances.

In short, sport culture is strong and an important influence on parents' emotions and decisions to participate in sport spaces. Therefore, it also played a role in the gendering of sport spaces as mothers and fathers made decisions in which the demands of driving and travel played a role. Differences were also found between mothers and fathers with respect to emotions and decisions regarding vehicle and home spaces.

Family Vehicle Space

Time spent traveling to practices and games in the family vehicle was important from virtually all of the parents' perspectives. The role of driver was critical to participation. Driving is a commitment that the parents must make to ensure the success of their children in their chosen sport. The positive aspects of travelling together had to do with communications and relationship building with the child. This happened irrespective of the amount of distance covered. There were many short in-town trips to practices and games, and fewer longer trips to tournaments. Only mothers talked about the negative aspects of driving or were concerned with long distances travelled for tournaments.

For fathers and mothers alike, the car represented a positive space. This was a place to have one-on-one time with the child. Many parents expressed their enjoyment in building and maintaining their relationship with the child while in the car. They had the child's undivided attention and the child had theirs. Many parents also valued the information they overheard while driving. Some felt that they learned a lot about what their children were concerned with, as the child apparently forgot that a parent was in the car when talking with team mates. One mother of three expressed her regret that she cannot always go to the tournaments with one or the other of her two competitive athletes. 'The kids [at home] aren't talking but in the car they talk and missing it means that I miss important information' (Sarah).

Fathers more often noted that they talked about the sport in the car. Claude, father of two competitive athletes, noted that he and his daughter talked about the game in the car. He had a chance to share his 'wisdom' as his daughter did not take losing well and he would help her to understand the loss and that losing was okay. Several men in this study related experiences such as Claude's, where they found that they had something to talk to their teenaged daughters about. This was a very positive aspect to sports for these fathers.

Time in the car was seen initially only from this positive angle. While carrying out the analysis, however, an unexpected trend emerged when investigating out-of-town hockey and volleyball travel. The attendance of mothers was noted to be more common for hockey tournaments than for volleyball. Hockey tournaments were, on average, 200 to 400 kilometres from home while the vast majority of those for volleyball were 500 kilometres away and sometimes as far away as 800 kilometres. Although a specific question was not asked about driving, 7 of the 12 mothers said that driving long distances was problematic or mentioned their dislike of driving. Some did not like driving on multi-lane expressways, while others did not like driving at night, and still others mentioned not wanting to drive without another adult in the vehicle.

Interviews with fathers did not reveal concern amongst them about driving to distant tournaments, regardless of the distance or whether their spouses attended or stayed behind. Concerns about the lack of female chaperones for some volleyball tournaments were voiced during some discussions and serve as another indication, and consequence, of fathers' dominance in driving when travelling long distances.

Fathers made comments that indicated they felt they were fulfilling a 'good sport father' expectation with the driving by getting their children where they were supposed to be. At the same time, they felt that they had better relationships with their children because of the discussions in the car. The sport, as noted previously, also gave them something in common that they could talk about. This again supports the statement of Coakley and Donnelly (2004) that sports are important to men and fathers in building relationships. Overall there were no negative comments from fathers about driving.

Mothers and fathers alike felt that attending distant tournaments was important. Jane and her husband opted to split tournaments. He would drive to one tournament and then she would look for a drive with another parent to the next one. Jane did not drive on four-lane highways. If she could not find a ride, Bob would go again. All mothers expressed their desire to see their daughters play their sports. Mothers, however, more frequently decided to stay at home to care for others rather than drive to distant tournaments. While this decision allowed mothers to fulfill their 'good mother' role at home it simultaneously induced stress and guilt for not attending the tournament and, therefore, mothers reported feeling they were not supporting their youth athlete. It seems that mothers felt guilt no matter which space they chose to occupy, but concerns about driving evidently played a significant role in the decision to stay or to go.

The family vehicle represents a critical resource for the sport space for getting children to their sport, but it is also an extension of the home space in that it provides opportunities for parenting. Therefore, it seems to represent the intersection of the two worlds. It serves dual roles since it is a place for preparing the athlete for the game or practice and for reviewing the time at play on the way home. It is also a place for parenting activities like demonstrating compassion, allowing communication to occur, and caring for the child as well as being a place to make sure nutritional needs are met. However, it is not a 'home' space in at least two important senses. Firstly, when the family vehicle is being used to ferry athletes to and from sporting events, only part of the family is expected to be present. Other family members are absent and in the case of long drives frequently one parent is missing and that parent is usually the mother. Secondly, the car or van is much more confined than the home and this often results in providing an environment that is more conducive to interaction and communication. However, it seems to be a space that is more often considered to be the domain of fathers, especially on long drives. Home spaces are traditionally seen to be the places where many more general parenting activities occur and are frequently the realm of mothers, and respondents' answers and comments support these statements.

Home Spaces

Home spaces are predominantly defined in this study as locations to provide meals and nutrition, and to nurture relationships within the family. The home space as it pertains to all of these things was a concern for most mothers. Some fathers expressed concern over relationships and time as a family, but only two fathers mentioned meals and none talked of nutritional concerns. Conflict for mothers over sport space and home space is evident in their comments.

Faith captured the feeling expressed by many mothers well, indicating the frustration that is felt. 'Who's available is critical. We end up with crappy meals. I'm stressed trying to get home from work and stressed getting them to practices and games on time and I feel bad about the poor meals.' Mothers spoke uniquely of nutritional concerns for themselves and their family. This supports the findings of Roy and Petitpas (2008) and those reported by Proudfoot (2009).

Time was the major factor in mothers' abilities to provide nutritious meals to their children as mothers often had three competing roles; wage earner, sports mom, and mom. Some participants commented on the frustration of coming home from work with little time before leaving for a game or practice. Two mothers told of times when they found that the athletes themselves had not eaten as instructed, leaving the mother feeling angry, nor was anything prepared for others in the household. The mothers were left with the difficulty of finding something light that would not make the athlete feel ill while playing and then felt guilt over leaving the other child/children in the household to fend for themselves. This presents a clear conflict of the 'good mother' providing a meal for the family and the 'good sports mother' getting the athlete ready, having eaten the correct food, and to the event on time. Virtually all mothers mentioned that this type of time pressure caused them a great deal of stress.

Only two fathers also commented about meal preparation and eating together. Mike said, 'The travelling is probably the easiest part and being the main driver for one, there is a fair amount of guilt about helping less with chores, etc., and I know how important it is to Chris [wife] to have dinner as a family . . .'. The other father simply noted that meals together are difficult. Fathers do not talk of the stress around sports with their daughters but some do mention that there is less time with their spouses and for themselves.

A second area of concern related to home space and the impact of the sports on this space is the area of relationships in the family. Half of the fathers mentioned concern about their relationship with other children in the family and virtually all mothers expressed negative feelings about other children left with less attention and/or more responsibility at home. Women also mentioned more frequently that an outcome of a lack of time was a compromised relationship with their spouses, with two respondents actually fearing what would happen to the relationship when the sports were over. As Janet commented, 'When we don't have hockey anymore, what will we say when we look across the table at each other?'

At the same time all fathers and a little over half of the mothers said that they had a better relationship with the athletes. Much of this relationship building is linked to time in the family vehicle mentioned above and the common interest of sport. As Mike said so well,

> 'Maybe one of the hardest things is due to the competitive nature of the sports—if it were a minor interest, there would be less interest from us parents. With the girls taking it so seriously, it becomes a major topic of conversation. Not just the sport, but the club politics, parent politics, team politics. It is almost all-consuming at times and really makes you want to give your head a shake. From the immediate bond with the daughter, it's another minor source of guilt that we have this thing in common, that Mum has a hard time entering'.

Most mothers and some fathers expressed feelings of guilt because they felt that they paid less attention to children at home. These feelings arose from not seeing as much of other children, knowing that more time and money was being spent on the athlete and remorse over not knowing the other children as well. As one mother expressed for her family, 'We split the kids down the middle because [the husband is the daughter's] coach so there's not a lot of time together . . . it's a little divisive' (Lucy).

Family relationships on the whole and in the formulation of the family were not seen to be positive because of demands of supporting youth in competitive sports. Over half of the respondents related a lesser quality in family life because of the time devoted to one or more child's competitive sports. Of these, the vast majority who mentioned lower quality of family life were mothers.

Mothers' concern about loss of family time and difficulty with maintaining relationships in the family is linked to the contradiction that exists between the different spaces within which mothers operate. They move from the home and school to the field, arena, and gym, and find that there are different expectations in the various locations. Marsha Marotta supports the significance of the expectations of behaviour by mothers in their various environments, saying,

> Mothers are classified as 'good' or 'bad' based on the practices they engage in and the spaces in which they do so. The idea of the 'good' mother is deployed through material and discursive spaces in order to mobilize subjectivities that are socially adapted and useful—keeping the attention of mothers focused on children and their needs, wants, and activities, which serve the needs, wants, and activities of Western culture with its hierarchies by sex, race, class, and so on. (Marotta 2005: 16)

The simple idea of bringing the incorrect food to a tournament for snacks left Lynn feeling as if she failed her daughter. An innocent comment was made by an athlete to her mother, saying, 'It's okay if you don't have a good snack for me, because [so-and-so's] mom will have one', left one mother with the burden of guilt.

As a 'good sport mother' a mother needs to focus attention on a single member of the family who has high demands. The athlete needs parents who ensure she can be at the practice or game at the appointed time, that she has eaten correctly, and has the appropriate equipment. The 'good sport mother' must pay the proper attention to her child playing the sport in the form of attending games and even practices. These roles put mothers in a position of conflict which manifests in stress and guilt. Stress comes from attempting to complete all of these tasks for family and athlete in limited time and often over long distances. Guilt is felt over not being there for the athlete or not being there for others in the family. Virtually all mothers spoke of both of these emotions whereas not one father mentioned them.

'Good sports fathers' likely are closer in definition to 'good fathers' as discussed earlier. The fathers all felt they had better relationships with their daughters. Several did mention less time for other children and loss of family time, but this did not bring the same level of emotional outpouring as it did for women. This generally more positive attitude towards the sport space is related to their greater affinity with sports as discussed earlier and their apparent comfort with driving. These combine to make being a 'good sport parent' more attainable for the men in this study.

Conclusions

There are, naturally, limitations to an exploratory study such as this. Firstly, there is inherent bias introduced by the researcher conducting the study. She came into the study with a pre-existing discomfort with respect to the norms associated with the sport cultures within which she was participating with her children. The realms of emotional geographies allow these emotions or passions to be recognized within geography rather than be hidden from view (Bondi et al. 2005: 1). These emotions are the reason that this research exists. Secondly, this research only explores three team sports and not the diversity of individual and team sports that are available today. Introduction of other sports would provide much broader information leading to better generalizations about the impacts of travel with competitive athletes on families. Next, this study focused on one suburban community in the west end of Ottawa, Ontario. This is a large-sized city with a population of 812,129 (2006 Census of Population). Research has indicated that size of city is connected to development of elite athletes (Cote et al. 2006). Finally, these were female teams only and there was little in terms of ethnic diversity represented on these teams. Even with these limitations, this investigation uncovered interesting links between supporting athletes and the creation and recreation of family geographies.

The findings are convincing that sports culture and emotional geography play significant roles in shaping the geographies of families with youth competitive athletes. The family geographies related to competitive sports consist of three spaces including sports, vehicle, and home spaces. While families make decisions based on their needs and the availability of mothers and fathers, gendering of the three spaces was observed when considering the combined input of the participant families.

One interesting finding was that different attitudes about driving athletes to distant tournaments exist between mothers and fathers. Mothers admitted to having concerns about long-distance driving, driving alone on multi-lane highways and in unfamiliar areas. Fathers expressed no such concerns, and were found to be disproportionately represented at tournaments that were more than 500 kilometres from home.

Sports culture and the desire of parents to be seen providing sporting opportunities to their youth athletes were important factors in shaping the behaviour of parents in the three spaces. Parents judging other parents' attendance at games and practices and proper pre-game nutrition were offered as examples to illustrate how parents expected each other to act within the sport culture.

Sports and sport culture reflect male interests in society at large (Coakley 2009; de Lench 2006; Lenskyj 2003) and this was supported in this investigation in two ways. Firstly, fathers did not report any stress or conflict between their roles as 'good father' and 'good sport father'. Secondly, mothers clearly indicated feeling conflicted between their dual roles of 'good mother' and 'good sports mother'. This was evident as they had to choose between occupying the three spaces and they reported feeling much anxiety and guilt regardless of their choice. The conflict that mothers reported feeling suggests that sport culture and its demands on parents is more conducive to society's expectations of men than women. This supports Marotta's (2005) argument that mothers are subject to far more limitations in terms of occupying space, or 'MotherSpace', than are fathers.

The hidden social courts of motherhood and fatherhood within sports culture and emotional geography allow for an understanding of parental decisions and the family geographies that result when parents support youth competitive athletes. These family geographies result in gendering of the sports, vehicle, and home spaces and strongly suggest that there is much overlap for fathers between their 'good father' and 'good sport father' roles.

Acknowledgements

The authors extend thanks to all the participants for their time and contributions. In addition, the authors wish to thank Ryan Bullock, Heather Hall, Nicole Yantzi, and Bonnie Hallman for insightful comments and suggestions on earlier drafts of this chapter. Nonetheless, any errors or omissions remain the responsibility of the authors.

Questions for Further Thought

1. Using experiences from your life and/or observations of other families, how do parenting behaviours and expectations of parents vary as families operate within their family geographies?
2. Using a feminist, structuration theory, or post-modern framework, offer an informed explanation of how and why sports culture impacts mothers and fathers differently.
3. Support or refute Marsha Marotta's (2005) concept of 'MotherSpace' using specific examples from your past or present family geography.
4. Using clear examples, explain ways that

family geographies can be modified and still offer the benefits associated with involving youth in activities like sports while drastically reducing the size of their family's ecological footprint.

Further Reading

Andrews, G.J., M.I. Sudwell, and A.C. Sparkes. 2005. 'Towards a geography of fitness: an ethnographic case study of the gym in British bodybuilding culture'. *Social Science & Medicine* 36, 4: 877–91.

Bale, J. 2003. *Sports Geography.* 2nd ed. New York: Taylor Francis Group, Routledge.

Tonts, M., and K. Atherley. 2005. 'Rural restructuring and the changing geography of competitive sport'. *Australian Geographer* 36, 2: 125–44.

Endnotes

1. Pearson, A. 2002. *I Don't Know How She Does It: The Life of Kate Reddy, Working Mother.* New York: Knopf.

References

Anderson, J.C., J.B. Funk, R. Elliot, and P. Hull Smith. 2003. 'Parental support and pressure and children's extracurricular activities: relationships with amount of involvement and affective experience of participation'. *Applied Developmental Psychology* 24: 241–57.

Andrews, G.J., M.I. Sudwell, and A.C. Sparkes. 2005. 'Towards a geography of fitness: an ethnographic case study of the gym in British bodybuilding culture'. *Social Sciences & Medicine* 60, 4: 877–91.

Bale, J. 2003. *Sports Geography.* 2nd ed. New York: Taylor Francis Group, Routledge.

Baxter, J.S., A.S.R. Manstead, S.G. Stradling, K.A. Campbell, J.T. Reason, and D. Parker. 1990. 'Social facilitation and driver behaviour'. *British Journal of Psychology* 81: 351–60.

Bernard, J. 1981. *The Female World.* New York: Free Press.

Bernhoft, I.M., and G. Carstensen. 2007. 'Preferences and behaviour of pedestrians and cyclists by age and gender'. *Transportation Research Part F* 11: 83–95.

Bergdahl, J. 2005. 'Sex Differences in Attitudes Toward Driving: A Survey'. *The Social Science Journal* 42: 595–601.

Bondi, L., J. Davidson, and M. Smith. 2005. 'Introduction: geography's "emotional turn"'. Pp. 1–16 in L. Bondi and M. Smith, eds. *Emotional Geographies.* Burlington, VT, US: Ashgate Publishing Company.

Coakley, J. 2009. *Sports in Society: Issues and controversies (10th ed.).* New York: McGraw Hill.

Coakley, J., and P. Donnelly. 2004. *Sports in Society: Issues and Controversies. 1st Canadian edn.* Toronto: McGraw-Hill Ryerson.

Côté, J., D.J. MacDonald, J. Baker, and B. Abernethy. 2006. 'When "where" is more important than "when": birthplace and birthdate effects on the achievement of sporting expertise'. *Journal of Sports Sciences* 24(10), October: 1065–73.

Côté, J., and J. Fraser-Thomas. 2007. 'Chapter 11: Youth involvement in sport'. Pp. 266–94 in P. Crocker, ed. *Sport Psychology: A Canadian Perspective.* Toronto: Pearson/Prentice Hall.

de Lench, B. 2006. *Home Team Advantage: The Critical Role of Mothers in Youth Sports.* New York: Harper Collins.

Donnelly, P., and J. Harvey. 1999. 'Class and gender: intersections in sport and physical activity'. Pp. 40–64 in P. White and K. Young, eds. *Sport and Gender in Canada.* Don Mills: Oxford University Press.

Doucet, A. 2006. *Do Men Mother?* Toronto: University of Toronto Press.

Duncan, S., and D. Smith. 2002. 'Family geographies and gender cultures'. *Social Policy & Society* 1, 1 : 21–34.

Hardy, S., and C. Wiedmer. 2005. 'Introduction: 'Spaces of motherhood'. Pp. 1–11 in S. Hardy and C. Wiedmer, eds. *Motherhood and Space: Configurations of the Maternal Through Politics, Home and the Body.* New York: Palgrave Macmillan.

Harré, N., J. Field, and B. Kirkwood. 1996. 'Gender differences and areas of common concern in the driving behaviors and attitudes of adolescents'. *Journal of Safety Research* 27, 3: 163–73.

Hayward, G., A. Diduck, and B. Mitchell. 2007. 'Social learning outcomes in the Red River Floodway Environmental Assessment'. *Environmental Practice* 9, 4: 239–50.

Heenan, C. 2005. 'Chapter 11: "Looking in the fridge for feelings": the gendered psychodynamics of consumer culture'. Pp.147–60 in L. Bondi and M. Smith, eds. *Emotional Geographies.* Burlington, VT, US: Ashgate Publishing Company.

Horn, T., and J. Horn. 2007. 'Family influences on childresn's sport and physical activity participation, behaviour, and psychological responses'. Pp. 685–711 in G. Tenenbaum and R.C. Eklund, eds. *Handbook of Sport Psychology 3rd ed.* Hoboken, New Jersey: John Wiley and Sons Inc.

Kostyniuk, L.P., and L.J. Molnar. 2008. 'Self-regulatory driving practices among older adults: health, age and sex effects'. *Journal of Accident Analysis and Prevention*: in press, retrieved on June 28, 2008.

Leckie, G.J. 1996. ' "They never trusted me to drive": farm girls and the gender relations of agricultural information'. *Gender, Place & Culture: A Journal of Feminist Geography* 3, 3: 309–28.

Lenskyj, H. 2003. *Out on the Field: Gender, Sport and Sexualities.* Toronto: The Women's Press.

Marrota, M. 2005. 'MotherSpace: disciplining through the material and discursive'. Pp. 15–33 in S. Hardy and C. Wiedmer, eds. *Motherhood and Space: Configurations of the Maternal Through Politics, Home, and the Body.* New York: Palgrave MacMillan.

Özkan, T., and T. Lajunen. 2006. 'What causes the differences in driving between young men and women? The effects of gender roles and sex on young drivers' behaviour and self-assessments of skills'. *Transportation Research Part F* 9: 269–77.

Proudfoot, S. 'Want good, healthy children? Then have supper with them: family dinners encourage better eating, and they aren't a thing of the past: study' *Ottawa Citizen*, May 11, 2009, retrieved May 12 on-line.

Rabinow, P. 1984. *The Foucault Reader*. New York: Pantheon Books.

Roy, B., and J. Petitpas. 2008. 'Rediscovering the family meal', *Contemporary Family Trends*, Ottawa: The Vanier Institute of the Family, June, 2008.

Shotwell, A. 2008. 'Gendering one another', *Colloquium Department of English, Laurentian University*, Sudbury, Ontario, January 25, 2008.

Stevens, D.E., K.L. Gammage, and L. Waddell. 2007. 'Female athletes and gender issues'. Pp. 315–43 in P. Crocker, ed. *Sport Psychology: A Canadian Perspective.* Toronto: Pearson-Prentice Hall.

Sundström, A. 2007. 'Self-assessment of driving skill—a review from measurement perspective', *Transportation Research Part F* 11:1–9.

Tonts, M., and K. Atherley. 2005. 'Rural restructuring and the changing geography of competitive sport'. *Australian Geographer* 36, 2: 125–44.

Tuan, Y.-F. 1974. *Topophilia: A Study of Environmental Perception, Attitudes, and Values.* Englewood Cliffs, NJ: Prentice Hall.

Valentine, G. 2001. *Social Geographies: Space & Society.* Harlow, England: Pearson Education Ltd.

Varpalota, A. 1995. 'Sport, leisure and the adolescent girl: single sex vs. co-ed?'. *Canadian Woman Studies* 15, 4:30–34.

Williams, D. 1987. *Perception of neighbourhood; gender differences in the mental mapping of urban and suburban communities.* Ottawa: Carleton University.

Williams, D., D. O'Brien, and R.E. Kramers. 2003. 'The Atlas of Canada web mapping: the user counts'. *Cartographic Perspectives* 44, (Winter): 8–28.

Wuerth, S., M.J. Lee, and D. Alfermann. 2003. 'Parental involvement and athletes' career in youth sport'. *Psychology of Sport and Exercise* 5: 21–33.

Yantzi, N., and M. Rosenberg. 2008. 'The contested meanings of home for women caring for children with long-term care needs in Ontario, Canada'. *Gender, Place and Culture* 15, 3: 301–15.

Chapter Three

'Mom, Dad, can I walk to school?': Road Risk in an Urban Family Context

Marie-Soleil Cloutier

Introduction

The various conceptualizations of childhood throughout history have led to different ways of studying children in their social and built environments. Starting from a purely biological definition (i.e., their age), the current debate about childhood can be viewed under two conceptual frameworks: the deterministic view of children as 'adults in the making' or the sociological view of children as 'social actors' (Holloway and Valentine 2000). While the first one considers childhood as a transitional phase where children have to reach some physical and social maturity to become fully integrated into the society, the second perspective regards children as members of a subculture, a minority or even a group structuring every society (Torres Michel 2007). In this context, the expanding area of research that is children's geography raises issues about the difference that place makes for children, and about the socio-spatial marginalization they can experience when confronted with a world made primarily for and by adults. The study of children's interactions within everyday life spaces becomes more and more important for geographers, and the application for this body of theoretical and empirical research is relevant to the definition and application of family geography as an analytical category.

Starting with the rather modern conceptualization of childhood as a social construct, the purpose of this chapter is to explore how family geography is influenced by child-pedestrian risk in an urban school context. As Ward had already written in 1978 (118), 'The street life of the city has been slowly whittled away to make more room for the motor car', which left pedestrian risk as one form of marginalization experienced by urban, and rural, children. To consider this issue as part of the reflection on family geography, this chapter is divided in three parts: first, the geographic nature of child-pedestrian safety in terms of factual and perceived risk will be overviewed; second, the results of a case study in Montreal, Quebec will be described; and finally, the implications of these results for family geography and prevention initiatives will be discussed.

Child Pedestrian Safety as a Geographical Problem

Everybody would agree on the rights of children to walk safely around their neighbourhood or to bike or walk to go to school. These daily activities should take place without parents' constant fear of traffic. But reality is never that simple and interactions with vehicles are frequent events in a child's routine outdoor experiences, especially in cities. That being said, major reductions have occurred in rates of injury and numbers of victims of road accidents over the past decades. For example, in the United States, the rate of traffic-related accidents for pedestrians aged 14 and under declined by 49 per cent for deaths and by 36 per cent for injuries from 1990 to 2000 (Hanley et al. 2002). Nonetheless, pedestrian injuries are still a leading cause of illness among children around the world, especially in industrialized countries. In Canada, between 1994 and 2003, pedestrian accidents were the first cause of injury-related death for children aged 5 to 9 years old (18 per cent) and the second one for children aged 10 to 14 years old (14 per cent) (SafeKids Canada 2007). In Montreal, for the same period, this public health concern touched more than 2,000 children, 12 per cent being severely injured, not to mention the physical and psychological consequences for the uninjured members of their families. When a child is injured, the whole family is affected.

But what are the main **risk factors** for child-pedestrian injuries? Most research on child pedestrians starts with a description of the demographic characteristics of the victims. Accordingly, the literature on risk factors for pedestrian injury has not changed much for the last twenty years. Boys are more at risk, as well as children between 5 and 14 years old, living in poor families (Adams et al. 2005; DiMaggio and Durkin 2002; Laflamme and Diderichsen 2000; Sunderland 1984). The experience of space is different for these families, which leaves them particularly vulnerable since their exposure to risk is twofold: children in poor families often walk more (parents are less motorized); and they tend to live in neighbourhoods with more busy streets (commercial and industrial zones) with higher traffic flows. In fact, the children who are most at risk from the presence of vehicles in the city are those who have least access to their benefits (Ward 1978).

In addition to the epidemiological description of the victims, research in psychology has tried to understand the cognitive processes involved in the different tasks required to be a 'safe' pedestrian. This research field has exposed the complexity of such tasks and related them to the cognitive and physical development of children at different ages. Crossing a street safely requires one to pay attention to traffic, to judge vehicle speed, and to make a decision to cross or not through an analysis of these components. While adults do these tasks without noticing, children cannot reliably do so, at least in their early childhood (i.e., before 12 years old). This type of study, first launched in the 1970s, profoundly changed the way child pedestrians were seen. From that point, it was established that young children, confronted with daily urban traffic, do not have all the cognitive abilities to choose the safest action to take (Assailly 1997; Connelly et al. 1998; Schieber and Thompson 1996; Stevenson and Sleet 1997; Underwood et al. 2007). Despite this new paradigm, the established approach to children in urban transport and environment planning relied on behaviour control and modification. These approaches see the child as an object to be manipulated to fit into the adult world (Davis and Jones 1996). Such positions make planners unable to respond to children's and families' needs in terms of daily movement throughout the urban landscape. In addition, none of the research done so far has demonstrated that these strategies, based mainly on education interventions, have a positive effect on injury rates, nor on **active transport** choices.

Other studies have also analyzed socio-economic and environmental risk factors for their direct influence on risk (traffic) or their indirect influence on children's **exposure to risk** (deprivation, in-house density, time spent on the streets, etc.). This group of studies is the most geographically driven one, as the environment is considered as a risk factor in itself. Accordingly, a number of the studies link higher levels of traffic, higher residential density, and specific street configurations to greater numbers of victims (Braddock et al. 1994; Yiannakoulias et al. 2002). For example, MacPherson et al. (1998) and Posner et al. (2002) established a strong positive relationship between higher exposure and injury risk. These authors measured exposure differently—as the number of streets children cross to go to school or by the outside playing routine—but they had the same result, in addition to findings pointing out that poor children are at higher risk. Indeed, MacPherson et al. (1998) found that children who attend more deprived schools cross, on average, 50 per cent more streets a day. Other studies, such as the one by Roberts et al. (1995), found a 14-fold increase in the injury risk on streets with the highest traffic flow compared to streets with the lowest. Similarly, LaScala et al. (2004) revealed that annual numbers of injuries were greater in areas with greater traffic flow, but their work also found more children injured in areas with higher youth population densities, more unemployment, and fewer high-income households. As well, several other studies specifically pointed out socio-economically disadvantaged areas as 'spots' of higher accident risk (Dougherty et al. 1990; Graham et al. 2005; Hewson 2004). Furthermore, Laflamme and Diderichsen (2000) recall the vulnerability of poor families within their exhaustive literature review: for most types of traffic injuries (pedestrian, cyclist, and car occupant), mortality and morbidity are often higher among children from families with lower social positions (e.g., parents with blue collar jobs) as well as in more deprived socio-economic areas. This last conclusion is explained by Carstairs (2000) in his work on the complex dynamic between poverty and health risks such as road injury. His work is markedly relevant to the present case and again proves the geographical nature of this public health concern.

> Socio-economic position does not, of itself, explain health state, rather it represents a complex of living experience, living and working conditions, attitudes and social orientation, income, wealth, and assets for the individual. As well as these personal attributes, 'poor places' provide socially adverse environments that strike at the health status of even the non-poor inhabitants. (64)

Arising from the existing literature, the main hypothesis followed in this chapter is that the growing increase in road-injury risk associated with urban life is modifying families' experiences of space. In fact, reductions in child-pedestrian injuries and deaths are increasingly related, among other factors, to a decrease in the frequency that children walk and, thus, a decrease in exposure to traffic, particularly on the way to school (DiGuiseppi et al. 1997; Dixey 1998; Joshi and MacLean 1995; Sonkin et al. 2006). Public health officials and researchers refer to this phenomenon with concern because of the known health problems associated with a sedentary lifestyle (obesity, etc.) and because walking has a positive effect on children's physiological and psychological development (Cooper et al. 2003; Crider and Hall 2006; Hillman and Adams 1992; Jutras 2003; Mackett et al. 2005; Prezza et al. 2005). Things such as freedom to explore, independence, or self-confidence can only take place when children explore the world their own way. Walking to school and playing outside are great opportunities to foster such psychological development and well-being. As Mullan (2003) pointed out, it seems that the official problem of road death among children is contributing to an invisible, unofficial problem of increased fear and worry amongst parents, resulting in restricted mobility for children.

But why don't children walk to school anymore? Is it because roads are seen by parents as increasingly less safe for their children? The literature partly answers this question, and several studies concerning road safety and injury prevention reveal that knowledge, beliefs, and risk perceptions have an influence on parental safety practices and behaviours, especially on questions related to their children, including walking to school (Hu et al. 1996; Joshi et al. 2001; Lam 2000; Rivara et al. 1989). Other teams of researchers found relations between high parental perceptions of traffic-related risks and the use of transport modes other than walking to school (Joshi and MacLean 1995; Pooley et al. 2005). Additionally, the research by Kerr et al. (2006) demonstrates that their parental-concerns scale was most strongly associated with child-active commuting while Merom et al. (2006) found that regular walking/cycling for school commuting was associated, among other factors, with parent-perceived safety. As Noland stresses, 'individuals are more likely to choose a given commute mode the safer they perceive it to be' (1995: 511).

Nevertheless, the origin of these modal choices is more complex than what Noland (1995) states. Changes in family structure and daily routine, media exposure, and perception of risk all contribute to the decision-making process of parents when it comes to letting their children walk or bike to school. For that reason, the family context is an important variable influencing every sub-group of potential risk factors. In fact, some of the family characteristics such as stress on family members (mostly mothers), in-house density, social support, or the neighbourhood in which they live can influence both the cognitive development of children and their exposure to traffic, creating inequalities amongst children when it comes to road safety (Christoffel et al. 1996; Taylor et al. 1997).

Conversely, the influence of the living environment of parents upon their knowledge, beliefs, and **perception of risk** concerning their children's safety is hardly addressed in the recent literature. From a geographical perspective, this is quite unexpected since we usually hypothesize that individual risk perceptions stem from actions undertaken in the context of everyday spaces (Silka 1996; Tsoukala 2001). In the small body of research found, there is no trace of the effect of such actions, but there are several conclusions arising from the study of urban environment characteristics. It seems that parents living in neighbourhoods with low socio-economic status and/or high-traffic volume have a lower understanding of child safety, higher levels of parental fear/anxiety, and more serious traffic concerns (Eichelberger and Gotschall 1990; Gärling et al. 1984; Lam 2001; Weir et al. 2006). For example, Gielen et al. (2004) came to the conclusion that parents from higher-risk neighbourhoods would report fewer safety practices, lower knowledge scores, and lower perceived risk. At the same time, families in lower-income neighbourhoods were more likely to report that their neighbourhood was not a nice place to walk and that they have concerns about drug dealers and crime. Such concerns might lead to restrictions of movement in space and change in family geography. The street seems to lose its meaning as a space where children meet friends, play soccer, or just wander around: they no longer have a place where they belong outside their home and their school. This rather 'growing post-modern assumption', as mentioned by Matthews et al. (2000), arises from the dangerousness attributed to public spaces including streets. This sentiment of "unsafeness" is partly created by the mass media coverage of tragic events, but could also be attributed to the lack of *geographical* knowledge of the neighbourhood that parents and children (no longer) explore.

Starting from these premises, the question is how to manage 'real' risk as well as parent perceived risk in a way that lets children walk to school. Part of the answer can be found in studies focusing both on the geography of pedestrian risk and on parental attitudes toward this same risk. The next section of this chapter presents an empirical example exploring these dynamics for several schools and families in Montreal, Quebec, following two objectives. The first one is to describe the geography of risk and the most important risk factors for child pedestrians around primary schools. The second objective is to compare parental knowledge, beliefs, and perceptions toward child-pedestrian risk according to the specific local environment surrounding each school. Both these objectives should enhance our knowledge of the parents' assessments of road safety in a school context, as well as our understanding of what shapes contemporary urban family geographies when it comes to school-related mobility.

Observed and Perceived Child-Pedestrian Risk around Primary Schools: A Case Study in Montreal, Quebec

The Geography of Risk For Child Pedestrians

The first step in this analysis was to explore the geography of risk for child pedestrians in a school context. To do so, the chosen study area was the island of Montreal, Canada's second largest city after Toronto, with 1.6 million people on the island, expanding to almost 3.6 million in the census metropolitan agglomeration (Statistics Canada 2006). Accident data were provided by the Quebec Public Automobile Insurance Society (*Société de l'assurance automobile du Québec* [SAAQ]) for all the territory, for children aged 5 to 14 years old and for the period 1995–2003. One of the main advantages of this unique, government-owned insurance company is the coverage of the database: every accident in the province has to be reported to SAAQ in order to get benefits (insurance payment). All accidents outside of school days were excluded from the original study database (summertime, holidays) to allow for a focus on elementary schools during the school year. This leads to the selection and mapping of 2,011 accidents (56 per cent of the total dataset).

Proportions of the total number of vehicular accidents involving children remained almost constant within the ten-year period: it fluctuated from a low of 176 victims (8.8 per cent of the total) in 2001 to a high of 228 victims (11.3 per cent) in 1994, with no significant decrease from the beginning to the end of the period. Boys (60.1 per cent) are more often involved than girls, as well as children in the 10- to 14-year-old age group; at 53.4 per cent when compared to the 5 to 9 year old group. The time of the accident was only available for the first five years (1994–99). Within this subset, periods of the day with the most accidents correspond to hours when children are not in class but are moving from or to school—in the morning (15.5 per cent of the total number of accidents between 1994 and 1999); at lunchtime (14.5 per cent); and in the afternoon after classes finish (30.5 per cent). Finally, a relative-severity indicator was available for every accident. This index, solely based on the on-site police officer's judgment, is well known to underestimate the real medical condition. Nevertheless, it reveals the relative impact of these accidents on children's health. Twelve per cent of accidents reported between 1994 and 2003 were considered severe (usually enough to go to the hospital) in addition to a total of fifteen deaths (almost 1 per cent) that occurred during school days. These numbers confirm child-pedestrian accidents as a public health concern, as well as the relevance of studying schools as a space of interest both for the concentration of nearby accidents and for the change in family geographies it could foster.

The spatial distribution of accident sites is generally clustered, with a **nearest neighbour index** value of 0.37 (*z-score: -54.2*). As Figure 3.1a illustrates, the "density of accident" is higher in the more densely populated boroughs of Montreal, around the central business district (CBD), in addition to three other concentrations of accidents in socio-economically disadvantaged areas: in Montréal-Nord (#8: see figure 3.1c), Côte-des-Neiges/Notre-Dame-de-Grâce (#3), and Sud-Ouest/Verdun (#16 and #17).

While the first two neighbourhoods are known for their significant immigrant populations, the last two are old working-class neighbourhoods. When the locations of accidents are compared to the locations of primary schools, the proximity is even more striking: a quarter of the accidents occur less than 200 metres from the closest school and the mean **network distance** between each accident and the closest school is around 350 metres. This last statistic reaffirms the importance of schools as points of interest for future child-road safety interventions.

In addition to this summary of the demographic and spatial characteristics of the dataset, two regression models were computed to verify the statistical relationship between the number of accidents around each school and socio-economic and environmental variables. Most of the variables introduced in our models were chosen for their known relation to child-pedestrian risk as stated in a previous section, while others were chosen to test such a relationship for the first time. Prior to the modelling, a first step was to allocate each accident location to a school. Data on primary schools were provided by the School Taxation Management Council (*Comité de gestion de la taxe scolaire de l'Île de Montréal*—[CGTSIM]). This dataset includes current addresses, the enrolment for the year 1999–2000 (September to June), and the linguistic affiliation of each school (English or French). Schools that were not open throughout the study period (1995–2003) were excluded, together with specialized ones (schools for disabled children or with specific educational programs) leaving a total of 331 schools included in the study (93.8 per cent of the original dataset). In order to allocate the accidents to a school, global image (raster) analysis and network distance were used to build proximity zones around schools. By creating such school 'catchment' zones, each accident location was spatially related to only one school—the closest to the accident if walking on the street network. All other variables were proportionally computed within these zones (a more detailed methodology can be found in Cloutier [2008]).

Results of both models were quite helpful in understanding the strength of each risk factor on the number of accidents associated with a school and its surrounding environment. The first model (linear) included seven variables and explained 36.5 per cent of the variance in the number of accidents for the period 1995–1999. Four parameters were statistically significant at the five per cent level: the number of accidents near a school was positively related to neighbourhood **social deprivation** (the strongest variable), to main road density and to entropy (land-use diversity index). The social deprivation index computed for this project included four variables: the percentage of single-parent families, the unemployment rate, the percentage adults (20-year-old and over) with less than a ninth grade education, and the percentage of low-income households. This information was extracted at the dissemination area level from the 2001 Canadian Population Census and standardized in order to have an index varying from one (low deprivation) to four (high deprivation). Finally, the number of accidents was also negatively correlated to school language (English, the language of a minority of schools). We introduced this last variable to take into account the bilingual status of the city population. The significance of this last factor can be explained by the fact that children attending English-speaking schools travel mostly by car or school buses since these schools

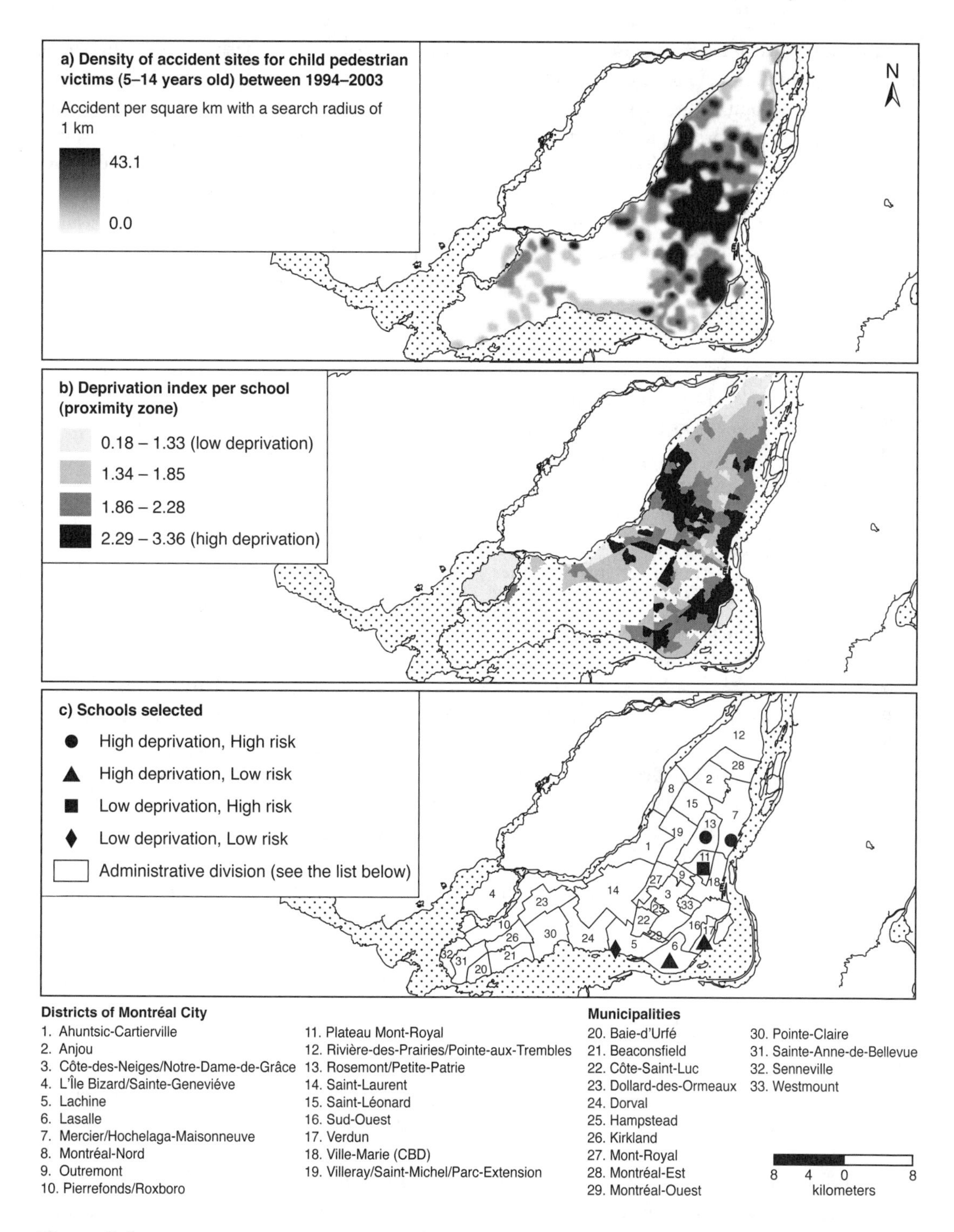

Figure 3.1 Source: MS Cloutier, 2009 (based on data from Statistics Canada, City of Montreal and SAAQ)

are less numerous and spread out over the territory. Non-significant factors were: proportion of children in the population, total road network density and school enrolment (Cloutier et al. 2007). The second model was based on a Poisson distribution for the accidents that happened between 1999 and 2003. Again, five out of nine variables were positively associated with pedestrian risk around schools: child population density (the strongest variable), neighbourhood social deprivation, residential density, land-use diversity, and number of crossing guards per school. This last non-intuitive positive relation can be explained by the selection of school crossing guards' intersections: their locations are usually chosen for their higher risk. Non-significant factors were total road and arterial network density, proportion of low-housing density, and proportion of children eligible for school bus transportation (Cloutier and Apparicio *in press*).

This geographic analysis confirms results of many studies previously done in Montreal and elsewhere in terms of factors associated with higher risk for accident. However, the most interesting implication from this first part of our analysis is in the attention drawn to the local nature of risk (Dougherty et al. 1990; Posner et al. 2002; Rao et al. 1997; Roberts et al. 1995). What we call the 'local risk' here is the one that makes a difference when studying the geography of families: it is part of what scares parents, part of what makes them forbid their children to walk to school. It does not matter if this risk is real or only perceived, the urban landscape between schools and homes can represent a barrier to the daily experience of space both for children and their parents. Indeed, the individual risk factors and therefore the global risk for children to be injured as pedestrians are not evenly distributed among primary schools, neither on the island nor the whole territory. For example, the pattern of higher risk of accidents seems to follow the pattern for higher deprivation, as shown by the mapping of the deprivation index for each school proximity zone (Figure 3.1b). These results indicate the need to raise questions related to health and social inequalities, and reinforce the idea that future prevention schemes should address local realities in their intervention planning, especially in poor neighbourhoods. That being said, there is a need to understand these local realities in order to reach children and families at risk. As well, any decrease in the risk related to child-pedestrian activities would benefit from a deeper exploration of the relationship between the observed—'real'—risk and parental knowledge of and attitudes toward such risk.

Parental Knowledge, Beliefs, and Perception of Child Pedestrian Risk/Safety

The objective of the second step of this case study was to compare parental knowledge, beliefs, and perceptions of child-pedestrian risk in different urban environments. This part of the chapter reports on analyses of a database of nearly 200 parents who answered a questionnaire that included 47 items in seven categories: demography, mobility (parents), child's way to school, map of the child's journey to school, road safety practices (parents with their children), risk perception, and knowledge and beliefs about road safety. Results and analysis from the two last categories are presented here.

The data collection process started with the selection of schools with the lowest residuals in the previously mentioned models. The selected schools were then divided into four types of neighbourhoods according to their deprivation index and their past number of accidents in the proximity zone of the school (1995–2003): high deprivation with high accidents, high deprivation with low accidents, low deprivation with high accidents, and low deprivation with low accidents.

In each school, the parents were reached either directly through the school board or via materials given to the child to take home. In the end, respondents were from six schools in all the different neighbourhoods (see Figure 3.1c). Because of the small sample in two of the former four subgroups (less than 30 observations or 5 per cent), we were able to compare our dataset only two by two: the deprivation index groups (high/low) were compared against each other, as well as the level of risk groups (high/low).

When looking at the demographic portrait of our dataset, it appears that respondents were mostly female (80 per cent) and the highest proportion of parents were in the 35- to 54-year-old category (81 per cent). Most of them were living with a partner in union (married or not: 81 per cent). A majority of respondents worked full time (60 per cent), as did their partners (70 per cent). The respondents in our sample are, however, relatively well educated and wealthy: 86 per cent had a degree superior to high school (compared to approximately 57 per cent for the population of Montreal) and 68 per cent had an annual family income of more than $40,000 compared to approximately 50 per cent for all of Montreal households (City of Montreal 2009). Respondents from the low deprivation index schools were the most educated (100 per cent of respondents had more than a high school diploma) and the wealthiest (87 per cent had a family income of more than $40,000). Respondents in the high accident risk schools had the lowest proportion for these two variables (78 per cent and 53 per cent respectively) and the highest proportion of single-parent families (35 per cent). The children referenced by the respondents were evenly distributed between boys and girls (48 per cent and 52 per cent), were mostly in the 6 to 9 years old age group (60 per cent) and were for the most part the oldest (39 per cent) or the youngest (29 per cent) child in the family. The families involved in the survey included a total of 385 children of whom 88 per cent were 12 years old or younger. For these last characteristics, no significant differences were noticed between sub-groups.

Knowledge

In order to test parents' knowledge concerning child pedestrians, several questions were asked. Results from the responses to two are presented here. To the question *in which order of severity would you place these sources of danger to your child*, only 11 per cent of all the respondents had the correct answer for the first source of danger according to the latest Canadian statistics (falls/injuries versus molestation/kidnapping, road traffic/accidents, cancer/severe diseases) (SafeKids Canada 2007; World Health Organization 2004). A significant proportion of respondents from high-deprivation schools (30.4 per cent) thought that *molestation/kidnapping* was the primary source of danger to their children compared to parents from low-deprivation schools (8.5 per cent). Also, the danger *road traffic/accident* was confirmed as a major preoccupation for parents since more than half (53 per cent) of all respondents put this source in first position. For this answer, a significant difference was noticed in the low-deprivation schools, which had a higher proportion of parents preoccupied by traffic (83.1 per cent versus 47 per cent for high-deprivation schools).

Again, only 11 per cent of respondents had the correct answer (*10–14 years old*) when asked *in which age group is one most likely to be a victim of a road accident as a pedestrian?* The other possible answers were: *less than 5 years old*; *6–9 years old*; and *15–24 years old*. Respondents from high-deprivation schools answered correctly in a higher proportion (15.3 per cent against 3.5 per cent) while respondents from low-deprivation schools answered in a higher proportion for the *6–9 year old* category (56.1 per cent against 39.5 per cent). We should mention here that, according to Canadian datasets on pedestrian injuries, hospitalisation and death rates are only higher for the 10- to 14-year-old group compared to the 6- to 9-years old since the year 2000 (Public Health Agency of Canada 2007). Indeed, if both proportions are added, 53 per cent of parents think children aged 6 to 14 are the primary victims of child-pedestrian accidents, which, after all, corresponds to the statistical reality.

Beliefs

One question related to beliefs is reported on here. Parents were asked to evaluate the level of danger (walking) along the route to their children's school using a **Likert scale** ranging from one (*never dangerous*) to seven (*always dangerous*) in two situations: a) for them and b) for their children. Overall, respondents rated their beliefs about the danger of walking to school 1.5 times higher for their children than for themselves. In addition, results show major differences between parents, depending on the deprivation status of the school surroundings. For example, respondents from low-deprivation schools considered the way to school *often dangerous* for their children in higher proportion than respondents from high-deprivation schools (respectively 17.7 per cent against 9.9 per cent). In contrast, parents from high-deprivation schools seemed to underestimate the danger for themselves and their children on the way to school: 26 per cent of them believed that it is *rarely* dangerous for their children and 15.3 per cent believed that it is *never* dangerous for them. This last result follows that of other work claiming that a higher risk environment, measured in terms of deprivation and/or past number of accidents, lowers the parent's perception of risk (Glik et al. 1991). The presence of a parental habituation phenomenon is the hypothesis followed in their work; when faced daily with environments where there is a fairly high potential for accidents, parents are in a way 'forced' to accept a certain level of risk. The risk becomes *invisible* to them under such circumstances (Glik et al. 1991).

Risk Perception

This final part of the analysis is based on the risk perception concept that can be defined as an intuitive assessment that most people will trust when faced with a particular danger (Slovic 2000). In the present case, our measure of risk perception was based on the work of Lam (2000, 2001). In his *Parental Road Environment Risk Perception Questions*, Lam asked parents to assess how dangerous they perceived six specific road environments and situations on a scale ranging from one (not at all dangerous) to four (very dangerous). Starting from the ideas found in Lam's questions, our final measure of risk perception integrated in the questionnaire included eight items describing risky situations (seven about *crossing*) for children as pedestrians on the streets:

1. playing on the sidewalk;
2. crossing an intersection without traffic lights;
3. crossing when the pedestrian signal is red and there are no cars coming;
4. crossing when the pedestrian signal is green;
5. crossing the street other than at an intersection;
6. crossing on a pedestrian crosswalk without traffic lights other than at an intersection;
7. crossing the street without looking;
8. crossing the street from between parked cars.

All these situations were illustrated by a drawing in the questionnaire to create the same mental image for every parent and reduce the optimistic bias involved when thinking about one's own child (for an example of the illustrations, see Cloutier and Apparicio [*in press*]). Again, parents were asked to rate the dangerousness of these situations on the same Likert scale (one to seven: *never dangerous* to *always dangerous*). In this case, the main advantage of the scale is the possibility to convert the answers into a score (numeric variable) by adding the values (points) on the scale and have one risk-perception measure for each respondent.

The measured risk perceptions remained relatively high no matter the type of urban environment parents and children experienced every day, and there is no significant difference according to the school surroundings' characteristics. Mean scores varied from 43.9 to 44.7 with a maximum of 56, and even if differences are not significant, high-risk schools have the lowest mean score. This last result is contradictory to the *belief* responses given by parents of high risk schools. Parents from these schools answered in a higher proportion that it was very *often dangerous* for them (7.8 per cent) and their children (15.7 per cent), compared to those of low-risk schools (respectively 1.4 per cent and 4.9 per cent for the same questions). This contrast might be explained by our methodology: faced with specific illustrated scenarios, parents from these schools seemed to underrate their perception of traffic danger compared to what they answered in previous general statement questions.

In addition to the comparison between the environments, variables from four different themes were introduced in a multiple regression model to observe their influence on parental road-risk perception as it was measured in the questionnaire. The first three themes were related to previous studies on risk perception (*demography*, *environment*, *cognitive elements*), while the fourth theme (*family mobility*) was added on the hypothesis suggested by Cutter (1993) arguing that risk perception cannot be grasped without an understanding of the daily life context. Results of this statistical analysis in turn confirm or invalidate certain of our hypotheses and provide us with information potentially relevant for family geography. This is why they are included here. The model, using a stepwise procedure, only retained three significant variables in two of the former themes (*demography* and *cognitive elements*): the parent's gender (*women* have higher risk perception); the leading source of danger (parents mentioning *traffic* have higher risk perception); and the sense of control (parents mentioning *lower sense of control* have higher risk perception).

In regard to demographic variables, it is noteworthy that none of the variables related to the child are found to be significant in our study, in contrast to some of the past studies where the age (younger) and sex (boy) of the reference children were positively related to parental risk perception (Gärling et al. 1984; Lam 2001; Sellstrom et al. 2000). However, the strong significance of the respondent's gender as a predictive variable, positively related to mothers, has implications for family geography. Since mothers are still playing a preponderant role in shaping how their children are raised and have a significant influence on their behaviour, their higher risk perception could modify, among other things, the walking and play territory of their children. It should also be mentioned that this result represents one of the challenges involved in education campaigns addressing parental accurate perception of risk: fathers and mothers should not be targeted in the same way.

Contrary to our initial hypothesis, none of the chosen variables related to the environment or to family mobility was linked to parental risk perceptions. In the case of the environment variables, the inclusion of only three observed variables (not reported by parents but measured in a geographic information system: socio-economic context, accident history, and presence of arterial roads) leaves little room for variability. Other elements of the urban landscape (graffiti, street configuration, etc.) and of the roads surrounding schools (road width, presence of signs, crossing guards, etc.) might also influence the perception of risk. In addition to these characteristics, we can hypothesize that the parental perception of the environment itself would be more correlated to risk perception, but this variable was not part of the present case study. In terms of mobility patterns, the only variable introduced in the final model, the frequency of the parents' use of the car (two categories: *only/often* or *sometimes/never*), was not significant. Even if this result was contrary to our hypothesis, it actually shows the need for more in-depth analysis of the link between environments, mobility, and risk through qualitative analysis of the route and the daily schedule followed by families. If we want to have a better idea of family geographies, we need to take a closer look at the routine between home, school, and work, as mentioned by another recent study done in Montreal, Quebec (*Groupe de recherche Ville et mobilité* 2008). This project revealed that children's mobility is tied up with their parents' mode of transport. Those children most likely to walk to school are those with at least one parent walking or taking public transit to go to work (*Groupe de recherche Ville et mobilité* 2008). Starting from these results, we can make the hypothesis that a given transportation mode is not only influenced by risk perception, but also by a range of variables including the mobility patterns of the family more broadly.

Finally, it is the cognitive variables that give us the most information about factors that influence parental risk perceptions since two of the three significant variables in the model are associated with this theme. Known variables such as sense of control (of the road risk) and beliefs about traffic were confirmed as important factors influencing risk perceptions. The importance of this dimension in our results underscores the need for action not only in regard to the real risk of accidents but also in regard to risk perceptions in order to encourage active modes of transport. As Bradshaw (1995: 19) said, 'As long as parents believe the roads are too dangerous for their children to go to school unaccompanied, there is a likelihood that an increasing number will choose to drive their child to school'.

Discussion

The previous sections of this chapter had two complementary objectives: to acknowledge the geographical nature of child-pedestrian safety and to look more closely, through an empirical example, at a dynamic that has been little studied yet—that of parents and their perceptions of risk for their children in an urban school context. Several implications can be drawn from the results of our investigation, informing both prevention initiatives and family geographies. This is what will be discussed in this last part of the chapter.

Implications for Injury Prevention Initiatives

Three conclusions can be highlighted from the empirical case study described here and they all have implications for prevention initiatives. First, accidents are concentrated around schools as well as in the central and southwest neighbourhoods of Montreal, contributing to the uneven spatial distribution of risk among child pedestrians. It means that besides being the main destination of children during weekdays, schools are in close proximity to several accident sites. This regrettable fact is creating opportunities to make schools, a major element in the infrastructure of our neighbourhoods (both physically and socially), a gathering point for key actors to get involved in interventions. Several road safety interventions are already taking place in schools every fall (i.e., at the beginning of the school year, such as police surveillance, educational pamphlets, etc.), but the idea of keeping road safety a priority all year long is very important for the efficacy of prevention measures. In short, there is a need to centre our initiatives on schools in order to be able to improve child-pedestrian safety and encourage active modes of transportation.

Second, the socio-economic status of the environment surrounding schools is an important factor to take into consideration when studying pedestrian risk. In addition to the significance of the deprivation variable in both models of road risk around elementary schools, the parental knowledge and beliefs about risk in the high-deprivation school areas were significantly different from the answers given by parents of low-deprivation schools. In fact, environments with a higher deprivation index accumulate multiple risk factors related to socio-economic status, family situation, and road design, which again increase inequity between people living in those neighbourhoods and the rest of the population. Misconceptions about the real risk, combined with higher traffic volume, make it even more important to undertake actions in those 'vulnerable' urban areas as a priority.

Finally, while risk perceptions remained relatively high no matter the type of urban environment parents and children experienced every day, it is the individual variables (motherhood and beliefs about traffic) that influence the most risk perceptions. Since these results open a gap between observed and perceived risk (half of the participating schools were located in low-risk environments), there is a need for action not only in regard to the real risk of accidents but also in regard to risk perceptions. Accordingly, the part played by families needs to be redefined since parents are often neglected in the ongoing strategies despite the important role they (should) play in injury prevention. In the child pedestrian context, parents actually play a double role: they are models in terms of road safety behaviour, but they are also the ultimate decision makers for their children when it comes to transportation. While both roles would always be important independently, it is the combination of the two that has the potential to change the geography of families.

However, parents alone will not be able to shift the broad propensity of our society toward vehicle-driven community design. In fact, the urban planning of our cities cannot be ignored when talking about child-pedestrian safety. Accordingly, recent studies reviewing the relationship between the built environment and traffic safety came to the general conclusion that dense urban areas appear to be safer than suburbs (Dumbaugh and Rae 2009; Ewing and Dumbaugh 2009). Still, this strong case against urban sprawl is not applicable in studies like ours that focuses on inner city neighbourhoods only. Saying so, cities with historical (grid street network) or conventional roadway designs (residential uses separated from arterial thoroughfares) should adopt child-centred approaches to the road environment if they really want to become 'family-friendly'. In its last report on children's road safety, the Organisation for Economic Co-operation and Development (OECD) confirmed that such interventions, including for example speed reduction (through traffic calming) or 'green district' creation, distinguished top-performing countries from those that did less well in terms of injuries reduction (OECD 2004). In a context where many authors agree now on the fact that parents, once convinced of the usefulness of their action, can serve as a powerful lobby group, there is a need for a shift of the parental preoccupation with traffic to the improvement of the overall urban environment (Bishai et al. 2003; Collins and Kearns 2001; DeFrancesco et al. 2003; Ziviani et al. 2004). Reconnecting parents with their own geography, in their neighbourhood, can be a first step in the understanding of real risk and of actions required to lower the different sources of urban insecurity. In addition to adults who want to teach children to be careful on the streets, we would like to see parents and neighbours attempt to improve the safety and walkability of neighbourhoods, including pathways to schools. This community involvement will benefit not only children, but, in fact, all neighbourhood citizens.

Contributions to the Notion of Family Geographies

Following these results, we can affirm that road risk around schools has the potential to change family geographies, and especially children's geographies. Right now, many urban children are paying for their parents' fears and beliefs. No matter if it is observed or perceived, child-pedestrian risk is altering mobility patterns and parental play supervision. The potential ramifications of this sense of risk has been family geographies that 'shrink' at the same pace as traffic flows! Parents restrain not only the walk to school, but also the play territory both in terms of distance (range from home) and the age children are allowed to play unsupervised (Moore and Young 1978; Valentine and McKendrick 1997). As said earlier, this is why the knowledge of the territory becomes important both for children and for parents.

The subject of pedestrian risk can also be viewed through the lens of 'modern' parenting, where parents take an increasingly active role in organizing their children's lives. The nature of childhood and of parenting is changing and it often means that children are brought up along more liberal lines. However, it can also imply increased supervision, home-centred play and high-parental involvement in organized leisure activities (Cheal 2008). The way these parenting practices are changing cannot always be considered a good thing for our children. As said earlier, playing and exploring the outside contributes to a child's development, to ties with the neighbours and families that live around them, and also to healthy behaviours now and as an adult. Walking to school is a good opportunity for such exploration, especially under the assumption that the freedom to walk around defines the limits of a child's world (Valentine and McKendrick 1997; Hillman, cited in Ward 1978). A family's geography, existing through walking, interacting, and playing in everyday spaces, contributes to a child's identity. Assuring parents that these spaces are safe is the goal we should all try to achieve, family geographers perhaps above all.

Questions for Further Thought

1. How can we take into consideration children and family needs when planning and modifying city neighbourhoods?
2. What role should be given to parents compared to school personnel (teachers, day care, school crossing guards) in terms of road safety education?
3. What role should government play in the issues of active transportation and pedestrian safety?
4. Are children considered, as Matthews and Limb (1999) argue, a 'neglected social grouping undergoing various forms of sociospatial marginalization'? Should our 'modern' conceptualization of the child be put under question?

Further Reading

Frank, L., P.O. Engelke, and T.L. Schmid. 2003. *Health and community design: the impact of the built environment on physical activity*. Washington: Island Press: 253.

Frumkin, H. 2006. *Safe and healthy school environments.*

Oxford: Oxford University Press: 462.

Gleeson, B., and N.G. Sipe. *Creating child friendly cities: reinstating kids in the city*. London: Routledge: 164.

Holloway, S.L., and G. Valentine. *Children's geographies: playing, living, learning*. Critical geographies; 8; London: Routledge.

Matthews, H., and M. Limb. 1999. 'Defining an agenda for the geography of children: review and prospect'. *Progress in Human Geography*, 23 (1), 61–90.

References

Adams, J., M. White, and P. Heywood. 2005. 'Time trends in socioeconomic inequalities in road traffic injuries to children, Northumberland and Tyne and Wear 1988–2003'. *Injury Prevention*, 11 (2), 125–26.

Assailly, J.P. 1997. 'Characterization and prevention of child pedestrian accidents: An overview', *Journal of Applied Developmental Psychology*, 18 (2), 257–62.

Bishai, D., P. Mahoney, S. DeFrancesco, B. Guyer, and A.C. Gielen. 2003. 'How willing are parents to improve pedestrian safety in their community?' *Journal of Epidemiology & Community Health*, 57 (12), 951–55.

Braddock, M., G. Lapidus, E. Cromley, R. Cromley, G. Burke, and L. Banco. 1994, 'Using a geographic information system to understand child pedestrian injury'. *American Journal of Public Health*, 84 (7), 1158–61.

Bradshaw, R. 1995, 'Why do parents drive their children to school?' *Traffic Engineering and Control*, 36 (1), 16–9.

Carstairs, V. 2000, 'Socio-economic factors at areal level and their relationship with health'. Pp. 51-67 in P. Elliot, et al., eds., *Spatial Epidemiology: Methods and Applications* Oxford Medical Publication; Oxford: Oxford University Press.

Cheal, D. 2008. *Families in Today's World: A Comparative Approach*. New York: Routledge. 172.

Christoffel, K. K., M. Donovan, J. Schofer, K. Wills, J.V. Lavigne, R.R. Tanz, M. Barthel, J. Jenq, C. Klinger, and P. McGuire. 1996. 'Psychosocial Factors In Childhood Pedestrian Injury: A Matched Case-Control Study'. *Pediatrics*, 97 (1), 33–42.

City of Montreal. 2009. *Socio-economic profile - Montreal Agglomeration*. [online database], (August 2009).

Cloutier, M.S. 2008. 'Les accidents de la route impliquant des enfants piétons: Analyse des risques potentiels et des risques perçus pour une meilleure prévention', PhD (Université de Montréal).

———. (*in press*). 'Predictors of parental risk perceptions: the case of child pedestrian injuries in school context'. *Risk Analysis*.

Cloutier, M.S., and P. Apparicio. (*in press*). 'L'environnement autour des écoles a-t-il un impact sur le risque routier impliquant des enfants piétons à Montréal? Apport de la régression de Poisson géographiquement pondérée'. *Territoire en Mouvement*.

Cloutier, M.S., P. Apparicio, and J.P. Thouez. 2007. 'GIS-based spatial analysis of child pedestrian accidents near primary schools in Montréal, Canada'. *Applied GIS*, 3 (4). http://arrow.monash.edu.au/hdl/1959.1/5120, accessed August 2009.

Collins, D., and R. Kearns. 2001. 'The safe journeys of an enterprising school: Negotiating landscapes of opportunity and risk'. *Health & Place*, 7 (4), 293–306.

Connelly, M.L., H.M. Conaglen, B.S. Parsonson, and R.B. Isler. 1998. 'Child pedestrians' crossing gap thresholds'. *Accident Analysis & Prevention*, 30 (4), 443–53.

Cooper, A.R., A.S. Page, L.J. Foster, and D. Qahwaji. 2003. 'Commuting to school: Are children who walk more physically active?' *American Journal of Preventive Medicine*, 25 (4), 273–76.

Crider, L.B., and A.K. Hall. 2006. 'Healthy school travel: Walking and bicycling'. Pp. 314–27 in H. Frumkin, ed., *Safe and healthy school environments*. Oxford: Oxford University Press.

Cutter, S.L. 1993. *Living with risk : The geography of technological hazards*. New York: E. Arnold, 214.

Davis, A., and L.J. Jones. 1996. 'Children in the urban environment: an issue for the new public health agenda'. *Health & Place*, 2 (2), 107–13.

DeFrancesco, S., A.C. Gielen, D. Bishai, P. Mahoney, S. Ho, and B. Guyer. 2003. 'Parents as advocates for child pedestrian injury prevention: What do they believe about the efficacy of prevention strategies and about how to create change?' *American Journal of Health Education*, 34 (5), S48–S53.

DiGuiseppi, C., I. Roberts, and L. Li. 1997. 'Influence of changing travel patterns on child death rates from injury: Trend analysis'. *BMJ*, 314 (7082), 710–13.

DiMaggio, C., and M. Durkin. 2002. 'Child pedestrian Injury in an urban setting: descriptive epidemiology'. *Academic Emergency Medicine*, 9 (1), 54–62.

Dixey, R. 1998. 'Improvements in child pedestrian safety: Have they been gained at the expense of other health goals?' *Health Education Journal*, 57 (1), 60–9.

Dougherty, G., I.B. Pless, and R. Wilkins. 1990. 'Social class and the occurrence of traffic injuries and deaths in urban children'. *Canadian Journal of Public Health/ Revue Canadienne de Sante Publique*, 81 (3), 204–09.

Dumbaugh, E., and R. Rae. 2009. 'Safe Urban Form: Revisiting the Relationship Between Community Design and Traffic Safety'. *Journal of the American Planning Association*, 75 (3), 309–29.

Eichelberger, M.R., and C.S. Gotschall. 1990. 'Parental attitude and knowledge of child safety'. *American Journal of Diseases of Children*, 244, 714–20.

Ewing, R., and E. Dumbaugh. 2009. 'The built environment and traffic safety: A review of empirical evidence'. *Journal of Planning Literature*, 23 (4), 347–67.

Gärling, T., A. Svensson-Gärling, and J. Valsiner. 1984. 'Parental concern about children's traffic safety in residential neighborhoods'. *Journal of Environmental Psychology*, 4 (3), 235–52.

Gielen, A.C., S. Defrancesco, D. Bishai, P. Mahoney, S. Ho, and B. Guyer. 2004. 'Child pedestrians: The role of parental beliefs and practices in promoting safe walking in urban neighborhoods'. *Journal of Urban Health*, 81 (4), 545–55.

Glik, D., J. Kronenfeld, and K. Jackson. 1991. 'Predictors of risk perceptions of childhood injury among parents of preschoolers'. *Health Education Quarterly*, 18 (3), 285–301.

Graham, D., S. Glaister, and R. Anderson. 2005. 'The effects of area deprivation on the incidence of child and adult pedestrian casualties in England'. *Accident Analysis & Prevention*, 37 (1), 125–35.

Groupe de recherche Ville et mobilité. 2008. 'Active travel and schools in Montreal and Trois-Rivières: An analysis of active travel by elementary school students in Quebec'. *Summary report.* Montréal: Université de Montréal, 40.

Hanley, M., B. Cody, A. Mickalide, C. Taft, and H. Paul. 2002. 'Report to the nation on child pedestrian safety'. Washington, DC: National Safe Kids Campaign, 8 pages.

Hewson, P. 2004. 'Deprived children or deprived neighbourhoods? A public health approach to the investigation of links between deprivation and injury risk with specific reference to child road safety in Devon County, UK'. *BMC Public Health*, 4 (15). www.biomedcentral.com/1471-2458/4/15, accessed August 2009.

Hillman, M., and J. Adams. 1992. 'Children's freedom and safety'. *Children's Environment*, 9 (2), 10–22.

Holloway, S.L., and G. Valentine. 2000. *Children's geographies: playing, living, learning.* Critical geographies; 8; London: Routledge, 275.

Hu, X., D. Wesson, P. Parkin, and I. Rootman. 1996. 'Pediatric injuries: parental knowledge, attitudes and needs'. *Canadian Journal of Public Health/Revue Canadienne de Santé Publique*, 87 (2), 101–05.

Joshi, M.S., and M. MacLean. 1995. 'Parental attitudes to children's journeys to school'. *World Transport Policy and Practice*, 1 (1), 29–36.

Joshi, M.S., V. Senior, and G.P. Smith. 2001. 'A diary study of the risk perceptions of road users'. *Health, Risk & Society*, 3 (3), 261–79.

Jutras, S. 2003. 'Allez jouer dehors! Contributions de l'environnement urbain au développement et au bien-être des enfants'. *Canadian Psychology*, 44 (3), 257–66.

Kerr, J., D. Rosenberg, J.F. Sallis, B. Saelens, L.D. Frank, and T.L. Conway. 2006. 'Active commuting to school: Association with environment and parental concerns'. *Medicine & Science in Sports & Exercise*, 38 (4), 787–94.

Laflamme, L., and F. Diderichsen. 2000. 'Social differences in traffic injury risks in childhood and youth: A literature review and a research agenda'. *Injury Prevention*, 6 (4), 293–98.

Lam, L.T. 2000. 'Factors associated with parental safe road behaviour as a pedestrian with young children in metropolitan New South Wales, Australia'. *Accident Analysis & Prevention*, 33 (2), 203–10.

———. 2001. 'Parental risk perceptions of childhood pedestrian road safety'. *Journal of Safety Research*, 32 (4), 465–78.

LaScala, E.A., P.J. Gruenewald, and F.W. Johnson. 2004. 'An ecological study of the locations of schools and child pedestrian injury collisions'. *Accident Analysis & Prevention*, 36 (4), 569–76.

Mackett, R.L., L. Lucas, J. Paskins, and J. Turbin. 2005. 'The therapeutic value of children's everyday travel'. *Transportation Research Part A: Policy and Practice*, 39 (2–3), 205–19.

Macpherson, A., I. Roberts, and I.B. Pless. 1998. 'Children's exposure to traffic and pedestrian injuries'. *American Journal of Public Health*, 88 (12), 1840–43.

Matthews, H., M. Limb, and M. Taylor. 2000. 'The "street as thirdspace"'. Pp. 63–79 in S.L. Holloway and G. Valentine, eds., *Children's geographies: playing, living, learning.* London: Routledge.

Merom, D., C. Tudor-Locke, A. Bauman, and C. Rissel. 2006. 'Active commuting to school among NSW primary school children: Implications for public health'. *Health & Place*, 12 (4), 678–87.

Moore, R., and D. Young. 1978. 'Childhood outdoors, toward a social ecology of the landscape'. Pp. 88–130 in I. Altman and J.F. Wohlwill, eds., *Children and the environment.* New York: Plenum Press.

Mullan, E. 2003. 'Do you think that your local area is a good place for young people to grow up? The effects of traffic and car parking on young people's views'. *Health & Place*, 9 (4), 351–60.

Noland, R.B. 1995. 'Perceived risk and modal choice: Risk compensation in transportation systems'. *Accident Analysis & Prevention*, 27 (4), 503–21.

OECD. 2004. 'Keeping children safe in traffic'. OECD Publishing, 130.

Pooley, C.G., J. Turnbull, and M. Adams. 2005. 'The journey to school in Britain since the 1940s: Continuity and change'. *Area*, 37 (1), 43–53.

Posner, J.C., E. Liao, F.K. Winston, A. Cnaan, K.N. Shaw, and D.R. Durbin. 2002. 'Exposure to traffic among urban children injured as pedestrians'. *Injury Prevention*, 8 (3), 231–35.

Prezza, M., F.R. Alparone, C. Cristallo, and S. Luigi. 2005. 'Parental perception of social risk and of positive potentiality of outdoor autonomy for children: The development of two instruments'. *Journal of Environmental Psychology*, 25 (4), 437–53.

Public Health Agency of Canada. 2007. 'Mortality and morbidity dataset, child pedestrian victims, 1994–2004'. *Unpublished data.*

Rao, R., M. Hawkins, and B. Guyer. 1997. 'Children's exposure to traffic and risk of pedestrian injury in an urban setting'. *Bulletin of the New York Academy of Medicine*, 74 (1), 65–80.

Rivara, F.P., A.B. Bergman, and C. Drake. 1989. 'Parental attitudes and practices toward children as pedestrians'. *Pediatrics*, 84 (6), 1017–21.

Roberts, I., R. Norton, R. Jackson, R. Dunn, and I. Hassall. 1995. 'Effect of environmental factors on risk of injury of child pedestrians by motor vehicles: A case-control study'. *BMJ*, 310 (6972), 91–4.

SafeKids Canada. 2007. 'Child and Youth Unintentional Injury: 10 Years in Review, 1994–2003'. Toronto: SafeKids Canada, 35.

Schieber, R.A., and N.J. Thompson. 1996. 'Developmental risk factors for childhood pedestrian injuries'. *Injury Prevention*, 2 (3), 228–36.

Sellstrom, E., S. Bremberg, A. Garling, and J. Olof Hornquist. 2000. 'Risk of childhood injury: predictors of mothers' perceptions'. *Scandinavian Journal of Public Health*, 28 (3), 188–93.

Silka, L. 1996. 'Transforming the psychology of risk: From social perception to the geography of communities'. In R.S. Feldman, ed., *The Psychology of Adversity.* Amherst: University of Massachusetts Press.

Slovic, P. 2000. *The Perception of Risk.* (Risk, society, and policy series). London: Earthscan Publications, 473.

Sonkin, B., P. Edwards, I. Roberts, and J. Green. 2006. 'Walking, cycling and transport safety: An analysis of child road deaths'. *Journal of the Royal Society of Medicine*, 99 (8), 402–05.

Statistics Canada. 2006. *Census of population.* [*online database*], (August 2009).

Stevenson, M.R., and D.A. Sleet. 1997. 'Which prevention strategies for child pedestrian injuries? A review of the literature'. *International Quarterly of Community Health Education*, 13 (3), 207–17.

Sunderland, R. 1984. 'Dying young in traffic'. *Archives of Disease in Childhood*, 59 (8), 754–57.

Taylor, S.E., R.L. Repetti, and T. Seeman. 1997. 'Health psychology: what is an unhealthy environment and how does it get under the skin?' *Annual Review of Psychology*, 48 (1), 411–47.

Torres Michel, J.J. 2007. 'La recherche par le projet d'aménagement : comprendre le vélo chez les enfants à travers les projets "Grandir en ville" de Montréal et de Guadalajara'. PhD (Université de Montréal).

Tsoukala, K. 2001. *L'image de la ville chez l'enfant.* Bibliothèque des formes; Paris: Anthropos, 191.

Underwood, J., G. Dillon, B. Farnsworth, and A. Twiner. 2007. 'Reading the road: The influence of age and sex on child pedestrians' perceptions of road risk'. *British Journal of Psychology*, 98 (1), 93–110.

Valentine, G., and J. McKendrick. 1997. 'Children's outdoor play: exploring parental concerns about children's safety and the changing nature of childhood'. *Geoforum*, 28 (2), 219–35.

Ward, C. 1978. *The Child in the City.* New York: Pantheon Books, 221.

Weir, L.A., D. Etelson, and D.A. Brand. 2006. 'Parents' perceptions of neighborhood safety and children's physical activity'. *Preventive Medicine*, 43 (3), 212–17.

World Health Organisation. 2004. 'World report on road traffic injury prevention'. Geneva: WHO, 178.

Yiannakoulias, N., K. Smoyer-Tomic, J. Hodgson, D. Spady, B. Rowe, and D. Voaklander. 2002. 'The spatial and temporal dimensions of child pedestrian injury in Edmonton'. *Canadian Journal of Public Health/Revue Canadienne de Santé Publique*, 93 (6), 447–51.

Ziviani, J., J. Scott, and D. Wadley. 2004. 'Walking to school: Incidental physical activity in the daily occupations of Australian children'. *Occupation Therapy International*, 11 (1), 1–11.

Chapter Four

Being Home: Family Spatialities of Teleworking Households

Jean Andrey and Laura C. Johnson

Introduction

The contemporary popular literature on **teleworking** from home includes a number of frivolous distractions pertaining to the ways costumes affect feelings of professionalism. Talk of 'working naked' (Froggatt 2001) or the pros and cons of fuzzy slippers versus dress-for-success attire do not capture the essential features of this alternative work arrangement that has the potential to alter the essential logistics of family life. In reality—as our research shows—the key ingredients for successfully combining home-based work and family life are: sufficient and appropriately designated household space, the latest in secure telecommunications equipment, a supervisor attentive to workers' needs for professional support and occasional social interaction with colleagues, temporal flexibility, and (in families with young children) a good childcare arrangement. Even with these in place, however, teleworking households engage in the re-negotiation of family practices and re-sculpting of family homes, thus altering family geographies in important, though often subtle, ways. This study illustrates various transformations of family geographies in analytically interesting ways. As well, it explores the implications of telework for social change—change related to workers' control over their lives and also to traditional gender roles in the family.

Over the past few decades, with the emergence of the **information society** and the possibility of anywhere-anytime work, the boundaries between paid employment and non-work life have become increasingly blurred. Indeed, as noted by Johnson et al. (2007: 144), 'Recent advances in telecommunications technology have dealt a blow to the "separate spheres" notion, liberating traditional office work from corporate offices—with much of it relocated to workers' homes'.

There is a growing appreciation that home-based work changes the ways in which the home is used, the routines and practices of its occupants, and their lived experiences of home (Michelson and Crouse 2002). As early as the 1990s, sociologists described two models by which homework can be accommodated in the home. One emphasizes the separation of employment activities from family life so as to allow non-work spaces to retain their sense of being a **haven** from the outside world. The other suggests a redefinition of home, as work and non-work activities are blended in time and space (Bulos and Chaker 1993; Huws et al. 1990; Miraftab 1996).

The ways in which public and private spheres are integrated through home-based telework, and their implications for family spatialities, are explored in this chapter based on insights gained through analysis of interviews with 52 home-based teleworkers. In the context of our study, telework refers to paid white-collar work that is facilitated by information technologies and is done by employees of corporations or institutions in locations other than 'the office'. That said, it is becoming increasingly difficult to differentiate the various arrangements of remote work: **e-work, mobile work, telecommuting**, and telework (Duxbury et al. 1992; Hardill and Green 2003; JALA, 2008), and increasingly apparent that the e-revolution in all its various forms is altering the nature of work and the very essence of home (Baines and Gelder 2003).

The interviews for the current study were conducted as part of a larger study funded by Canada's Social Sciences and Humanities Research Council on the implications of home-based telework for individual workers, their families, and society at large. Publications from this study, based on subsets of the sample set, have focused alternatively on leisure (Shaw et al. 2004), environmental sustainability (Moos et al. 2006), the experiences of workers (Johnson et al. 2007), and mothering (Hilbrecht et al. 2008). In this paper, we draw on the full set of interview transcripts to explore family spatialities with particular attention being given to the ways in which work and family life intersect and blend.

The Telework Phenomenon

The last two decades have been associated with major changes in the nature of work in most western economies. In Canada, national surveys and large-scale workplace studies reveal that employment demands have increased, and associated with this trend are growing levels of time stress and **work–life conflict** (Duxbury and Higgins 2002; Johnson et al. 2001). At the same time, a 'quiet revolution' has been occurring (Gartner Inc. 2005), whereby an increasing number of employees working in **knowledge-intensive sectors** are working from home—all or part of the time, something that has been identified as one possible way to reduce work–life tensions (Duxbury et al. 1998; Hilbrecht et al. 2008; Pratt 1999).

While telework adoption was fairly slow throughout the 1980s and 1990s, it has grown considerably over the past decade. Gartner Inc., an information technology research and advisory company, in their report published in 2005, estimated that by 2008, 41 million corporate employees worldwide would spend at least one day a week teleworking and 100 million would work from home at least one day each month, with the highest penetration of this work form in the United States. They went on to highlight that telework adoption in Canada has been 'cautious' relative to the United States, but that it was, nevertheless, still expected to total 1.2 million Canadian employees working from home at least one day per week by 2008, which translates into 8.4 per cent of the employee population.

The Canadian Telework Association identifies some of the reasons why telework is growing in popularity in Canada: a high per-capita ratio of high-tech, government, and other information workers; high rates of computer ownership and broadband penetration; and the expansive geography of Canada. As well, it is generally understood that there are potential cost savings for employers. For example, fewer parking spots are needed and space costs may be reduced as dedicated offices are replaced with a smaller number of hotelling workstations or meeting rooms. There is also the promise of higher worker productivity, due largely to fewer interruptions and less employee absenteeism. There can also be advantages for individual employees in the form of time savings because of the foregone commute, time flexibility leading to more optimal scheduling of activities, and advantages associated with 'being present' at home—themes that will be explored in detail in this chapter.

Women, in particular, have often emphasized the importance of home-based work for managing their multiple roles as employee and as mother/wife/homemaker/caregiver. In a recent study of Australian women, for example, Kelly and Kelly (2008: 1) found that most women prefer home-based work over conventional away-from-home options, and that the strongest influence on their expressed preferences was a 'perceived conflict between a woman's career and her family life.' This finding is not surprising, given increased demands of the workplace, continuing gendered division of responsibilities in most homes, and new ideologies of parenting that value more structured childhoods, increasing parents' workload through added activities that require planning, scheduling, coordinating, and chauffeuring (Arendell 2001; Turcotte 2007).

Telework and Spatialities of the Family

The spatiality of social life has been a central theme in human geography over the past few decades. Only recently, however, have geographers begun to focus on home as the 'house in which we live'—in other words, to explore the 'space of the house' as a complement to our understandings of 'a place called home' (Gregson and Lowe 1995: 225–26). Spatial arrangements within homes reflect and reinforce socio-cultural values, including power relations and hierarchy. They get at some of the smaller scales of geographic production—providing opportunities to reflect on how rituals, behaviours, symbols, and practices contribute to shared social meanings.

Geographers have looked variously at the home as a source of meaning and identity in everyday life. Humanist geography, as it developed in the 1970s, investigated the human experience of important places such as the home, which were imbued with meanings based on experiences of living. According to humanistic geographers, those significant places with profound and personal meaning, in turn, grew to become part of human identity (Rose et al. 1997). Cultural geographers subsequently addressed the ways that dominant meanings of place were dictated more by culturally prescribed dominant ideas and ideals, rather than by the actual lived experiences of place identity (Rose et al. 1997). The home might be portrayed in media as a warm, welcoming, place of leisure and nurturance, in spite of the contradictory subjective experiences of residents. Feminist geographers took a different approach in understanding the influence of place on identity and considered the different ways men and women experienced the home and other significant places within the geographies of human experience.

Looking historically from the nineteenth century at the home as a site of spatial division of labour among men and women, the home was associated with women's identity and role. For women as homemakers, the home was the locus of their family and household work roles. Later in the twentieth century, as women left the home for the paid labour force, research focused on the dual roles of women, and their struggle to balance the competing responsibilities of domestic and paid labour (Gregson et al. 1997).

The topic of paid work in the home has been of special interest to feminist scholars, spawned by historical studies of industrial homeworkers (Boris 1994). While it is well known that the industrial revolution saw the shift of most manufacturing into factories, there is less focus on the overflow of production that continued to be done in workers' homes. Work of that kind was often characterized by substandard and exploitative conditions. As documented by muckraking American social reformers such as photojournalists Lewis Hine and Jacob Riis (Boris 1994), homework around the early twentieth century was typically women's work and often involved child labour. Working long days for extended hours in cramped, poorly ventilated conditions, women and children engaged in manufacturing were paid on a piece rate basis at rates typically below those paid to factory workers (Domosh and Seager 2001; Johnson and Johnson 1981). While employment standards, health and safety regulations, and child labour legislation were enacted to control the worst abuses of homework production, the practice of home-based production persists. In the privacy of homes, protected from public scrutiny and government regulation, 'sweatshop' production persists.

Recent developments in telecommunications technology have expanded the range of home production. In today's knowledge economy, office work has largely replaced factory work as the main type of home-based employment. Researchers examining the conditions of employment of contemporary homeworkers focus on the ways they blur boundaries between home and work. Feminist geographers Domosh and Seager (2001: 55) describe that boundary problem: 'for homeworkers, there is no time or place when they are not (or could not be) "at work," and the waged homework often expands to fill all available time and space. As feminist geographers have shown, the microgeography of the home is seldom designed to accommodate wage-earning work'.

Contemporary studies of home-based activities raise a number of important questions, from the home being a space which is critical to the gender constitution of society, to different definitions and understandings of woman (Gregson and Lowe 1995). And these studies most certainly confront the issue of the **emotional work** around family. By emotional work, we mean participation in and commitment to those behaviours and activities that maintain relationships among family members and improve their psychological well-being. As expressed by England and Farkas (1986: 91), emotional work is to 'understand others, to have empathy with their situation, to feel their feelings as part of one's own'. Only recently have researchers begun to explore in detail emotional work, the ways in which it intersects with gender and family circumstances, and its implications for well-being of individuals and relationships (Duncombe and Marsden 2004; Erickson 2005; Strazdins and Broom 2004). The matter of family circumstance is particularly relevant to teleworking households where physical adjustments to the house and the physical presence of a parent alter opportunities for emotional work and care. Victoria Lawson (2007), in her presidential address to the American Association of Geographers in Chicago, spoke of geographies of care and responsibility, arguing that geography and geographers have a responsibility to care, especially now because of '. . . the relentless extension of market relations into almost everything'. She went on to reflect that 'care ethics involves values of empathy, responsiveness, attentiveness, and responsibility—values most readily mobilized in places with which we are most familiar (our homes, communities, perhaps our nation)' (6). Teleworking households provide a unique opportunity to explore the ethic of care as mobilized in the most familiar of all spaces—the family home.

Writings on the electronic home/cottage and imaginings of home designs also inform this research as they explore the possibilities associated with a future society that is primarily home-based and technology-mediated. They raise questions around how different work environments potentially affect the meaning of the home and reinvent or reinforce social roles. As noted by Gurstein (2001: 148) in her book, *Wired to the World, Chained to the Home*, the live/work trend has major implications for dwelling design and raises serious questions about the meanings of home—meanings related to 'control, security, refuge, orientation, comfort, entertainment, solitude, memories, accomplishment, family, children, space, personalization and expression, responsibility, investment, and seclusion'. Recent studies provide abundant fodder for considering the manner in which telework alters work and family life. Holloway's (2007) work on boundary issues and the manner in which teleworking households attempt to take some control over the placement of new technologies and uses in the home is particularly relevant. As well, publications related to the current study provide insights into boundary construction and crossing, gender issues, and work–life balance (Hilbrecht et al. 2008; Johnson et al. 2007).

The Case Study

This research is based on a case study of a large cross-Canada financial firm. We first approached this employer in the year 2000, and entered into an informal partnership to explore the implications of home-based work for their employees. The employer was interested in learning more about two topics: best practices as they related to management of remote employees and the particular challenges faced by their off-site workers. We were interested in learning how teleworking affects individual workers, their families, and the community at large.

Over the five years that followed, we met regularly with the case study firm's management and went about recruiting individual workers to participate in the study. Participants were recruited in one of three ways: through a list of teleworkers provided by the company's human resources department based on input from line managers; a teleconference arranged by the employer attended by managers and workers from specific units where telework was practiced, where we described the objectives of our research; and through snowballing where contacts were provided by study participants. All potential participants were contacted by electronic communication and provided with an outline of the study. In total, we contacted just over 100 employees and had participation from 74 of them. Of the 74, 52 completed an interview—either in person or by phone—and it is these interviews that are used here to deepen our understanding of home-based work as it intersects with family life.

The interviews were semi-structured, and lasted from 60 to 90 minutes. Interviews consisted of open-ended questions organized around three themes: work-life balance, home workspaces, and general experiences with home-based work. With the participants' permission, we tape recorded the interviews, transcribed them, and coded and analyzed them in NVivo computer software to identify key themes.

Forty-one participants also provided a detailed time diary for a reference week, photo documentation of home workspaces, and answers to an on-line questionnaire that administered several standardized measures of quality of work and life, including time pressure and job satisfaction. While these materials enabled us to know our participants better, the focus in this chapter is on the insights gained from the narratives in which people describe and explain their actions and experiences.

We are committed to protecting the confidentiality of all those who participated in the study, and as a result we provide few details on the nature of the work performed by specific employees; similarly, we do not identify the communities in which individual participants live. We also use pseudonyms when providing quotations; these pseudonyms are consistent with those used in a previous publication based on a subset of these same interviews (Hilbrecht et al. 2008), when the same interviewee is being referred to.

The sample of 52 participants comprises 46 females and six males, reflecting the largely female composition of the labour force of this company and sector. The ages of participants ranged from early 30s to late 50s. Most participants reported that they currently were or had previously been married, and all but a few participants had children. Living arrangements varied. A few were living on their own and others with only a partner, but most were living in family arrangements: traditional families, blended families, single-parent families and, in one instance, an extended family. They lived in communities of all sizes, from Canada's largest cities to rural regions of the country. All had professional occupations, and their annual gross household incomes varied from about $50,000 to over $100,000, with a median category of $80,000 to $99,000. Almost all had post-secondary training and most had one or more university degrees.

It is important to point out that, despite the broad cross-section of family types and residential locations, our participants largely represent middle-class Canada, a slice of society that is more likely to be white and less likely to be of recent immigrant status. We, as researchers, share this social heritage, and it is through this shared cultural lens that interpretations of 'home' and 'domesticity' are presented. This must be borne in mind when extending the findings to other settings.

As a way of entering into the case study, we begin by profiling two of the teleworkers in our sample set. Meet Zlata and Pam.

Profile 1: Zlata

Some people, Zlata among them, would consider Zlata's work-life balance to be just about perfect. She and her husband have what is sometimes termed a million dollar family, with a son aged seven and a daughter aged five. Their son is in grade two and their daughter attends kindergarten in the mornings. Zlata works from home for three extended (10-hour) days per week, usually Tuesday, Wednesday, and Thursday. On work days, her nanny arrives at 10 a.m., does some housekeeping, picks the children up from school, and serves them lunch. After three o'clock, the nanny packs up the little girl and together they pick the son up from school, and go over to the nanny's house, where they spend the rest of the afternoon in a playgroup.

When Zlata's employer summoned her with the news that she was being relocated to a home office, she describes her initial reaction as 'horror': 'It really was not my choice . . . When they first told us they were moving us into home offices, I was really horrified by the idea of the two things I do—my home life and my work life—being entwined'. She explains, 'I'm . . . somebody who really likes to work when I'm working and, when I'm not, I don't want to work. So I was really worried when I first came home that I would be working all the time because I have a tendency that way anyway'.

But now, after three years of homeworking, she finds it an excellent fit with her needs. She credits her nanny—who helps with child care, housekeeping, the laundry, and some meal preparation—with much of the success of the arrangement. 'I can't believe how lucky I am. It's amazing.' But credit for the Mary Poppins nanny aside, Zlata also attributes her positive experience with home-based work to good planning in terms of the physical arrangement of her home workspace (she and her husband renovated the basement to create a home office) and the careful training of those near and dear to her about what it really means to work at home.

Sometimes those nearest and dearest need a bit of a refresher course in the parameters of her working from home. She recalls an awkward occasion when one spring, after they had re-sodded their lawn, her husband asked if she might take care of watering the lawn. He was asking what he thought was a relatively simple task, to just remember to move the sprinkler every half hour or so. Zlata declined, explaining that her work required her full concentration: 'I put him in his place pretty fast. I resent that, because it's almost like I'm not working'. In order to protect against distractions during her working times she even had to train her mother not to drop by. 'I told her, like I tell other people: "I've got to work this many hours every day and if I don't do it, I'm in trouble." My mother did not want me to be in trouble'.

The price tag on the basement home office renovation was $5,000. Zlata points out that her office isn't luxurious, but they decided to build a room with a door that can be closed securely. The employer gave a standard allowance of only $500 to defray the costs of setting up a home office. Zlata resents the discrepancy, particularly, as she points out, since the teleworking arrangement was initiated by the employer. The employer, she observes, realizes significant benefits from this arrangement.

For starters, Zlata is convinced that she is much more productive working from home, compared with her previous experience doing the same job from head office. 'I'm billing way more hours working from home than I ever did working from the office.' At home, she explains, there are fewer interruptions than at the office. By sending Zlata and her colleagues home, the company, in her opinion, gets a good deal. The employer saves on office overhead and gets efficient and highly motivated workers.

From Zlata's perspective, the only downside is her own occasional need for social support from those who understand the challenges of her job. Working in her home office, she finds that she is lonely sometimes. She says, '. . . when I was in the office, when you were having one of those days where some wacko was just yanking your chain, you could go into your boss's office or co-worker's office and sit down and say, "Oh, man, let's go get a cup of coffee." And I really miss that, that kind of camaraderie, and I think we learned a lot from each other by talking. And we, one co-worker and I, still do that over the phone—but it's just not the same . . . I'm lonely for other professional contact'.

Zlata also talks about how telework has affected some aspects of family life, mostly in positive ways. For one, the family is eating more home-cooked dinners. In addition, there is more time for doing things with the kids—like having tea parties with her daughter in the late afternoon and skiing with her children on Friday afternoons. She sums it by saying, 'I really like this. I really like the working at home'.

Profile 2: Pam

Being part of her community is a definite advantage of homeworking for Pam. She is getting to know more of her neighbours and local shopkeepers, and is grateful that she no longer wastes up to 90 minutes each day commuting to the company's head office.

While some of her teleworking colleagues were 'sent home' by the employer, Pam had requested that work arrangement. On her own initiative, she began working from home one day a week. When the employer subsequently relocated her whole department to home offices, she was more than willing to comply. Homework enthusiasts, she says, 'are like a cult; once you join you never want to leave'. She is unequivocal in stating that she would not take a job that did not allow her to work from home.

That said, Pam's house is not ideally configured for home-based work. Her relatively small house and growing teenagers prevent her from allocating an entire room to the office function. Her home office is part of her bedroom, which she acknowledges is not ideal. In fact she has thought a lot about the home office that she would like, 'So, I'd rip the carport off . . . and I would put in a two-storey addition. Then on the main floor, the lovely office for me, and it would look onto my yard, and it would be really nice. I have it all designed in my head'. But for now, she has chosen a space that provides the kind of corporate privacy that is needed, and she has spent the employer's contribution on a 'really old oak desk' that she likes.

While she is a conscientious and capable worker, Pam emphasizes that she is no workaholic. She is attentive to the balance between work and the rest of her life. She acknowledges pressure to allocate too much time to work, but is clear that she monitors that allocation and turns off the switch if she is devoting excessive time to work. Telework, she believes, helps her to manage everything. 'It helps so much. I mean, I have to take my dog to the vet and that's not a problem; on Friday I'll do that. Or if I have an electrician coming, then I can schedule for a day that I'm going be home all day, so I think it helps so much. And then my kids, and just a million other things. I think it's the greatest thing ever!'

Pam's teenaged children have mostly got the drill down pat, and know the etiquette of the corporate home office: no interrupting phone calls, no playing with the office computer. But she does have some interference from the animal kingdom. Her dog, slower than her children at learning the rules against acoustic interruption, often barks when she is on the telephone. Barking blows her cover, forcing her to explain to clients that she is working from home 'for the day'.

Despite Pam's enthusiasm for homeworking, the arrangement does have its downside. Since the company has outsourced technical support, that has proved to be problematic for her. The offshore tech support firm is less able, she finds, to solve the problems that arise. Not an enthusiast of domestic chores, Pam finds that housework rarely lures her away from her paid work. In no way does she wish she had more time for household tasks. 'If it was zero I'd be thrilled.' Her only regret is that she can only afford to hire someone to come in and clean for her on a monthly basis.

Pam acknowledges that home-based work is not for everyone. There are some who crave the companionship and social interaction of an office. She does not. Perhaps, she considers, her age makes her comfortable being by herself without feeling lonely. She does schedule lunch meetings and other social contacts with colleagues and friends. She also credits her successful home-based work arrangement with a particularly capable supervisor, who regularly convenes the virtual team. The result is that colleagues know one another and can work together effectively, despite generally being separated by considerable distance.

Commentary on the Two Profiles

The profiles of these two teleworkers provide an introduction to various opportunities and tensions associated with home-based work. While both women are enthusiastic about their work arrangements, there are a number of contradictions in their respective lived experiences. For example, Zlata feels professionally isolated whereas Pam is content operating at a distance with only occasional contact with co-workers. Zlata also feels some pressure to take on traditional female roles—meal preparation and other domestic responsibilities—whereas Pam, who is no longer married, seems to pick and choose what she will do and for which services she will pay. Zlata also displays some of the signs of **intensive parenting** with her involvement in the children's soccer and skiing outings. Pam's children, on the other hand, are nearly grown and her concern seems to be more about just being there. Pam faces the challenge of having her home office in the most private part of her home—her personal bedroom—whereas Zlata has adopted an arrangement that deliberately attempts to separate home and work.

These glimpses into two women's lives provide an entry point into a fuller discussion about the implications of telework for spatialities of the family.

Integration and Separation

There is a growing appreciation that the geography of the home changes when a residence doubles as a workplace. Where do work-related activities occur, and in what ways are they integrated with or separated from 'normal' family life? Under what conditions are home offices secured and separated from the mainstream areas of the home? In a qualitative study of Swedish home-based teleworkers, Wikström et al. (1998: 203) contrast the 'place in the sun' arrangements where workers—typically females—opportunistically seek out available areas of peace and tranquility within the home, with the physically bounded workplaces of 'splendid isolation' more typically used by male teleworkers. But are these patterns reproduced in the current study, and what experiences and meanings are attached to the various spatial arrangements?

Work into Home

The very definition of home-based work is that remunerative activities occur within the confines of one's residence. The employees interviewed for the current study talked at length about their relocation to home offices. For some employees, the opportunity to work at home was sought out, and in those cases the creation of a suitable workspace usually was not problematic. These employees spoke of the spaciousness of their home workspace relative to the cubicle they occupied at the office, and how it incorporated most of the functionality of the corporate office while offering a more homey atmosphere in which to work. But for others, the move home was involuntary and/or sudden, and the physicality of the home offered limited options for office work. In these cases the work space was seen as confining or inadequate, or the family space was seen as being compromised, or both.

Several participants described the ways in which a work space was inserted into or carved out of domestic spaces, how these initial arrangements were sometimes inadequate, and the ways in which the allocation of former family spaces to corporate activities sometimes required sacrifice on the part of selected family members.

> '. . . when I started working from home the first time, it [the office] was in a spare bedroom, and I worked in the closet.' (Sarah)

> 'It [the home office] needs to be cleaned up because it used to be a laundry room.' (Karen)

> 'Way back ten years ago when I started a home office . . . I was in an apartment . . . and I must admit that was tough . . . [with] my children being very young.' (Len)

> 'I had to put my two boys in the same room—which they hate!' (Loretta)

> 'When I bought this house, it was just me and my three kids. It's only about 1,020 square feet. Then I got sent home to work. We had to convert my living room to my office . . . it pretty much took up the whole thing.' (Kirsten)

> 'She [my daughter who is away at university] was grievously saddened that I was sacrificing her bedroom. I literally gave away the furniture because I didn't have a place to keep it because I had to bring in these big filing cabinets.' (Laleh)

Participants also spoke of intrusions of work activities into shared-use or primarily private spaces, including the teleworker's bedroom. As explained by Cathy, a full-time homeworker with four children, this arrangement was problematic. 'So my office went in my bedroom . . . and believe me, it was brutal, really horrid.' Alice, who works at home two or three afternoons per week, spoke of similar concerns. She was, nevertheless, considering moving her office into the bedroom she shares with her spouse because of her need for a workspace behind a door that can be locked. 'We're contemplating putting the desk in the bedroom, but quite frankly, I don't want it in our bedroom. You know, the last thing you want to be doing is looking at a computer at night.'

For some full-time homeworkers the intrusion of work into home concretized in the form of supervisory audits. Coreen, for example, perceived these visits as 'a real invasion of privacy' particularly given her husband's deteriorating health.

> 'I felt like, to a certain extent because we have a home office for a large corporation, there is a sense of "we own you and we own your home". So they could say that they wanted to come and you didn't have any choice about it. So, if my husband was ill or something and had to sleep all day, I didn't have the privileges of saying, "I am sorry but you can't come to my home".'

Sheila, too, expressed discomfort with home visits by her supervisor. 'So, it was like, come to the front door, walk up to my office, don't look anywhere else.'

In terms of the location of the office within the home, participants expressed a strong desire to maintain separation between paid work and other home-based activities, consistent with emerging theory on '**boundary work**' (Holloway 2007; Nippert-Eng 1995) and somewhat at odds with studies that purport women who work at home may be rather nomadic. The importance of a door, and preferably a door that locks, was partly related to the confidentiality of work materials. But there was also a sense that an open office, particularly one located near the epicenter of family life, made it too difficult to focus on work in some instances, and alternatively turn work off in others. In probing what advice participants would give prospective teleworkers, most offered the opinion that the workspace should be as far away from where people 'live' as possible. As stated emphatically by Karen, 'I would keep that strictly office. I like separation. I know we call it work-life integration, and yes, to some degree I have that, but I would say for the most part I try to separate the two as best I can'.

In addition to helping workers turn off, others talked about how physical separation affects turning work on, as described by Lloyd. 'Once I get into the seat in the office . . . it actually kind of feels like flipping a switch, like a light switch or something like that.' Participants said that being motivated to work is typically not a problem, but sometimes they feel distracted for short periods and need self-discipline to re-enter their mental work zone. Dedicated and separated workspaces help to provide that discipline.

Caring for Children and the Extended Family

The desire for work-family separation was not absolute, however, and for many parents the metaphor of a two-way mirror seems appropriate. Participants, especially mothers, spoke of being busy in their offices and deliberately not engaging with their children. There was, however, a commitment to being vigilant and 'available should anything happen' (Tina). Some spoke of how older children were mandated to come directly home after school or how younger children felt protected by parental presence regardless of where in the house the teleworker was located. This 'boundary work carried out by teleworkers at the micro level' (Holloway 2007: 34) was frequently described as being one of the main advantages of home-based work—that, and the foregone commute. As summarized by Nancy, a mother with two young children who works part-time and exclusively from home, 'Oh yeah, I feel it's more difficult working from home. I find it more rushed, but I'm home for the kids'. As in other studies, the main concerns were being there during times of children's sickness as well as after school, and the main advantages were opportunities for oversight and closeness facilitated by the parent's physical presence. This phenomenon of being home, physically and symbolically, speaks to the very identity of many of these working mothers, and confronts in a direct way the reality of being both an employed person and a responsible parent.

New spatialities of the home and family were also found to intersect with intensive parenting, an ethos adopted by most participating parents, seemingly without question. Most mothers were involved in child-centred activities during the work day, such as taking part in school fieldtrips or taking work breaks to chauffeur their children to 'enriching' activities. Virtually all arranged their work activities so that they could be home during critical family times. Their movement in and out of the family sphere, in terms of both spatiality and the rhythm of activities, is the most shared characteristic of teleworking mothers who were part of our study. Their priorities are consistent with a national trend whereby working mothers are now spending more time with family (Turcotte 2007).

Styles of fatherhood also appear to be shaped or at least influenced by telework. Much less has been written about models of fatherhood in the context of home-based work. Lupton (1998) describes how home-based work has the potential to change **gender dichotomies**, and Marsh and Musson (2008: 35) explore fathers' experiences of not only 'inhabiting a gendered space, but also an emotional context which has previously been held to be largely for women only'. They then profiled three fathers—each of whom was used to illustrate a different approach to home-based work. These approaches include 'privileging a parental identity', 'privileging a professional identity', and 'an attempt to have it all'. In the current study, only five of the 52 interviewees were fathers. Lloyd, who works full-time at home with his spouse on maternity leave and his infant son always nearby, comes closest to the parental identity category. Lloyd sees telework as having a positive impact on his home life and as something that will provide him with opportunities to be a good father and good husband.

> '. . . That's what a lot of fathers miss out on is that whole, that scene through every moment of development and you know, I'd say that there's less a chance now that like, when he [son] takes his first steps, or says his first words or whatever, there's more of a chance that I'm going to be around because I'm going to be right here for work as well, so that seems pretty exciting to me . . . I'm really looking forward to spell[ing] my wife off when she's had a particularly bad day, maybe, or a bad evening. And just say, 'okay sweetie you have a few hours sleep with the baby' and maybe I can take a few hours off in the morning to allow her to get some sleep and substitute that with some evening work. I'm looking forward to being able to help out . . .'

Of the remaining four fathers, Len appears to be in transition from a father who tried to have it all to one who yearns to spend more time with his children, who are eleven and seven. The other three fathers appear optimistic that they can have it all; they talked about seeing more of their children since relocating to home offices, but there was little indication that these fathers were taking on serious caregiving roles. Instead, most of the responsibility remained with the mothers, or it was diverted to childcare workers and relatives who were available on fixed schedules.

Participants also talked about how telework allows them to provide care to extended family members. Sometimes these family members are aging parents, where the emphasis is on maintaining relationships and providing support. Other times, the family members are nieces and nephews, and the focus is on building relationships and providing opportunity.

> 'On Thursday, Sue's dad dropped in. It was great, I hadn't seen him for a while, so we had a gab and a cup of coffee.' (Zach)

> '. . . when my first child was born, my husband . . . wanted to move to the country, back closer to his parents and my parents. And because I worked in the home, it really didn't matter where I lived.' (Eileen)

> 'And I do feel better as my parents are aging, that if there is an emergency or anything that I can drop what I am doing and go and help them. I can meet my mom for lunch or I can pick up my nephews if I have to . . . I'm after grandmother on the [emergency contact] list so if there is something with one of my nephews . . . they are like my pseudo kids.' (Sheila)

> 'Like this week, my niece, she's 12, and I signed her up for volleyball camp and it's from 9 to 12, so I have the flexibility. When she sleeps over here, I drive her, come back, work, and then I pick her up.' (Sharon)

Regardless of the individual circumstances, participants spoke passionately about how the co-location or proximity of work and family altered family spatialities, and a central message was one of new opportunities for care.

Incubation and Isolation

Telework also affects home life in other important ways. Without any specific prompting, many of the respondents provided statements as to how home-based work had positive implications for their health and overall well-being. It was almost as if the family home acted as an incubator for those aspects of personal well-being in need of healing. Some of the comments referred in specific ways to the inferior work environment in corporate offices.

> 'Oh, and I'm so claustrophobic I couldn't stand . . . a lot of cubicles were three or four feet . . . you're not even near a window, you know? And I'm like, Oh my god! I used to get up to go down to the fax machine just to get away from it. But, in this one, I've got all the light I want . . . it's perfect.' (Ranjana)

> 'When I was working at the office . . . I was burnt out and I was taking mental health days just to recoup because of all the running back and forth, . . . it [telework] helps me manage . . .' (Betty)

> 'I had breast cancer two years ago. . . . I have the opportunity to stay more composed here at home . . . I did this from a lifestyle decision and I'm very pleased . . .' (Jennifer)

Susan, who alternated between working at home and at corporate headquarters, drew a sharp contrast between her experiences in these two settings.

> 'I ate better when I was at home . . . I ate very healthy, I always had my cereal . . . and I was more comfortable. I would go to the washroom on a more regular basis, I didn't have to wait . . . or I was more comfortable using my own washroom than say the washroom at

work . . . Now [that I am back in the corporate office] I am not as healthy. I have gained weight because you're always eating junk food, . . . my neck hurts and I am constipated all the time.'

Others spoke about being 'healthier, happier' (Abby), of having 'more energy' (Jody) since they started working from home, or of developing new routines and better habits.

But some home-based workers had a very different experience. Despite individual and corporate strategies for replacing day-to-day, face-to-face communications at work with a variety of remote connections or off-site meetings (i.e., touch-point meetings with supervisors, instant messaging with co-workers, and scheduled conference calls and lunches with virtual team members) some participants felt isolated, as if they had become 'office orphans'. They talked about missing the social contact at the office, that sense of community that can transform a workplace into more than a place to work. Several spoke of home-based work as being isolating. 'I think it's just isolating 'cause even in the summertime I was inside. You could get weird.' (Jennifer)

Laura described it as 'a big feeling of exclusion', but regardless of the phrasing it was clear that many felt they had lost part of their social network, their extended family of sorts. In Kirsten's case, she found the move home seriously harmful to her well-being, and she ended up seeking alternate employment in order to be able to return to a normal office environment.

> 'I spent entire days by myself, and the move home corresponded to a personal crisis that I had a hard time getting over. The personal crisis led to depression, which eventually I got help for, but I found it hard to get over it working the whole time at home. It's a whole mindset change when you walk out the door and go to the office. I never had that, and I think that made it more difficult for me to get over things.'

While questions remain unanswered about the circumstances under which telework becomes detrimental to both work performance (Golden et al. 2008) and personal well-being, what is clear is that home-based work changes social contact and networking in fundamental ways. The implications of these changes for family geographies are not immediately obvious, but some participants did talk about being starved for positive, personal interaction. Jennifer was one of the few to connect this need to family life, noting that her husband gets 'more venting' than he used to. Most others, however, seemed to derive satisfaction from having their families as the focal point of their social interactions, allowing for a more intense experience of family.

Home into Work

Regardless of the spatial arrangement of the home office and despite attempts to keep work and family spheres mostly separate, sometimes the workspace is a 'leaky capsule' to use the phrase of Adams (1995) in his description of how telecommunications technology transforms a dwelling. Leakages occur in both directions. Home creeps into the office, and office work permeates the home. Where there were repeated leakages of family life into work, work that requires concentration and confidentiality, new rituals or practices were introduced into family life. For example, Linda explained that during the telephone interview with us, she had her office door closed and a sign on the door saying 'In a meeting, be quiet'. Signage, scheduling of work activities around the presence/absence of others, and the introduction of new house rules around noise and visitors were key strategies used to contain unwanted family noises and distractions from leaking into work-spaces.

But not all were successful at achieving adequate separation, as reflected in the following comments about interruptions from children:

> 'Well, we put French doors on it (the office) . . . And I think from an office perspective, that was probably a mistake because the kids do come to the basement because it's our one unfinished place in the house and they can ride their bikes down there and play with their scooters down there and stuff . . . and it kind of bugs me, and it bugs them a little bit too, I think.' (Zlata)

> 'When my kids get home early on Fridays I can hear my daughter playing guitar upstairs. I know that they are around. Sometimes at night, if I'm trying to work at night, it's a problem because they've got the big screen TV blasting and the rec-room is in the basement, so they are just on the other side of the wall.' (Charlotte)

> 'Not to mention the fact, that if my kids are at home . . . it's ultra-distracting, because I mean . . . my two year old stands at the bottom of the gate, looking up, 'Mom, mom, mom, mom!' You know, so there's no working.' (Alice)

> '. . . summertime is crazy . . . you're not only dealing with your own children, you're dealing with their friends, their families.' (Rochelle)

In fact, boundary issues, the circumventing of walls intended to keep spheres separate, were portrayed as challenges in various ways. Sometimes tensions arose around the inability of the teleworker to negotiate a private space for confidential work, as illustrated by Kirsten's struggle to keep her mother out of her home office. 'My mom is very old. When she was here I know she was going in this office . . . .' Other times it had to do with the expectations of family members.

> 'But now they [children] come home at 3:30 or 3:45, and they expect that I just drop everything because I'm here. They have an adjustment issue.' (Cathy)

> 'Oh yeah, if my husband comes home early, which he often does if I'm at home, it is somewhat distracting . . . it's sort of like he's finished his day, maybe I should be too.' (Jennifer)

Regardless of the personal circumstances, the reconfiguration of the home and the redefinition of boundaries that telework entails have implications for family life. The very intersection of these two spheres, work and family, which Western society had grown accustomed to having physically separated, is contributing to new geographies of the family, geographies that involve mixing and blurring of functions and roles in varied and dynamic ways.

Time Over Space

In some cases, the very presence of work in the house changes the sense of home, such that it loses its feeling of sanctuary or haven. Instead, the co-location of work and non-work sometimes creates a sense of perpetual obligation to the job, a virtual ball and chain (Gurstein 2001). Jennifer expressed in words what has been altered.

> 'I must say I have a different sense coming in the driveway now I know that there are messages waiting for me. If I come in off the road,

> the first thing I do is go down to the office as opposed to going into my room or fixing tea or something so yeah, I do see it differently. Not really bad but I can't put a word on it but it is different . . . You have to be extremely disciplined to say I'm not working now, I've worked enough.' (Jennifer)

Others also talked about the need to resist the temptation to work constantly.

> '. . . because the kind of work I do, it's never finished, you're never caught up ever, and . . . somebody will phone at the last minute, they need this and that, and you are there.' (Jiqisha)

> 'I can get so into my work [at home] that I don't even eat.' (Sahira)

> 'It's probably more insidious, you know it's really easy to run in here at 11 o'clock at night and think oh I'll just see if so-and-so has e-mailed me back. Or I better get that letter [off], oh I'll do it right now so I am not stuck with doing that first thing in the morning. And so from that perspective it sucks me back in, whereas before it was, well, I am not driving 45 minutes back to the office to check e-mail.' (Coreen)

Still, in many cases the opportunity to work at home was seen as a chance to take control of one's life, to prioritize things of importance, in essence to let time conquer space. This is the crux of the telework **utopia** argument or illusion. Rochelle talked about this phenomenon in terms of empowerment. 'There is an empowerment aspect where you are your own boss.' Most others spoke specifically of the temporal flexibility afforded by home-based work. There is the opportunity to arrange work hours around the unexpected, for example, a chance to sleep in if a child didn't sleep through the night, or to engage in valued activities that were often not possible or at best inconvenienced by rigid office-based schedules.

Some participants expressed their satisfaction with telework in terms of their having more discretionary time. Ruth described teleworking as giving her just 'a bit of down time', while Benita and Geoff expressed the benefits of home-based work in terms of managing the double juggle.

> 'And because I work so far away, I want to be able to have a life at home, and yet, be able to get my work done as well. So, to be able to work from home allows me to do that, and it really allows me to do it quite well.' (Benita)

> 'So, the home office has kind of worked out that way in terms of managing the family, and the daycare, stuff like that.' (Geoff)

The physical removal of workers from the corporate office typically weakened employees' ties to workplace routines and cultures, allowing more compatibility between work obligations and domestic life. Modifications in the spatiality of domestic life took various forms: parents driving or walking children to and from school, parents taking the opportunity to attend a child's basketball game with laptop in hand, parent and child standing around the kitchen after school debriefing about the day, or sick children being accommodated on cots beside their teleworking parents' desks. Some of our participants, as well as some of the business-oriented literature on telework, would argue that this intensified sense of time control is the upside of telework, the aspect of change that allows parents, mothers in particular, to navigate the difficult terrain of the double juggle. Others would caution that one should not overstate the benefits of time sovereignty (cf.

Baines and Gelder 2003), but focus instead on '... wider systems of social support for caring and to provide a better work-life balance for men as well as women' (Greenhill and Wilson 2006: 387).

Gender Roles

Our study shows that telework also has implications for other family dynamics, including gender roles. Many of the female participants spoke about their roles in **homemaking**, and, as one would expect given national data on time use differences between men and women (Turcotte 2007), the majority of mothers indicated that they have primary responsibility for child care, meal preparation, cleaning up, and other domestic chores. However, in many cases, respondents indicated that this arrangement had nothing to do with their homework location. As stated summarily by Susan, '... my husband never helped anyways. (Laughs) Nobody ever helped, so nothing changed'.

But in some cases, there were changed expectations around domestic responsibilities. It was not uncommon for teleworking women to report that they felt that they should be taking on more responsibilities around the home, and there was guilt and anxiety associated with not conforming to this self-imposed stereotype.

> 'When I first started, I had a lot of guilt around not doing housework, not doing the laundry, not doing those kind of things and it took me a while to say, "No, this is my work time and I can't do it all" ... When I first started, I felt like it had to be clean, it had to be ... Like I was home, you know, why isn't the dinner ready, why isn't the house clean?' (Kate)

> 'And my biggest issue with that was, well, you work from home, so you must do your laundry, cook amazing dinners, have the cleanest house, you know, have a beautiful garden, do all these fun things, watch TV in the afternoon, and so I always felt pressure that I had to perform at or above everyone else to prove that I wasn't goofing off.' (Sarah)

Other times, it was husbands who had changed expectations of their teleworking wives, but with varying outcomes.

> 'There's been the assumption that I could be doing more and rightly so. I said, "I do have extra time".' (Jody)

> 'My husband sometimes says, "Can you run into the Zehr's [supermarket] and do that errand?" "Well, I'll have to see how my work goes".'(Jackie)

> 'I have heard him [husband] say to other people, "It's so nice that she works from home. She's really on top of things. She can get so much [housework] done working from home" and I feel like saying, "Well that's just because I am efficient and organized, and that means I go go go, I don't take a break and just sit and relax. At lunchtime others enjoy lunch with their friends or read a book at lunch, stuff like that. I never do that".' (Eileen)

For these women, home-based telework was associated with pressure to re-embrace traditional female roles, rather than renegotiate social norms. For the most part, however, the homeworkers in this study resisted any pressures associated with their home location and remained work-focused during designated hours and not easily distracted by household tasks, except for an occasional load of laundry or putting something in the oven or crock pot to get a start on dinner preparations. And there were several instances where domestic responsibilities were shared between husband and wife in more equitable ways, regardless of their work locations.

Conclusions

This case study of teleworkers provides insights into how the temporal and spatial rhythm of family life is renegotiated and fundamentally altered by this alternative work arrangement. While the introduction of telework may appear on the surface to be uncomplicated, it often acts as a catalyst and facilitator of fundamental changes in lifestyle and life choices. Such changes are sometimes stressful, creating ongoing challenges in household, workplace, and family dynamics, and other times empowering in terms of adjusting one's priorities or adopting new routines.

Many of the challenges are associated with tensions related to the integration versus separation of paid work and personal/family life. The insertion of work activities into one's house sometimes becomes intrusive in one's home, generating a range of strategies for containing unwanted 'leaks'. At the same time, the very presence of domestic spaces and family members can create distractions during work hours and interfere with people's ability to let go of work at the end of the day. As well, the physical separation of a teleworker from nodes of corporate activity can create 'office orphans', i.e., employees who on the surface have all the resources needed to perform the job, but who lack the nurturing offered by regular face-to-face contact.

A strong theme to emerge relates to the ethic of care, and the way in which home-based telework is valued, sometimes to the extreme. Parents—both mothers and fathers—spoke appreciatively of how working from home creates opportunities for being more present and more involved in the lives of their children. And the ethic of care was not restricted to parenting, but in some instances extended to elder care and extended-family members, and also affected care of self. Interestingly, for the most part, telework does not appear to be a catalyst for rethinking gender roles, although there were instances where teleworking women resisted taking on more domestic responsibilities and others where fathers took on more of the emotional work of the family. It remains unclear, however, as to the degree to which telework is an agent of change in society versus the extent to which it allows individuals and families to live out current societal norms in less stressful ways. Rather, our participants provide mixed messages about whether telework liberates them as workers and as members of a highly gendered society.

In summary, our study reaffirms the complexity of the links between physical setting and social life, and re-articulates the importance of physical separation between work and family domains even as home-based workers foster particular types of leakage between work and family life. Our research also supports assertions that an intensified sense of time control, which is highly valued, is experienced by most home-based workers. The findings, however, challenge earlier work that purports that men and women prefer different kinds of home workspaces. The research also raises questions about the ways in which family spatialities may be influenced by the sometimes involuntary nature of home-based telework, and it suggests that this work form, which is becoming increasingly common, has broad implications for society writ large including parenting styles and the ethic of care in extended families.

Acknowledgements

While we cannot identify the case study firm, we are most grateful to its management for providing us with access to its teleworking employees. We are particularly grateful to study participants for allowing us into their lives, and often homes, and for the arduous task of documenting their time use patterns over a reference week. The research was funded, in part, by a Standard Research Grant from the Social Sciences and Humanities Research Council of Canada, and by research grants from the University of Waterloo. For their roles in conducting this research, we wish to thank colleagues Professor Susan M. Shaw and Dr. Margo Hilbrecht, as well as the many graduate students who contributed to data collection and analysis.

Questions for Further Thought

1. What are the implications of home-based telework, and virtual work more generally, for family life in different cultural contexts?
2. If telework leads to higher levels of worker productivity, will this lead to increased employer expectations and worker exploitation?
3. If part-time telework becomes the new norm, will this change housing preferences, both in terms of location and housing style?
4. Are teleworkers 'better' parents than non-teleworkers?

Further Reading

Baines, S., and U. Gelder. 2003. 'What is family friendly about the workplace in the home? The case of self-employed parents and their children'. *New Technology, Work and Employment* 18, 3: 223–34.

Castells, M. 2000. *The Rise of the Network Society*. 2nd Edition. Oxford: Blackwell.

Gurstein, P. 2001. *Wired to the World, Chained to the Home: Telework in Daily Life*. Vancouver: UBC Press.

Hill, E.J., A.J. Hawkins, and B.C. Miller. 1996. 'Work and family in the virtual office: perceived influences of mobile telework'. *Family Relations* 45: 293–301.

Holloway, D 2007. 'Gender, telework and the reconfiguration of the Australian family home'. *Continuua: Journal of Media and Cultural Studies* 21,1: 33–44.

References

Adams, P. 1995. 'Bringing globalization home: a homeworker in the Information Age'. *Urban Geography* 20, 4: 356–76.

Arendell, T. 2001. 'The new care work of middle class mothers: managing childrearing, employment, and time'. Pp. 163–204 in K.J. Daly, ed. *Minding the Time in Family Experience: Emerging Perspectives and Issues*. Oxford: Elsevier Science.

Baines, S., and U. Gelder. 2003. 'What is family friendly about the workplace in the home? The case of self-employed parents and their children'. *New Technology, Work and Employment* 18, 3: 223–34.

Boris, E. 1994. *Home to Work: Motherhood and the Politics of Industrial Homework in the United States*. New York: Cambridge University Press.

Bulos, M., and W. Chaker. 1993. 'Home based workers studies in the adaptation of space'. Pp. 55–79 in M. Bulos and N. Teymur, eds. *Housing: Design, Research, Education*. Aldershot: Avebury.

Domosh, M., and J. Seager. 2001. *Putting Women in Place: Feminist Geographers Make Sense of the World*. New York and London: The Guilford Press.

Duncombe, J., and D. Marsden. 1993. 'Love and intimacy: the gender division of emotion and "emotional work": a neglected aspect of sociological discussion of heterosexual relationships'. *Sociology* 27, 221–41.

Duxbury, L., and C. Higgins. 2002. *The 2001 National Work-Life Conflict Study: Final Report*. Ottawa: Health Canada.

Duxbury, L., C. Higgins, and S. Mills. 1992. 'After-hours telecommuting and work-family conflict: a comparative analysis'. *Information Systems Research* 3, 2: 173–90.

Duxbury, L., C. Higgins, and D. Neufeld. 1998. 'Telework and the balance between work and family: is telework part of the problem or part of the solution?'. Pp. 218–55 in M. Igbaria and M. Tan, eds. *The Virtual Workplace*. Hershey, PA and London: Idea Group Publishing.

England, P., and G. Farkas. 1986. *Households, Employment, and Gender: A social, Economic, and Demographic View*. New York: Aldine.

Erickson, R.J. 2005. 'Why emotion work matters: sex, gender, and the division of household labor'. *Journal of Marriage and Family* 67: 337–51.

Froggatt, C. 2001. *Work Naked: Eight Principles for Peak Performance in the Virtual Workplace*. San Francisco: John Wiley, Jossey-Bass.

Gartner Inc. 2005. *Teleworking: The Quiet Revolution*. www.gartner.com/DisplayDocument?doc_cd=122284

Golden, T.D., J.F. Veiga, and R.N. Dino. 2008. 'The impact of professional isolation on teleworker job performance and turnover intentions: does time spent teleworking, interacting face-to-face, or having access to communication-enhancing technology matter?'. *Journal of Applied Psychology* 93, 6: 1412–021.

Greenhill, A., and M. Wilson. 2006. 'Haven or hell? Telework, flexibility and family in the e-society: a Marxist analysis'. *European Journal of Information Systems* 15: 379–88.

Gregson, N., and M. Lowe. 1995. ' "Home"-making: on the spatiality of daily social reproduction in contemporary middle-class Britain'. *Transactions of the Institute of British Geographers* 20, 2: 224–35.

Gregson, N., U. Kothari, and J. Cream, et al. 1997. 'Gender in feminist geography'. Pp. 49–85 in N. Gregson, G. Rose, J. Foord, S. Bowlby, et al., eds. *Feminist Geographies: Explorations in Diversity and Difference*. Women and Geography Study Group of the Royal Geographical Society, with the Institute of British Geographers. Harlow, England: Longman.

Gurstein, P. 2001. *Wired to the World, Chained to the Home: Telework in Daily Life*. Vancouver: UBC Press.

Hardill, I., and A. Green. 2003. 'Remote working—altering the spatial contours of work and home in the new economy'. *New Technology, Work and Employment* 18, 3: 212–22.

Hilbrecht, M., S.M. Shaw, L.C. Johnson, and J. Andrey. 2008. '"I'm home for the kids": contradictory implications for work-life balance of teleworking mothers'. *Gender, Work and Organization* 15, 5: 454–76.

Holloway, D. 2007. 'Gender, telework and the reconfiguration of the Australian family home'. *Continuua: Journal of Media and Cultural Studies* 21, 1: 33–44.

Huws, U., W.B. Korte, and S. Robinson. 1990. *Telework: towards the elusive office*. Chichester: John Wiley and Sons.

JALA International. 2008. *Just What is Telework?* www.jala.com

Johnson, K.L., D.S. Dero, and J.A. Rooney. 2001. *Work-Life Compendium 2001*. Centre for Families, Work and Well-Being, University of Guelph, Ontario, Canada. http://dsp-psd.communication.gc.ca/Collection/RH64-3-2001E.pdf

Johnson, L.C., J. Andrey, and S.M. Shaw. 2007. 'Mr. Dithers comes to dinner: telework and the merging of women's work and home domains in Canada'. *Gender, Place and Culture* 14, 2: 141–61.

Johnson, L.C., and R.E. Johnson. 1981. *The Seam Allowance: Industrial Home Sewing in Canada*. Toronto: The Women's Press.

Kelly, S., and C. Kelly. 2008. *Women's Work Preferences: The Importance of Home-Based Work*. http://ssrn.com/abstract=1322386

Lawson, V. 2007. 'Geographies of care and responsibility'. *Annals of the Association of American Geographers* 97, 1: 1–11.

Lupton, D. 1998. *The Emotional Self: a Sociocultural Exploration*. London: Sage.

Marsh, K., and G. Musson. 2008. 'Men at work and at home: managing emotion in telework'. *Gender, Work and Organization* 15, 1: 31–48.

Michelson, W., and D. Crouse. 2002. 'Changing demands on the time-use analysis of family, work, and personal outcomes in late-century: implications of trend analysis for time pressure'. Paper presented to the Conference on Time Pressure, Work-Family Interface, and Parent-Child Relationships: Social and Health Implications of Time Use, University of Waterloo, March 22.

Miraftab, F.K. 1996. 'Space, gender, and work'. Pp. 63–80 in E. Boris and E. Prügl, eds. *Homeworkers in Global Perspective*. New York and London: Routledge.

Moos, M., J. Whitfield, L.C. Johnson, and J. Andrey. 2006. 'Does design matter? The ecological footprint as a planning tool at the local level'. *Journal of Urban Design* 11, 2: 195–224.

Nippert-Eng, C.E. 1995. *Home and Work: Negotiating Boundaries through Everyday Life*. Chicago: University of Chicago Press.

Pratt, J.H. 1999. *Telework America National Survey: Cost-benefit of teleworking to manage work/life responsibilities*. Prepared for the International Telework Association and Council. Washington, DC: ITAC.

Rose, G., N. Gregson, J. Foord, et al. 1997. 'Introduction'. Pp. 1–12 in N. Gregson, G. Rose, J. Foord, S. Bowlby, et al., eds. *Feminist Geographies: Explorations in Diversity and Difference*. Women and Geography Study Group of the Royal Geographical Society, with the Institute of British Geographers. Harlow, England: Longman.

Shaw, S.M., J. Andrey, and L.C. Johnson. 2004. 'The struggle for balance: work, family and leisure in the lives of women teleworkers'. *World Leisure Research* 45: 15–29.

Strazdins, L., and D.H. Broom. 2004. 'Acts of love (and work): gender imbalances in emotional work and women's psychological distress'. *Journal of Family Issues* 25: 356–78.

Turcotte, M. 2007. 'Time spent with family during a typical workday, 1986 to 2005'. *Canadian Social Trends* Catalogue No. 11–008. Ottawa: Statistics Canada.

Wikström, T., K. Palm Lindén, and W. Michelseon. 1998. *Hub of Events or Splendid Isolation: The home as a context for teleworking*. Sweden: Lund Universit, School of Architecture.

Chapter Five

Eggs, Raspberries, and Domestic Space: The Spatial Construction of Family in 1920s–30s New Brunswick

Robert Summerby-Murray

Introduction

This chapter focuses on the connection between changes in women's mobility and the dynamics of family geographies. Most recent research on women's mobility in North America has focused on the journey to work for paid employment. Applying a time-geographic approach, primarily to contemporary urban and suburban settings and in the context of dramatic increases in female participation rates in the waged workforce since the 1960s, various insights have been gained into the renegotiation of traditional gender roles, the relationship of waged work to domestic practices, and the blurring of traditional views of home, work, family, and **spaces of production and reproduction**. In contrast, little attention has been paid to the origins of these processes in the 1920s and 1930s as automobile use changed individual mobility and created new tensions within the geography of the family. Even less consideration has been given to rural Canadian families and the changing roles of women between the two world wars. This chapter analyzes the diaries of two farm women in 1920s and 1930s New Brunswick and finds dramatic differences in spatial mobility within the family unit. Eggs and raspberries become metaphors for both the constraints on women's mobility, as well as crucial connections to the world beyond the farm gate. In reflecting on and negotiating their mobility in a male-dominated economy, these women provided important examples of the renegotiation and reconstruction of the farm family and its spatial dynamics in a time of socio-economic adjustment.

Gender, Space, Mobility, and Family: Intersections in the Literature

The relationship between **gendered mobility** and the spatial construction of family life requires further exploration, despite significant work in this area over the past two decades. The advent of explicitly feminist epistemologies in the social sciences and the investigation of gendered spaces in disciplines such as geography, anthropology, history, and sociology have given rise to a profound reassessment of the social, economic, and political relationships that underpin and condition women's lives and daily activities. This reassessment has implications for the construction of concepts of family and family geographies, as well as the particular epistemologies that condition our analysis. Several authors have challenged what Duncan describes as the 'unmarked masculinist subjectivity' of traditional geographical approaches (Duncan et al. 2004: 363), arguing instead for greater incorporation of women's voices and understanding. For example, Rose (1993) deconstructs various approaches within geography ('sparing no-one in her critique, including herself', as Domosh puts it, in Duncan et al. 2004: 366) while Duncan constructively suggests that we must be aware of the 'myriad examples of geographers' exclusions and erasures of women and others who happen to vary from a naturalized, unstated white, masculinist, bourgeois, heterosexual norm' (Duncan et al. 2004: 363). Our current understanding of the geography of the family gains much from this discussion; rather than being a new or sudden analytical view, the concept of family geographies flows from earlier identification of the erasures highlighted by Rose, Duncan, and others. The current chapter attempts to address some (although certainly not all) elements of erasure, particularly the exclusion of women's voices from our understanding of change in rural farm economies. The chapter suggests that gendered mobility provides a lens through which the shifting dynamics of family geographies can be viewed.

Strongly influenced by studies in time geography, women's mobility and its influence on the geography of the family has been analyzed largely through the journey to work or the **spatial entrapment thesis** (England 1993). This analysis relies on Marxist interpretations of productive and reproductive work such as those so handsomely and elegantly developed by Massey in her now-classic discussion of the spatial divisions of labour (Massey 1984). Applying such analysis to the dramatic increase in female participation in the waged workforce in the post-1945 era (and particularly since the 1960s), there is now a preponderance of discussion on contemporary urban and suburban settings. In some cases, this work has an explicit focus on the shifting dynamics of family and home (for example, Dyck 1990, 2005; MacKenzie 1989; MacKenzie and Rose 1983); in other cases, the focus has been more explicitly time-geographic and considers the mobility constraints of women and their fluid negotiation of productive and reproductive dualisms (Christensen 1993; Dyck 1989; England 1993; Hanson and Pratt 1991; Kwan 1999, 2000; Laws 1997; Pratt and Hanson 1993; Rosenbloom 1993). Linda McDowell sums this up with a suitably gendered domestic metaphor derived from Foucault's earlier discussion:

> While we are all affected by the radical transformation of local and global relations outlined above, by the power of multinational capital and global telecommunications, there are radical inequalities in the spatial spread of individuals' lives. For some, the network of points or skein referred to by Foucault above is a tightly constrained local pattern,[;] the skein, with its wonderfully woolly metaphor, is a trap, whereas for others the interstices of the network are separated by enormous distance and the connections are paths to greater freedom, an internet in cyberspace perhaps, rather than the homely skein of knitting wool. (McDowell, in Duncan ed. 1996: 31)

Much of the discussion outlined above is derived from explorations of contemporary urban life and notions of the mid-to-late twentieth century nuclear family. Indeed, there is a significant and valuable contribution from contemporary feminist political economy literature that has sought to explore the spatialities of women's work and to situate this in the context of class and gender struggle (Maroney and Luxton 1987). The contributions of historical geography to our understanding of the skein of gendered mobilities and the construction of the network interstices that help to construct family spatial relationships need further exploration. Particularly, far more attention must be paid to these 'radical inequalities in the spatial spread of individuals' lives' and their historical precedents.

In this regard, the examples in this chapter are informed by historical work on the late nineteenth- and early twentieth-century industrial nuclear family (Bradbury 1990; Franzoi Bari 2000; Miller 1982). Bradbury's analysis of pressures within the industrial families of two nineteenth-century working class suburbs of Montreal reveals the significant roles of 'working wives' (as well as child labour) in supporting a process of urban industrialization. Similarly, analyses of women working beyond the home (such as the industrial 'working girl' whose experiences have a surprising relevance for one of the central characters in this chapter) provide various examples of dramatic changes (and ambiguities) in the spatialities of women's work (Boyer 2004; Domosh and Seager 2001; Strange 1988). For rural women and families particularly, the historical precedents for the post-1960s changes in women's mobility need further investigation, especially in relation to the rise of **automobility** in the first decades of the twentieth century (Scharff 1991). In an earlier assessment, Kinnear (1988) analyzed the experiences of rural women in 1920s Manitoba, noting the significance of women's work and the social conditions of farm life (as evidenced by a survey carried out by the United Farm Women of Manitoba). While this survey and Kinnear's analysis considered perceptions of comparative isolation, they neglected the new dynamics of family geographies that were being reshaped by the use of the automobile in the 1920s. Similarly, examples from this period within Conrad et al.'s (1988) analysis of diaries and letters of Nova Scotian women are strong on daily activities and seasonal rhythms but give less attention to the changing dynamics of women's mobility, particularly in relation to economic change. While differences in gendered mobilities are implied in these studies, there is little explicit recognition of their impact on the operation of family life. The concept of family geographies requires a much better developed understanding of these mobilities.

In the analysis which follows, I explore the day-to-day spatial activity patterns of two rural women in 1920s–30s New Brunswick, drawing upon the diaries of Alice (Etter) Bulmer and Ella Blanche Anderson to consider the impact of shifting gendered mobilities on their family geographies. In many ways the stories of Alice and Ella are unremarkable—and that is precisely the point. The words of these ordinary women, who are bound up in local agricultural economies dominated by conceptions of men's productive work, provide representative examples of the connections between changing gendered mobilities, the nature of women's reproductive work in the home, and the spatial construction of families. By relying heavily on these women's own words, the chapter attempts to challenge the **masculinist normativity** of earlier analyse of gendered mobility and family geographies and to allow something of Alice and Ella's own worlds to intervene in our understanding. In addition, the chapter indicates that the literature on feminist political economy can be further informed by the family geographies of these women and their lived experience. Particularly, I argue that the enhanced mobility of rural women in the 1920s and 1930s (albeit at dramatically different rates than for men) challenged the former domestic order within family geographies and opened up new roles for women as economic contributors beyond the farm gate. It is clear, however, that Alice and Ella represent a particular circumstance of women's experiences and family geographies. This experience is predominantly white, Anglophone, and of a rising middle class. Crucially, all three of these factors contribute in the first instance to the availability of Alice's and Ella's diaries. Simply put, the diaries and letters of women of colour, Francophone women in this region, and working class women rarely make it to the archives. Tellingly also, the primary reason for the availability and accessibility of Alice and Ella's diaries is their inclusion in the archival fonds of

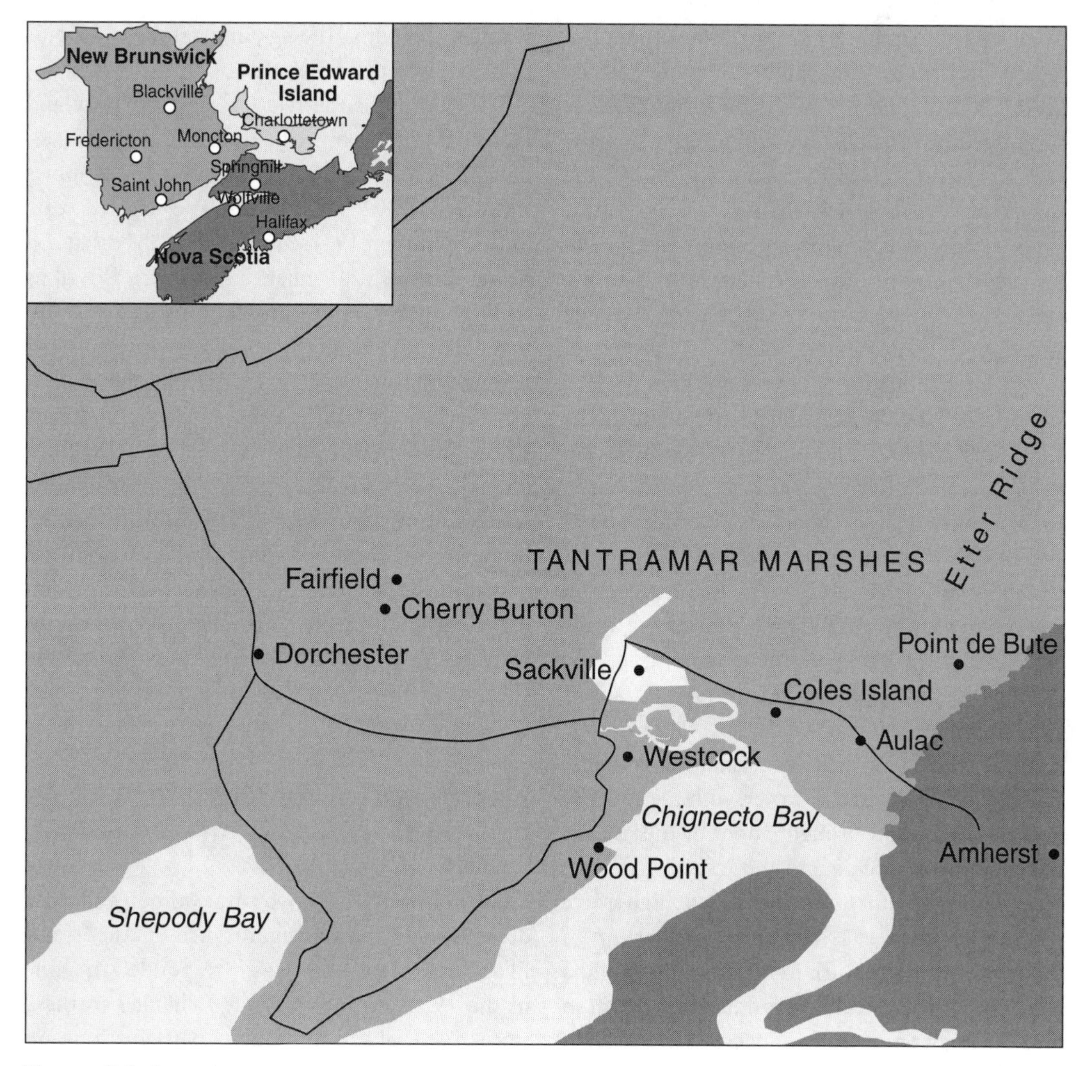

Figure 5.1 Location map

their spouses and families rather than via inclusion on their own right. Conrad et al. (1988) refer to the potential of women's diaries and letters to 'turn history inside out', to build a social history based on everyday realities rather than great events alone. Even in the examples which follow, with their attempt to redress the erasure of women's voices, the reader is cautioned that 'turning history inside out' still requires us to ask 'whose history' and 'for whom'.

Economy, Society, and Space in the Tantramar Marshes in the 1920s and 30s

Alice Bulmer's and Ella Anderson's diaries record the myriad details of farming life on the Tantramar Marshes at the head of the Bay of Fundy, New Brunswick, in the 1920s and 1930s and record numerous connections to small towns and villages locally and regionally (Figure 5.1). Both diaries tell of the mixed-farming economy of this productive region, particularly the farming of cattle and the production of marsh hay, an important export crop. Building (often literally) on the seventeenth- and eighteenth-century Acadian dyking of the coastal marshes and river margins, an intensified export-oriented agricultural economy emerged in the later eighteenth and early nineteenth centuries, benefiting from urban growth in Maritime Canada and from the demands for foodstuffs and hay in the logging industry (Summerby-Murray 1999; Wynn 1985). By the mid-nineteenth century, the Tantramar Marshes had become the centre of a hay economy, aided by new railway connections in 1872, exporting hay as far afield as Newfoundland and Boston to supply horse feed for both industrial uses and urban transport systems. In 1921, hay production covered 72 per cent of the available agricultural land with the remaining acreage devoted to oats, potatoes, and turnips (Bezanson and Summerby-Murray 1996).

Alice's and Ella's lives were changed dramatically shortly thereafter, however, with the collapse of the hay economy brought about by the sudden rise of the automobile and the resulting decline of demand for horse feed. From a high price of $28 per ton in 1920, hay fell to $7 per ton in 1938, prompting a return to mixed farming and a search for other agricultural products. This shift in focus brought large-scale changes in land use and agricultural practices, particularly for marsh agriculture in the 1930s (Ennals, Summerby-Murray, and Ennals 2004; Summerby-Murray 2003). As the value of hay declined, farmers were less able and willing to invest in the maintenance of drainage and dyking and large areas were abandoned. For farming families, new crops and livestock had to be managed, creating new labour demands. Significantly, other forms of enterprise were engaged to generate off-farm income and diversify the family economy. This chapter notes that many of these additional enterprises were undertaken by women, including the egg and raspberry operations referred to in the title.

The decline of the regional hay economy and increased automobility had positive but differential effects on Alice's and Ella's family geographies through the 1920s and 1930s. The selected years of diary entries (1931–32 and 1937–38) discussed below demonstrate enhanced mobility as well as rising socio-economic status for those farmers who managed to diversify beyond the hay economy through the 1930s. It is clear from the diaries, however, that this diversification also created new stresses in domestic and family life, ranging from increased pressure on women to contribute to sources of off-farm income to frustrations at the limits on daily mobility. Key findings from the diaries include significant differences in the reporting of male and female activities, clear divisions between male and female mobility, and resulting differences in the geographies of family life.

Narratives of Family Life: Economy and Mobility

We turn now to these two rural women and their diaries, analyzing the significance of Alice's and Ella's changing mobility as it responded to shifts in the 1930s economy, created changes in these women's social and community participation, and produced new pressures on the geography of the family.

Our local diarists were very much aware that they were chronicling a narrative that was significant beyond their immediate selves. Alice Bulmer entitled her diary, *The Diary of the Bulmers, written by Mrs. Walter J. Bulmer of Aulac, New Brunswick.* There was an implicit, public element to this narrative function, even if the audience was less clearly defined. It is unlikely that either woman anticipated that her diary would be preserved (let alone studied in an academic setting) but there was a strong imperative in each case to create a faithful record of daily occurrences that affected family, farming, and community. These diaries are not personal daily reflections, for the most part; indeed, Alice and Ella often appeared to be at pains to avoid having their personal opinions intrude on what was a more public document. Instead, each diary reflects a strong sense of responsibility in recording the activities of the family, the weather, the daily round of economic functions, and the social connections of the district. Matched with farm accounts (which are available in the case of the Anderson family), the diaries provide a humanizing commentary on changes in the farm economy of the 1920s and 1930s. Yet, while it is clear that these two women diarists hardly saw themselves as carrying out anything remotely exotic, let alone romantic, in the recording of daily weather conditions or who was mowing hay on the marsh, both diaries display elements of tension in the domestic sphere and the family as new economic geographies emerged in the 1930s. Alice Bulmer's selection of a notebook with an exotic 'Gypsy Girl' cover might be considered to be her own personal statement (Photo 5.1), while Ella Anderson was frequently resentful of the limits of her world. These tensions hint at dissent with the metanarratives of domesticity and masculinist economic materialism that dominated the rationales for the diaries in the first place. The diary narratives thus provide a remarkable source for informing our understanding of the gendered mobilities of family life in early twentieth-century rural New Brunswick.

Photo 5.1 Cover of Alice Etter Bulmer Diary 1920

Photo: Mount Allison Archives. Alice Etter Bulmer Fonds. 8822

Alice Bulmer (1870–1956)

Alice Bulmer lived with her husband Walter and three children Mary, Gordon, and Charlie in the small settlement of Aulac on the New Brunswick–Nova Scotia border (see Figure 5.1). Both the Bulmer and Etter families (Alice's maiden name) had farmed in the area through the nineteenth century and the area is still referred to as the Etter Ridge. Alice's diary entries follow a consistent pattern from the early 1920s, listing in priority order the weather, the men's work of the day, any visitors, and any other noteworthy events (Photo 5.2). The original diaries noting activities from September 1920 to November 1932 are held in the Mount Allison University Archives; one representative year (1931–32) has been selected for most of the analysis in this chapter but frequent reference is made to annual patterns throughout the 1920s.[1]

Analyzing these diary entries from a spatial perspective produces a highly gendered geography reflecting the new dynamics of family relationships in this period. As one would expect of a rural New Brunswick family in the 1920s and early 1930s, the men's work was concentrated outside the home, largely on the adjacent marshlands, and involved

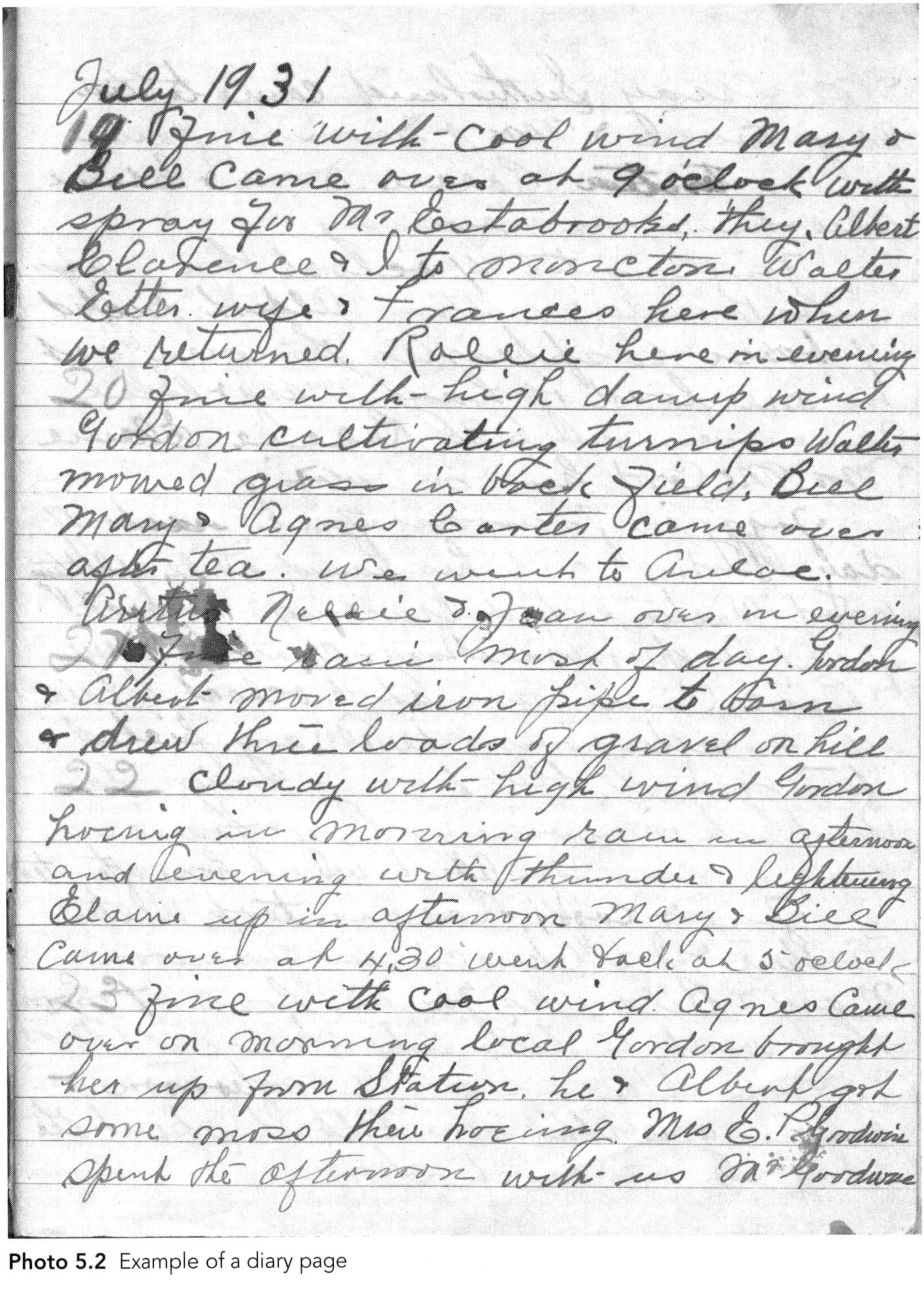

July 1931
19 Fine with cool wind Mary &
Bill came over at 9 o'clock with
spray for Mr Estabrooks, they, Albert
Clarence & I to Moncton. Walter
Etter wife & Frances here when
we returned. Rollie here in evening
20 Fine with high damp wind
Gordon cultivating turnips Walter
mowed grass in back field. Bill
Mary & Agnes Carter came over
after tea. We went to Aulac.
Arthur Nessie & Joan over in evening
21 Fine rain most of day. Gordon
& Albert moved iron pipe to barn
& drew three loads of gravel on hill
22 Cloudy with high wind Gordon
hoeing in morning rain in afternoon
and evening with thunder & lightning
Elaine up in afternoon Mary & Bill
came over at 4.30 went back at 5 o'clock
23 Fine with cool wind. Agnes came
over on morning local Gordon brought
her up from Station, he & Albert got
some moss their hoeing. Mrs E. P. Goodwin
spent the afternoon with us Mr Goodwin

Photo: Mount Allison Archives. Alice Etter Bulmer Fonds. 8822

Photo 5.2 Example of a diary page

making hay, hauling manure, and the planting and harvesting of potatoes, carrots, and turnips. There was further seasonal work on the marsh dykes and ditches in the spring and work in the forested woodlots in the winter. As well, the men planted strawberries, tended raspberries, and carried out a variety of farm maintenance tasks. As an example, Alice Bulmer reported on the activities of seven men on the Bulmer farm in September 1931 (her husband Walter, son Gordon, and several other farm workers). Just under half (49 per cent) of the diary entries recording men's activities were for work in the fields, 20 per cent for general chores, just over 6 per cent for maintenance work on dykes and ditches, a similar percentage for harvesting raspberries, and just under 5 per cent each for work in the family greenhouse and collecting firewood in preparation for winter. A further 9 per cent of entries noted off-farm social events, primarily church-related or attending funerals—although it is worth noting that attendance at these appear from the diaries to have been initiated by Alice, with men in accompanying and chauffeuring roles. The work of the three women (Alice, daughter Mary, and another domestic worker, Elaine) is grossly under-reported, representing only 20 per cent of the total entries during September 1931 and ignoring much of Alice's daily routine of cooking, cleaning, and maintaining the household. Only 6 per cent of the entries referring to women's activities involved cooking and cleaning. Instead, Alice's recording of her own activities highlighted primarily special events outside the house. Alice and Mary were more active than the men in the family greenhouse during September (38 per cent of entries), but it was the limited off-farm activities that generated the most comment in the diaries: attendance at church (including funerals) dominated (37 per cent), with attending W.A. (Women's Auxiliary of the Anglican Church)[2] meetings taking up 12.5 per cent of women's activities. Other socializing (including dances and tea parties) accounted for 6 per cent of the recorded activity.

While there was an important seasonality to the work of both men and women noted in Alice's diary (from men making hay to Alice spring-cleaning), the annual pattern of activity was remarkably consistent with that of September 1931 (with the addition of more time on an annual basis devoted to the men harvesting wood and the appearance of occasional winter recreational events such as skating). Notable, however, was the division of labour between seasonal tasks and particular individuals. Alice's son, Gordon, was taking on a larger amount of the work associated with working the marsh (making hay, cultivating crops, managing animals) and harvesting wood, in part because of the usual intergenerational transfer of responsibilities but also because Alice's husband, Walter, was by this time increasingly ill. Throughout 1931–32, Walter travelled to Montreal, Quebec, for medical treatment every two or three months although Alice notes 'no improvement in his condition' (Alice Bulmer Diary, 30 June 1932). When able, Walter's work during these years was mostly confined to the household's raspberry operation. Gordon's increasing work responsibilities reflected the Bulmer's expanded mixed-farming operation as the family sought to diversify their operation away from its reliance on hay. His activities were recorded on a daily basis, ranging from weeding carrots and hoeing turnips, to ditching, making hay, hauling lumber, ploughing, and sowing oats. Other operations were also carefully divided: farm worker Rollie mostly assisted in the marshes and provided his services as a driver (including driving Alice to church and various social events on numerous occasions); off-farm labourer Bill (who may also have become Mary's husband, Alice's son-in-law) was hired almost entirely to work in the farm's small greenhouse operation and travelled from Amherst, Nova Scotia, on most days.[3] The tasks of Alice, daughter Mary, and farm worker Elaine were recorded in less detail but demonstrate work in the greenhouse operation and support in the production of raspberries and eggs (particularly during the 1920s). Mary was allowed to contribute to some of the men's work on the marsh; the 4 per cent of fieldwork recorded for women's activities for 1931–32 was entirely the result of her assisting with haymaking.

The comparative geographies of men and women were highly distinctive, reinforcing the argument of separate spatial spheres within the family with attendant gendered mobilities. Alice recorded men's movement in great detail, as part of the chronicling of the operation of the farm economy. We learn where men were headed in the morning, sometimes down to the exact field:

> 11 May 1932: Fine with southwest wind. Albert here helping in greenhouse. Bill came over in car in morning. Gordon sowed 22 bags oats in upper field. Uncle Art took down hall stove pipe, polished pipe. Rollie got team to plough his garden.

> 18 May 1932: Fine. Gordon took load logs to mill in morning, choring and leveling ditch, in after tea. Bill came over in morning. Gordon and Rollie took cattle to north field after tea. Percy got up 90 lbs local potatoes.

On occasion, Alice even added intriguing but favourable commentary on son Gordon's increased number of visits to a Miss Palmer in Woodpoint:

> 30 August 1931: Miss Palmer here to tea Gordon drove her to Dorchester in Rollie's car.

> 27 May 1932: he [Gordon] and Miss Palmer to Amherst to dance in the evening.

> 5 June 1932: Gordon to Wood Point in afternoon brought Miss Palmer here to tea.

By comparison, Alice's work and movement, when it even makes it to her diary, is limited to the geography of the home or to domestic enterprises such as raspberries and egg production which provided important sources of off-farm income. For example, on 18 May 1932, when Alice records a considerable amount of men's activity and mobility (see above), she notes for Mary and herself simply, 'Mary over in evening and helped me clean. I cleaned spare room'. Also, there is only one instance in the diary of Alice benefiting at the domestic level from new technology: the arrival of a second-hand washing machine which was either a gift or on loan. One week after its arrival, Alice recorded, 'We used our electric washer very pleased with it' (12 October 1931).

Alice's references, particularly through the 1920s, to the work of managing poultry, collecting and selling eggs, and harvesting seasonal berry fruits such as raspberries, blueberries, and cranberries provide rare examples of breaking the boundaries of the domestic sphere with enhanced mobility. For example, annual trips to Fairfield (approximately 19 kilometres) to harvest blueberries and to Blackville (over 100 kilometres) to pick cranberries were significant instances of off-farm travel. Such occasions for travel were described in comparative detail, including one case of a journey to Point de Bute, a distance of a little over six kilometres, to vote in an election; the diary entry is complete with the names and parties of candidates and the number of votes gained by each (Alice Bulmer Diaries, 13 October 1931). Similarly, a shopping trip to Moncton (approximately 75 kilometres distant) elicited a long and excited recitation of the items purchased.

The sale of eggs in the nearby town of Sackville in the early 1920s provided an interesting case of Alice's domestic enterprise and its expansion beyond the farm. The diaries record monthly totals of eggs laid by Alice's pullets (over 4,000 eggs per year). Even allowing for the egg-rich cooking of the 1920s, this was a considerable surplus for a household. More than just a further (and welcomed) cash input to the household economy, the sale of eggs was an enterprise controlled by Alice. Ironically, it is the journeys of Walter and the boys to Sackville to sell Alice's eggs that are chronicled in the diary entries here rather than Alice's work or her physical connection to a market economy. Alice's geographic mobility was severely constrained by the local and domestic economy and her role within it, even to the point of her diary entries concentrating on male mobility, perhaps in an attempt to travel vicariously but more likely simply reflecting the dominance of men's mobility within the construction of a family geography.

The development of the Bulmer's raspberry operation was a further example of the intersection of male and female labour but with constrained female mobility in terms of family geographies. By 1931, the production of raspberries was a significant seasonal activity contributing to farm income (along with an expanded greenhouse operation which provided cut flowers for funerals and some lettuce production). Alice controlled the accounts for the raspberry operation, based on the notes in her diary (and various account records in the Alice Etter Bulmer fonds, Accession 8822/2). Although some of the men (such as hired help Percy, Bill, and Albert) assisted at various times through the season (including Walter increasingly), raspberries were clearly women's work and women's space, particularly at harvest time. Alice's diary contains numerous references to 'the girls picking raspberries' (particularly Mary). A diary entry during the height of the harvest of August 1931 records the picking of berries by 'the girls' and the role of men in providing most of the subsequent mobility for the transportation and sale of the raspberries:

> Girls here picking raspberries. Gordon and Albert took one crate to Sackville at noon in Rollie's car. Mary over at 1:30 took crate back to Amherst with her. She Bill and Mr/Mrs Jetts [spelling uncertain] over at 6:30. Mary, Albert and I to Sackville in evening took raspberries 12 boxes to Mrs JM Oulton also 12 boxes to Mrs AC Anderson. (Alice Bulmer Diary, 4 August, 1931)

Male mobility beyond the farm in Aulac is represented extensively in the diaries, part of the regular geography of a farming family. The men travelled constantly, largely between Amherst, Nova Scotia, Sackville, and Moncton. Even social visiting that may have been initiated by Alice involved Walter through the 1920s and later, as Walter's health declined, Rollie—presumably because Alice did not drive. Alice's diaries also indicate that such social visiting (for quilting and afternoon tea, for example) was not a priority for males: when women friends visited Alice in the 1920s particularly, a number of diary entries noted that Walter was 'working outside', present and recorded as part of Alice's narrative responsibility but not part of the feminized domestic spaces of the house.

Comparing frequency of off-farm visits to particular destinations for the period 1920–32 supports the gendered and economically dependent nature of Alice's mobility and provides a useful snapshot of the spatial connections of the family. Based on the entries in her diaries, Alice travelled from Aulac to Sackville on average once per week (a return distance of approximately 10 kilometres), to nearby Point de Bute and more distant Westcock (20 kilometres return) once a month, to Fairfield and Moncton (approximately 150 kilometres return) once a year, for an annualized average travel distance of 1,100 kilometres. In contrast, the men of the family were in Sackville, Point de Bute, Amherst, Cherry Burton, and Dorchester once a week and Moncton once a month. This travel was primarily by Walter, Gordon, and Bill, yielding an annual average travel distance of nearly 4,800 kilometres for male travel.[4]

Daughter Mary's mobility raises important questions also. She travelled almost as often as the males in the household: to Amherst, Point de Bute, and Westcock at least once a week and to Dorchester and Moncton once per month, with an annual average of 2,760 kilometres. In part, this was because in the 1920s she appeared to have been attending school or working in Amherst. Alice was clearly fascinated with the heightened mobility of her daughter's generation and went to great lengths to document it, noting whether Mary travelled to Amherst by car or on the 'local' or 'No. 17' train service between Aulac Station and Amherst. In the early 1930s, Mary and Bill travelled many times per week from Amherst to the Bulmer farm, either on the train or, more frequently, in their own car. For example, 'Mary got a car at Smiths garage and took Bill to Sackville to deliver flowers at Mrs Arthur Smiths then came over for Bill in their own car at 5 o'clock both went home after tea' (Alice Bulmer Diary, 13 November 1931).

In summary, Alice's diary provides a remarkable articulation of the spatial underpinning for the daily round of rural family life in 1920s and 1930s New Brunswick. Her own sense of space and mobility, at least as recorded in the diary, was almost entirely domestic and limited to the geography of the home, farm economy, and immediate community. It was predominantly men's work that was accorded priority; apparently she considered her own work so routine that it rarely made it to her diary. Generally, Alice's view of the materialist economic structures around her conformed to the highly gender-differentiated spaces that dominated late nineteenth- and early twentieth century-Canadian conceptions of the domestic and productive sphere: men were productive, with the spaces of production involving travel and mobility—even simply mobility around the farm from woodlot to hayfield. These productive actions by males were deemed worthy of record in Alice's diary while Alice's own work and mobility was rarely mentioned, reinforcing the central notion within the study of this family geography: that women's spaces were those of the reproductive domestic sphere, limited to spaces of the home and the maintenance of family life.

Ella Blanche Anderson (1880–1955)

Ella Anderson's diary shares Alice Bulmer's preoccupation with eggs and raspberries, demonstrating a similar constrained mobility. There are indications of increased economic and travel independence, occasioned by greater economic security and better access to automobiles in the late 1930s. As her diary records, however, Ella is less content with the gendered limits to her mobility and provides more evidence of tension within the family structure.

Ella Anderson was born at Coles Island, currently the location of the Radio Canada International shortwave broadcasting towers that dominate this section of the Tantramar marshlands. She completed school in Sackville and attended Acadia University in Wolfville, Nova Scotia, for one year. Unlike Alice Bulmer, Ella Anderson had early experience of a surprisingly frequent cause of mobility for young Maritime women: leaving the region to find work in the industrial towns and cities of New England (see, for example, Blewett 1990; Dublin 1979, 1993; and Norkunas 2002). Ella worked in a watch factory in Waltham, Massachusetts for a number of years before returning to live on the family farm at Coles Island (Photo 5.3) in the1930s. Her experiences as an industrial working girl (Boyer 2004; Strange 1988), with an income and independence, undoubtedly influenced her later attitudes towards life in a rural setting and prompted a number of resentful comments in her diary.[5]

As with Alice Bulmer's narrative, Ella Anderson's diary reflected the operations of the mixed-farm economy of the 1930s: the spring ploughing,

Photo: Mount Allison Archives, Alice Etter Bulmer Fonds, 8317

Photo 5.3 Patterson family and Anderson family homes on Coles' Island Road, 1930

manuring of fields, and moving of cattle on the marsh. With less detail than Alice, Ella recorded the elements of the daily round, the binding of hay, the shingling of roofs, the cleaning of closets. Sometimes, the entries were brief and to the point: 'men went to marsh while Self tended to chickens'; at other times, more detail was given or Ella gave way to personal comments on her own situation. The division of labour within the Anderson household demonstrated distinctive male and female patterns similar to those of the Bulmer family earlier in the decade. In the 1937–38 year, for example, 45 per cent of the men's recorded activity was for work in the fields with a further 27 per cent of the entries noting their general chores. Harvesting and haymaking took up 7 per cent and planting approximately 2 per cent. Entertainment, mostly off-farm, was recorded in 7 per cent of the entries. Ella also took note of illnesses within the family, listing these as 3 per cent of the men's annual activity. The work and whereabouts of Ed and Albert in particular were noted almost daily in Ella's diary, with these two men carrying out the bulk of the work on the farm. Indeed, Ella's diary provides a significant example of the activities of marshland farming and the financial success of the Anderson family in diversifying their operation.[6]

The activities of the women of the household were reported in far greater detail than in Alice Bulmer's diary, perhaps reflecting Ella's earlier independent work experience and a stronger sense of the value of domestic work in sustaining the family and its farming operation. It is clear also that Ella's recording of her work was often deliberately resentful (see below). The activities of women in the household were dominated by general housework (29 per cent) and food preparation (26 per cent). Further work outside the house was noted (5 per cent) as well as contributions to harvest (1 per cent). Significantly, Ella records a much larger proportion of her activity engaged in recreation (27 per cent) than the men in the family and considerably more than Alice Bulmer. It is argued below that much of this increased recreation activity resulted from Ella's enhanced mobility, but the extensive reporting of it in the diary also represented Ella's dissatisfaction with the constraints of her family geographies, especially domestic responsibilities and the spatial limits imposed on her by the operation of the family farm. Her recording of her daily work activities, as well as those of her aunt, Cassie Patterson, was frequently resentful, as the following excerpts demonstrate:

> Still cleaning hope to finish soon (28 May 1937).
>
> Self trying to cook worked all day and accomplished nothing ready for bed (15 October 1937).
>
> Cass doing some washing, self doing the work (25 January 1938).
>
> Cass and self manage to be busy most of the day (18 March 1938).
> (Ella Anderson Diaries, MTAA)

Turning now to gendered mobility within the family's geography, Ella recorded in considerable detail the men's off-farm travel. Much of this was the result of the men's successful participation (with selected livestock) in various agricultural fairs, field days, and competitions in Canada's Maritime provinces, involving travel to Prince Edward Island, Halifax, Kentville, Saint John, and Fredericton during 1937–38, as well as more locally in Sackville, Port Elgin, and Amherst. Ella recorded the logistics of moving cattle from one exhibition to another and listed with some pride the results of these competitions, including any prizes ('didn't get as many 1st prizes in Saint John as in PEI—chickens did well' [9 September 1937], 'Albert arrived home with $122 in prizes from Halifax Exhibition' [4 October 1937]). She was particularly pleased when she was also involved:

> A wonderful day for the exhibition, got away with the cattle in good time took 27 and got 16 first prizes and a few seconds on them. Didn't get anything on my bread but got a second on [undecipherable] (22 September 1937).
>
> Had a big day at the fair. Ed and William went to Amherst early. The big fair is open at 10:00 a.m. Self went to town, Clarry came for me, got home about 5. Men got back at six. Won the grand championships, carried off the honours in the beef classes (5 November 1937).

Male mobility was clearly rewarded in the family dynamic, both in terms of greater freedom from the home spaces of the farm and in terms of external status through recognition at agricultural exhibitions—which in themselves tended to celebrate male achievement and the male productive spaces of agricultural enterprise. This is not to say that women's domestic work was not also celebrated in agricultural exhibitions and fairs but the winners of prizes for the best raspberry conserve were implicitly confirming the boundaries of women's domestic and reproductive work, mobility and geography. And although Ella would never have put it in these terms, these differences in mobility within the family unit reflected the 'radical inequalities in the spatial spread of individuals' lives' suggested by McDowell (in Duncan ed. 1996, 31), reinforcing a particular political economy of the family.

Opportunities for Ella to increase her mobility and extend her geography beyond the domestic sphere were recorded in great detail. Most of this travel also involved males of the family as drivers and companions but in some cases Ella was able to move beyond the constraints of the house and farm gate and participate in the sort of mobility enjoyed by men. Ella's ability to travel locally was much greater than that of her earlier neighbour, Alice Bulmer. The variety of reasons for travelling—the presence of a considerable number of social events and calls, and access to enhanced recreation (including movies and shopping trips)—all combined with greater access to cars, suggests that Ella's mobility was benefiting from the improved financial circumstances in the rural economy as the Anderson family members diversified their farming in the later 1930s. Her diary records numerous instances of going 'to town'(meaning Sackville), Amherst, and Moncton for 'shopping', 'a dress fitting', 'church', and 'to go to a movie' (usually then followed by a brief comment of what she thought of the movie). This level of recreational travel was completely unknown (or at least unrecorded) by Alice Bulmer, just a few years earlier. Ella is explicit that some of her travel is a luxury—in some cases for Sunday picnics or simply joyriding. For example,

> Went for a ride after dinner had Nellie, Ruth, Jean and Barbara, Clarry did the driving. Called at the greenhouse,[7] museum + ship-railway which isn't . . . (30 May 1937). [8]

> Neals has a new car Pontiac. Went back with Fred then went to Springhill and the farm trying the new car out (1 July 1937).

> Got our lunch up and left for Amherst shore before 11, Clarence, Barbara, Jean, Flo, Mrs Morton and Self had a great day (18 July 1937).

There was also frequent reference to increased mobility related to egg and poultry sales, a significant contributor to family income, as was the case for Alice Bulmer a few years earlier.

> Clarry and self went to town took eleven dozen eggs, did some shopping (March 17, 1938).

> Ed and self went to town this afternoon took 6 dozen eggs (April 1, 1938).

> Fred over for eggs he took 7 dozen (16 April 1938).

> Ed and Cass went to Amherst delivered the turkeys (21 December 1937) [the 8 turkeys dressed the previous day].

> Ed and Cass went to town, took turkey and chicken (23 December 1937).

> Dressed chickens Ed helped (20 January 1938).

In Ella's diary, women had direct involvement with these sales processes, including delivery, demonstrating increased mobility—although not yet automobility as they were still driven by the men of the household. Despite these increases in mobility as compared to Alice Bulmer, there remained significant differences within the Anderson family in terms of off-farm travel geographies. Based on an analysis of one volume of Ella's diary (for the period May 1937 to July 1938), men were engaged in almost one and a half times as much off-farm travel as women, with a concentration of activity during the fall as a result of attending the various exhibitions and fairs noted above. June and July also showed an increased number of days of off-farm travel, particularly related to recreational activities; indeed, women's travel, as reported in the diary, either equaled or exceeded men's in these two months. August had fewer days for both men and women as a result of harvest. While the general seasonal pattern applied to both men and women, there were significant differences in some months; in May 1937, for example, little off-farm travel was recorded for women, just one day compared to the men's twelve.

Further, while Ella does not mention travelling beyond Sackville, Amherst, Springhill,

Table 5.1 Travel distances and frequencies, Anderson Family, May–December 1937

Destination	Return distance from Coles Island (km)	Ella	Cass	Ed	Albert	William	Clarry
Sackville	8	25	11	24	15	8	3
Dorchester	30	0	0	0	1	1	0
Amherst	11	10	9	18	5	5	6
Moncton	106	1	1	1	0	0	0
Fredericton	600	0	0	0	1	0	0
Saint John	560	0	0	0	1	0	0
Halifax	600	0	0	0	1	0	0
Total distance (km)		416	293	496	1965	149	90
Average per week (km)		17	12	21	82	6	4

Source: Ella Blanche Anderson Diary, May 1937–June 1941, Mount Allison University Archives, Anderson Family Fonds, Accession 7832/2/4/2

the Northumberland Strait, and Moncton, the male members of the household (and particularly Albert) were making much longer trips to Shediac, Prince Edward Island, Kentville, Saint John, Fredericton, and Halifax. Table 5.1 shows the average distances travelled by each member of the Anderson household recorded in Ella's diary for the twenty-four week period May to December 1937 as well as the frequency of these journeys. While Ella was not travelling long distances, her trips to Sackville and Amherst were significant—and her mobility increased during 1938, particularly in terms of the social outings noted earlier. A more affluent economy was emerging and the mixed-farming enterprise of the Anderson family was clearly doing well. Further sources of funds came from land sales to the federal government to establish the broadcasting site for Radio Canada International. Between October 1937 and February 1938, Ella recorded several visits of men 'from Moncton and Ottawa to see about deeds', including the presence in her kitchen of an 'Ottawa lawyer in to get warm a good talker', and 'H. Freil of Moncton came in paid a second option on the Patterson property where the radio station is to be built'. Ella was later involved in 'visits to the lawyers' to manage parts of this transaction.

However, Ella provides clear signs in her diary of a disruptive tension in the family's geography, particularly her frustration with her limited mobility and the gendered differences among family members. In most cases, this was recorded simply as resentment at being left alone at home with no means or justifiable rationale to leave the farm. It doesn't appear from the diary that Ella ever drove; she was always driven by a man—although there is one entry about bicycling. There is also one instance of her returning to the farm from Sackville independently by public bus. Ella expresses frequent displeasure at the restrictions on her travel, despite what appear to be significant increases in her mobility by comparison to the situation of her neighbour Alice just a few years earlier

> All away tonight even the dog isn't here for company (29 May 1937).

> Alone most of the day. Another dull day. Cass not coming home this morning (6 June 1937).

All away in the afternoon, self alone (4 July 1937).

It is hard to imagine who Ella intended to have read these somewhat self-pitying entries but they speak to the loneliness and isolation of rural women (as Kinnear [1988] notes also), even in the midst of rising affluence and subsequent positive changes to family mobility.

Analysis of the activity and travel patterns of the Anderson family members shows a much wider geography than the Bulmer family in the earlier section of this chapter. Based on Ella's diary, there is even the suggestion that the gender differences in mobility are declining, in response to greater affluence, further altering the family dynamic. But it is also clear that these changes are insufficient from Ella's perspective and that she continues to resent the considerable inequalities in the geographic mobility of family members. The spatiality of family life, with its juxtaposition of individual differences based on gender, was clearly a continuing source of tension within the Anderson family.

Conclusions: Gendered Mobility and the Construction of Family Geographies

Eggs and raspberries function here as metaphors for the spatial fixities of domestic life that constrained the mobility of rural women in New Brunswick in the 1920s and 1930s and were significant determinants of family geographies. Just as Kwan (2000) argued for women in 1990s Columbus, Ohio, Alice Bulmer and Ella Anderson encountered (and constructed) various 'binding effects' that conditioned the geographies of family life, even at a time of increasing automobility. This chapter has argued also that it is imperative to view perspectives on mobility through women's eyes, through their own words recorded in their diaries, in order that we develop greater understanding of the historical geographies of gender and the construction of family in earlier twentieth-century rural Canada. Both Alice Bulmer and Ella Anderson shared domestic geographies that were dramatically constrained in comparison to the present day and yet the expansion of mobility for these women in the 1920s and 1930s suggests that some elements of the simple productive/reproductive space dichotomy outlined at the beginning of this chapter were being challenged. Crucial here is the extent to which these women were in control of their domestic spaces and their mobility. The nineteenth-century conception of gendered spaces, translated from middle class urban-industrial settings and mediated by a Victorian sense of domestic femininity within the family structure and the cult of domesticity in a rural setting, reflected the idea that men were highly mobile while women were relatively static. This was a significant determinant of the spatialities of family life in both cases analyzed in this chapter. In Alice Bulmer's case, domestic tasks referred to in her diaries reinforced traditional conceptions of male (productive) and female (reproductive) space. And yet, the diaries show that overall mobility was changing with some family members gaining more than others, including the suggestion that gender was not the only factor. While Ella Anderson demonstrated a more extensive social life and expanded travel distances, including those relating to her more extensive involvement in the mechanics of the agricultural economy, it is clear that she remained heavily dependent on males of the household (and beyond the household) for her actual mobility, reflecting dominant structures of family power, that is, masculinist power, as Rose (1993) would suggest. She was not herself automobile, generally, and thus remained in a state of dependency, not unlike that of Alice Bulmer. This dependent relationship reinforced stereotypical mobility roles within the family unit but it is clear from Ella Anderson's diary that increased affluence provided some women, at least, with greater opportunities for social travel. While these roles were to be altered dramatically during the coming war (including by the dynamic of women working in wartime industries), the gendered mobility and family structures of the rural economy in the 1920s and 1930s provided an important precedent for the suburban constraints on mobility that later authors have analyzed in urban-industrial settings in the immediate post-1945 era.

Finally, what do we learn here of the spatialities of family geographies as an object of study? This chapter has attempted to shed light on women's spatial constraints, arguing that inequalities of mobility have significant effects on the operation of family life. By relying heavily on Alice and Ella's own words, the chapter gives voice to women's perceptions and understandings of their family dynamic and counters a masculinist normativity that influences even the way questions are asked by the researcher and observer, frequently to the point of erasing women's voices, even if unintentionally. It is remarkable, for example, that traditional analyses of rural economies in Maritime Canada have made little use of women's diaries. The present chapter attempts to interpret these formerly silenced voices. To do this in the context of an exploration of the spatialities of family life suggests that a far more rigorous and nuanced understanding can be gained from the inclusion of women's voices. These voices themselves help to fill in the interstices of the network noted by McDowell (1996) at the outset of the chapter. And while it is simplistic to view the skein only as a spatial trap, after Foucault, it is clear that the homely skein unravels in different ways within the family, alternately reinforcing and challenging family power structures and spatial inequalities.

Acknowledgements

The preliminary research assistance of Lucie Orlando and Ryan MacDonald, students in GEOG 3301 Historical Geography of North America, is gratefully acknowledged. Particular thanks are due to Cheryl Ennals, former Mount Allison University Archivist, for guiding me towards the diaries and the significance of the many entries on eggs and raspberries; and to Donna Beale of the Mount Allison University Archives for assistance with related documents; and to Rhianna Edwards, Mount Allison University Archivist for the assistance with images and copyright permission. Detailed research assistance and contributions to subsequent analysis were provided by Mallory Baxter and Christina Tardif. Christina also produced the map.

Questions for Further Thought

1. What social conventions of the 1930s may have limited women's capacity to resist the gendered mobility constraints discussed in this chapter? In what ways do these conventions structure and shape family geographies?
2. Do the diaries analyzed in this chapter provide a full picture of the geographies of farming families in 1930s New Brunswick? What other types of data would be useful in helping us assess the variety of factors affecting these families?
3. What major changes in Canadian society since the 1960s in particular have affected the mobility of women in farming communities? How have farm families (and their geographies) been transformed as a result? Do the examples of Alice and Ella provide us with useful clues to the origins of these social changes and their resulting geographies?

Further Reading

Dyck, I. 1989. 'Integrating home and wage workplace: women's daily lives in a Canadian suburb'. *The Canadian Geographer* 33(4), 329–41.

England, K. 1993. 'Suburban pink collar ghettos: the spatial entrapment of women?' *Annals of the Association of American Geographers* 83(2), 225–42.

Kwan, M.P. 2000. 'Gender differences in space-time constraints'. *Area* 32(2), 145–56.

Miller, R. 1982. 'Household activity patterns in nineteenth century suburbs: a time-geographic exploration'. *Annals of the Association of American Geographers* 72, 355–71.

Endnotes

1. Throughout this chapter, various excerpts from each diary are quoted. Spelling and punctuation have been left as close as possible to the original diary entry.
2. The Women's Auxiliary, still a vibrant force in rural New Brunswick, was renamed the Anglican Church Women in 1967, a little over a decade after the Church of England in Canada became the Anglican Church of Canada (1955). At least one of Alice Bulmer's descendants is currently a member of the ACW in the Parish of Sackville (K. Stockall 2008, pers. comm.).
3. There were clearly some development issues with the family's expanding greenhouse operation in the 1930s. As an example, Alice records, 'Bill came over on morning local [train]. Fumigated greenhouse with sulpher [sic] with disastrous results killed most of the plants. He went home on the night local . . .' (Alice Bulmer Diaries, MTAA, 5 February 1932).
4. Walter's trips to Montreal for medical treatment have been excluded from this analysis as they are not representative of the general pattern of male mobility within the family between 1920 and 1932.
5. Ella's aunt, Catherine (Cassie) Patterson had followed a similar path to Waltham, recording her experiences in a diary and leaving us with a detailed record of a seasonally mobile female industrial worker from Maritime Canada contributing to the 'Lowell' industrial system.
6. The author was fortunate to interview Albert Anderson in 1997, just a few years prior to his death in 2004 at 92 years of age. His memories of haymaking and the operation of the family farm were particularly vivid. See Anderson (1997).
7. This was probably the Alice and Walter Bulmer family greenhouse. Both the Ella Anderson diary and the Alice Bulmer diary contain references to what appear to be the same people, either family members (for example, the 'Albert' of the Bulmer diary appears to be Albert Anderson) or other workers (Clarence [sometimes 'Clarry'] appears to the be same person, a farm worker or relative, in both diaries).
8. The ship-railway referred to here was the site of a proposed seventeen mile overland route by which small ships would be carried on a rail system across the Chignecto Isthmus, from the Bay of Fundy to the Northumberland Strait. Supported by Amherst, Nova Scotia, business interests (particularly Henry Ketchum) and with federal assistance, the project was begun in 1888 but was abandoned in 1896 as further federal funds dried up. See Underwood (1995) and University of New Brunswick Archives.

References

Anderson, A. 1997. Typescript of interview with P. Ennals, R. Summerby-Murray and C. Ennals. Albert Anderson Family Fonds, Mount Allison University Archives, Accession 8317/10/2. Audiotape reference accessible at www.mta.ca/marshland/pages/archivalsrc/anderson_audio.htm

Anderson, E.B. Diary, May 1937–June 1941. Mount Allison University Archives, Albert Anderson Family Fonds, Accession 7832/2/4/2

Bezanson, N., and R. Summerby-Murray. 1996. *Background Report: a historical geography of Tantramar's Heritage Landscapes*. Fredericton: Department of Municipalities, Culture and Housing, New Brunswick.

Blewett, M., ed. 1990. *The Last Generation: Work and Life in the Textile Mills of Lowell, Massachusetts, 1910–1960.* Amherst: University of Massachusetts Press.

Boyer, K. 2004. 'Miss Remington goes to work: gender, space and technology at the dawn of the information age'. *Professional Geographer* 56(2), 201–12.

Bradbury, B. 1990. 'The family economy and work in an industrializing city: Montreal in the 1870s'. Pp. 124–51 in G. Stelter, ed. *Cities and Urbanization: Canadian Historical Perspectives*. Toronto: Copp Clark Pitman.

Bulmer, A. Diaries 1920–32. Alice Bulmer Fonds, Mount Allison University Archives, Accession 8822.

Crang, M. 1998. *Cultural Geography*. London and New York: Routledge.

Christensen, K. 1993. 'Eliminating the journey to work: home-based work across the life course of women in the United States'. Pp. 55–87 in C. Katz and J. Monk, eds. *Full Circles: Geographies of women over the life course*. London and New York: Routledge.

Conrad, M., T. Laidlaw, and D. Smyth. 1988. *No Place like Home: diaries and letters of Nova Scotia Women 1771–1938*. Halifax: Formac.

Domosh, M., and J. Seager. 2001. *Putting Women in Place: Feminist geographers make sense of the world*. New York & London: Guilford Press.

Dublin, T. 1979. *Women at Work: The Transformation of Work and Community in Lowell, Massachusetts, 1826–1860*. New York: Columbia University Press.

———. 1993. *Farm to Factory: Women's Letters, 1830–1860*. New York: Columbia University Press.

Duncan, N., ed. 1996. *Bodyspace: destabilizing geographies of gender and sexuality*. London & New York: Routledge.

Duncan, N., M. Domosh, and G. Rose. 2004. 'Classics in human geography revisited: Rose, G. 1993: Feminism and geography: the limits of geographical knowledge'. *Progress in Human Geography* 28(3), 363–68.

Dyck, I. 1989. 'Integrating home and wage workplace: women's daily lives in a Canadian suburb'. *The Canadian Geographer* 33(4), 329–41.

———. 1990. 'Space, time and renegotiating motherhood'. *Environment and Planning D: Society and Space, 1*, 459–83.

———. 2005. 'Feminist geography, the 'everyday', and local-global relations: hidden spaces of place-making'. *The Canadian Geographer* 49(3), 233–43.

England, K. 1993. 'Suburban pink collar ghettos: the spatial entrapment of women?' *Annals of the Association of American Geographers* 83(2), 225–42.

Ennals, P., R. Summerby-Murray, and C. Ennals. 2004. *Marshlands: Records of Life on the Tantramar* www.mta.ca/marshland/topic7_marsheconomy/marsheconomy.htm (Accessed 17 October 2008).

Franzoi Bari, B. 2000. '. . . with the wolf always at the door . . . Women's work in domestic industry in Britain and Germany'. Pp. 164–73 in M. Boxer and J. Quataert, eds. *Connecting Spheres: European Women in a Globalizing World, 1500 to the present*. New York & Oxford: Oxford University Press.

Hanson, S., and G. Pratt. 1991. 'Job search and the occupational segregation of women'. *Annals of the Association of American Geographers*, 81(2), 229–53.

Kinnear, M. 1988. 'Do you want your daughter to marry a farmer?: women's work on the farm'. Pp. 137–53 in D. Akenson, ed. *Canadian Papers in Rural History*, Vol. VI. Gananoque, Ontario: Langdale Press.

Kwan, M. P. 1999. 'Gender, the home-work link, and space-time patterns of non-employment activities'. *Economic Geography*, 75(4), 370–94.

———. 2000. 'Gender differences in space-time constraints'. *Area*, 32(2), 145–56.

Laws, G. 1997. 'Women's life course, spatial mobility and state policies'. In J.P. Jones III, H. Nast, and S. Roberts, eds. *Thresholds in Feminist Geography*. New York: Rowman and Littlefield.

Mackenzie, S. 1989. 'Restructuring the relations of work and life: women as environmental actors, feminism as geographic analysis'. Pp. 40–61 in A. Kobayashi and S. Mackenzie, eds. *Remaking Human Geography*. Boston: Unwin Hyman.

Mackenzie, S., and D. Rose. 1983. 'Industrial change, the domestic economy and home life'. In J. Anderson, S. Duncan, and R. Hudson, eds. *Redundant Spaces and Industrial Decline in Cities and Regions*. London: Academic Press.

Maroney, H., and M. Laxton. 1987. 'From feminism and political economy to feminist political economy'. Pp. 1–28 in H. Maroney and M. Laxton, eds. *Feminism and Political Economy: Women's Work, Women's Struggles*. Toronto: Methuen.

Massey, D. 1984. Spatial Divisions of *Labour: Social Structures and the Geography of Production*. London: MacMillan.

McDowell, L. 1996. 'Spatializing feminism: geographic perspectives'. Pp. 28–44 in N. Duncan, ed. *Bodyspace: destabilizing geographies of gender and sexuality*. London & New York: Routledge.

McDowell, L. 1999. *Gender, Identity and Place: Understanding Feminist Geographies*. Minneapolis: University of Minnesota Press.

Miller, R. 1982. 'Household activity patterns in nineteenth century suburbs: a time-geographic exploration'. *Annals of the Association of American Geographers* 72, 355–71.

Norkunas, M. 2002. *Monuments and Memory: History and Representation in Lowell, Massachusetts*. Washington and London: Smithsonian Institution.

Norton, W. 2006. *Cultural Geography: Environments, Landscapes, Identities, Inequalities*. 2nd edition. Oxford and Toronto: Oxford University Press.

Pratt, G., and S. Hanson. 1993. 'Women and work across the life course: moving beyond essentialism'. Pp. 27–54 in C. Katz and J. Monk, eds. *Full Circles: Geographies of women over the life course*. London and New York: Routledge.

Rose, G. 1993. *Feminism and geography: the limits of geographical knowledge*. Cambridge: Polity Press.

Rosenbloom, S. 1993. 'Women's travel patterns at various stages of their lives'. Pp. 208–42 in C. Katz and J. Monk, eds. *Full Circles: Geographies of women over the life course*. London and New York: Routledge.

Scharff, V. 1991. *Taking the Wheel: Women and the coming of the Motor Age*. Albuquerque: University of New Mexico Press.

Stockall, K. 2008. Personal communication regarding the history of the Women's Auxiliary and Anglican Church Women in the Parish of Sackville.

Strange, C. 1988. 'From modern Babylon to a city upon a hill: the Toronto Social Survey Commission of 1915 and the search for sexual order in the city'. Pp. 255–77 in R. Hall, W. Westfall, and L.S. MacDowell, eds. *Patterns of the Past: Interpreting Ontario's History*. Toronto and Oxford: Dundurn Press.

Summerby-Murray, R. 1999. 'Interpreting cultural landscapes: a historical geography of human settlement on the Tantramar marshes, New Brunswick'. Pp. 157–74 in C. Stadel, ed. *Themes and Issues of Canadian Geography III*. Salzburger Geographische Arbeiten, Vol. 34. Salzburg, Austria: Institute of Geography and Geoinformation, University of Salzburg.

Summerby-Murray, R. 2003. 'Commissioners of Sewers and the intensification of agriculture in the Tantramar marshlands of New Brunswick'. *The North American Geographer* Vol. 5, No. 1–2, 183–204.

Underwood, J. 1995. *Ketchum's Folly*. Hantsport, N.S: Lancelot Press.

University of New Brunswick Archives www.lib.unb.ca/archives/finding/ketchum/chignecto_railway.html (accessed 8 May 2009).

Wynn, G. 1985. 'Late eighteenth century agriculture on the Bay of Fundy marshlands'. Pp. 44–53 in P. Buckner and D. Frank, eds. *The Acadiensis Reader: Volume 1: Atlantic Canada before Confederation*. Fredericton: Acadiensis Press.

Chapter Six

Schools and Youth Networks: Challenges for Newfoundland's Seasonal Migrant Families

Joan Marshall

Introduction

The concerns about the devastation of Canada's east coast fishery, particularly the collapse of the cod stocks that led to the moratorium in 1992, have tended to focus on the unemployment of thousands of people living in the coastal villages and 'outports' of Newfoundland. Debates about retraining, the **migration** of villagers to larger urban centres, and the provincial strategies for economic development, have focused on solutions for families who stay in Newfoundland year-round. There have been few studies that have explored the outcomes and experiences of families who become seasonal or permanent migrants. In telling the stories of these families the importance of spatiality for family relationships is highlighted. Family geographies are seen through the trajectory of the migration experience.

This chapter draws upon over a decade of ethnographic research on Grand Manan Island, New Brunswick, and specifically intensive interviews with Newfoundland families who migrated to this small, island fishing community in the period between 1995 and 2004. They were initially encouraged by the recruiting efforts of a fish broker on Grand Manan who wanted steady labour for his lobster pound. Later, migrant families responded in larger numbers to active recruitment in 1999 by the manager of the Seal Cove fish packing plant on Grand Manan, (owned by Connors Bros.) during a period of restructuring by the parent company in Toronto, Weston Corp. The Seal Cove plant was adding an extra shift in order to increase capacity in the region with the result that new labour had to be hired.[1] In the narrative that follows, the trajectories of different migratory choices by the Newfoundland families are examined, with particular focus on the importance of the school system and youth networks for determining (1) the assessment by the Newfoundland families of the merits of their decisions, especially in relation to their children and overall effects on their families, and (2) the relative success of the migration experience. For both the community of Grand Manan and for the approximately 35 Newfoundland families who migrated to the island during the period of study, there were almost insurmountable adjustments that continue to reverberate in 2009.

While migration flows have tended to be seen in relation to labour markets, there is an 'increasing awareness that migration needs to be understood not only in terms of economic causation but also as a social process' (Ogden 2000: 505). The move itself expresses a particular worldview, infusing it with meaning as 'an extremely cultural event' (Fielding 1992). To a significant extent, the migration literature has neglected the contextual interaction of routines and structured forms of behaviour upon which the flow of daily life depends. The 'emphasis on the stresses—the "pushes" and "pulls" of the origin and destination—caused by the environment neglects the way in which the individual (and, by extension, the family unit) formulates and deals with these stresses' (Halfacree and Boyle 1993: 335). Consequently, there has been a lack of attention paid to the problematic aspects of migration, and notably to the role of social networks and institutions within both sending and destination communities. Additional attention also needs to be paid to the diversity of the experiences and meanings of migration for migrants, both individually and as family groups, and for the receiving communities. The importance of social networks has been explored by Boyd (1989) who saw the domestic unit (the family) as an important component of the migration process. Domestic units are sustenance units and therefore have their own structural characteristics such as the age and sex structure of the family household and the stages of the family life cycle. Insofar as migration is related to a household's past and present, it is an action in time with implications for its predicted and projected future. As socializing agents, families transmit cultural values and norms which influence who migrates and why, as well as the impacts upon the family, its individual members, and the receiving communities. Although social networks can provide migrants with an adaptive and anchoring function in receiving societies, the homogenizing effect of these networks, as we shall see, can be socially problematic. Pohjola (1991) notes that family integration tends to maintain cultural and normative continuity when moving from traditional communities to a different cultural sphere. Nonetheless, family forms are deeply influenced by new local conditions that are structured within a context of historical meaning. Constraints and choices have a distinct geography (Smith 2008).

The Context: Grand Manan Island

Grand Manan is an island situated in the Bay of Fundy off the coast of the Canadian province of New Brunswick. It is accessible via a 35 kilometre ferry trip, which takes approximately 90 minutes. The size of the island (26 kilometres north-south and 11 kilometres east-west), and the distribution of the population in five villages along the east coast, are important factors affecting the relationships amongst islanders, and between islanders and any newcomers. Historically the single road down the east coast provided only limited access between villages until it was paved in 1948. As a result, distinctive village identities and perceptions of belonging have evolved, rooted in the location of schools and churches. While gradually becoming more muted, this history continues to inform community sensibilities and personal identities.

The population of Grand Manan had been steady at about 2,600 people since Confederation, until tragic events on the island combined to influence the plans of young families. The result has been that by 2006 the island population had fallen by almost 6 per cent to 2,460, and sources suggest that by 2009 even more islanders had moved away.[2] With an economy based on a diverse wild fishery, which includes herring, lobster, scallops, sea urchins, groundfish, halibut, clams, and dulse, Grand Manan has enjoyed a sense of security; if one fishery was down, others would be up. Overlapping seasons provided fishers with an annual rhythm of activities requiring varied skills and levels of capitalization. Until the mid-1990s Grand Manan had experienced the vagaries of a resource-based economy, but one that nevertheless

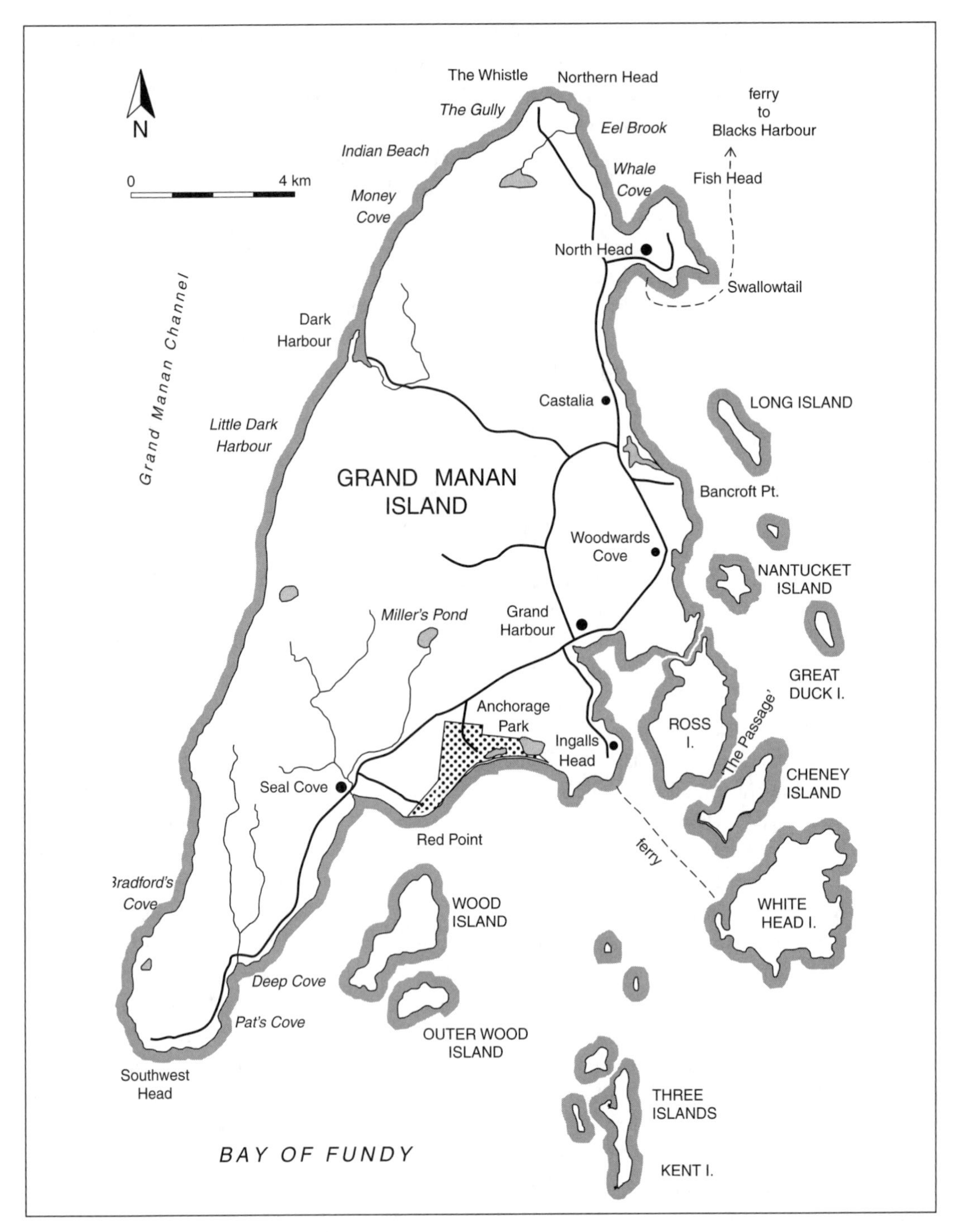

Figure 6.1 Grand Manan Island

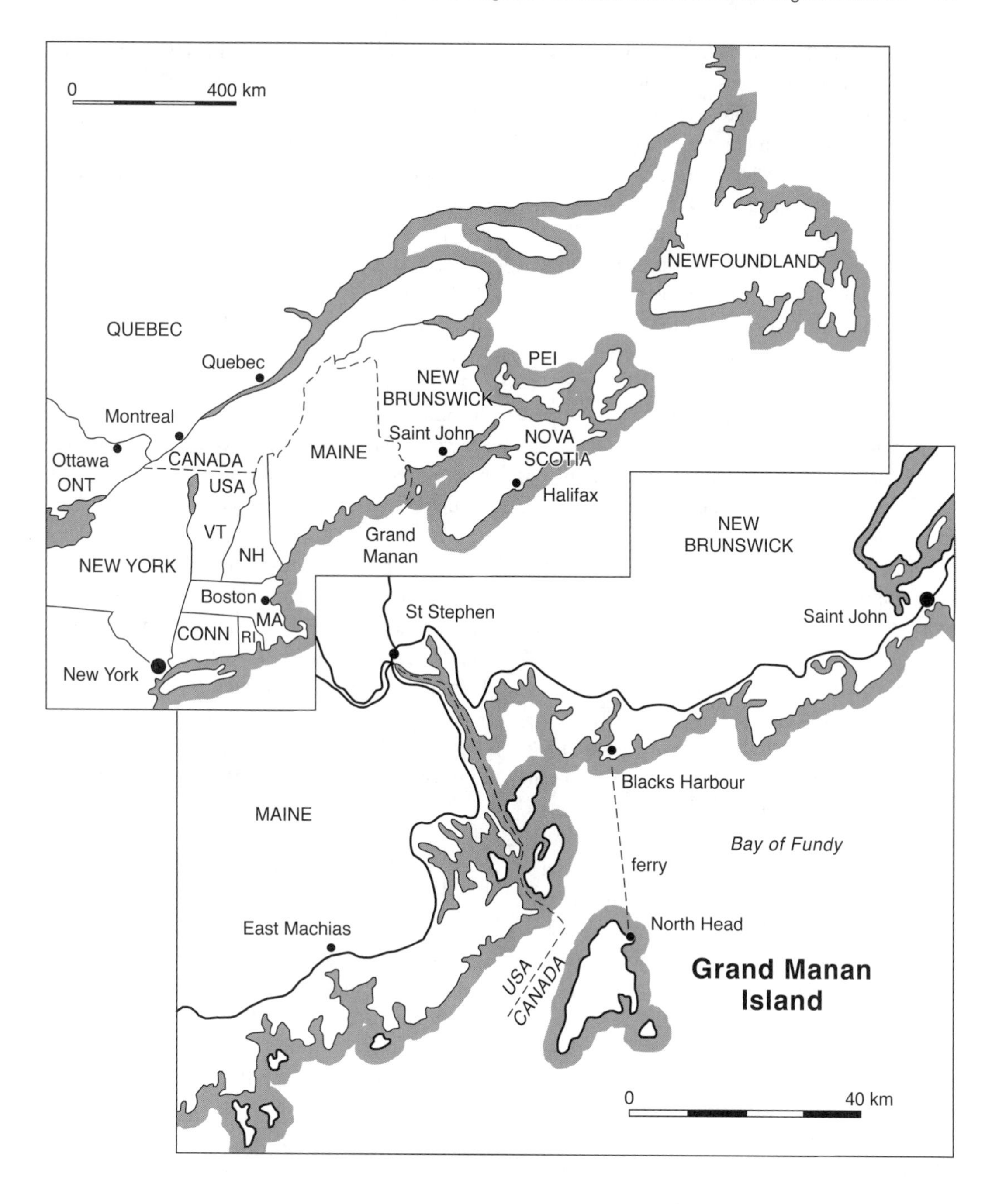

Figure 6.2 Grand Manan Island

provided good livelihoods for anyone willing to work. There is a strong sense of stable community identity rooted in family lineages that extend back some 200 years, and in the shared struggles of the fishery and an unforgiving ocean with its tides, currents, and storms. Scarcely a family has been untouched by the tragedy of loss at sea.

While summer residents and tourists had visited the island for decades beginning at the end of the nineteenth century, for the most part the population history of the community is remarkable for the low level of in-migration and the degree to which Grand Mananers, even if they went away for a short period of education, returned to the island. Although the situation is changing, as noted earlier, until 2002 net in-migration was about 18 persons per year, a rate of about 5 per cent (Marshall 2009: 23). In 2006, 87 per cent of the population had lived at the same address for more than five years, compared to the Canadian average of only 25 per cent. Moreover, until 2003 most high school graduates stayed on the island, and of those who did leave for further studies most would return within two years. The remarkably high rate of Grand Mananers staying on the island (over 50 per cent overall) became an especially entrenched part of the demography at the time of the Canadian recession at the beginning of the 1970s (Marshall 2009: 339). With the international extension in 1977 of the 200-mile exclusive economic zone (EEZ) of marine jurisdiction and the initiative by the federal government beginning after 1978 to grant subsidies for larger boats, fishers were encouraged to expand their operations, thereby making their island home much more economically attractive. The result for the larger community was even less **mobility** and out-migration. At the same time, until 1990 when a new ferry doubled vehicle capacity, there was little encouragement for newcomers to migrate to the island, either as retirees or as young families looking for work. Thus, until the 1990s, the community was one that was intrinsically defined by family lineage and some 200 years of shared experience of the wild fishery.

For Grand Mananers, being one of the multi-generational families has been a key marker of belonging, which includes not only acceptance but also access to certain jobs. Family roots and the 'ascriptive nature of family characteristics' (Cohen 1985: 104) provide the touchstone for youth relationships in school and amongst their youth peers. The notion of belonging, as Foucault (1980) points out, is not only a question of who is 'in' but importantly is also related to exclusionary practices. These practices have impacted various newcomers to the island, including female spouses brought to the island by Grand Manan men, new residents 'from away', and migrant workers and their families. Without the shared stories of generations and intermingled families a sense of belonging, even for those who have moved permanently to the island, is inevitably diminished. This 'unbelonging' permeates all of the shared collectivities, rights and values of the community (Rogoff 2000: 5), and is an important factor in the (limited) capacity for adaptation and change.

Grand Manan Identity: Challenges to Established 'Webs of Significance'

Several forces of change almost overwhelmed Grand Manan culture in the 1990s, causing a context of resistance, a lack of resilience, and ultimately a collision of identities between islanders and the new Newfoundlanders. As George Herbert Mead posited back in the 1930s in his classic discussion of the formation of self, identity formation occurs within a dynamic social network of interactive relationships, a process that is inherently unstable and dependent upon relations of difference (Mead 1934). Layers of economic, social, and political relationships within particular contexts of history, migration and mobility create 'webs of significance' and **lifeworlds** of meaning that are constantly being created and recreated through social interaction (Geertz 1973). 'Identity', said Hall (1987: 44), 'is formed at the unstable point where the unspeakable stories of subjectivity meet the narratives of history, of a culture'. During periods of migration, for both the movers and for the receiving community, identities are necessarily being renegotiated and reshaped through complex networks of social relations and institutions that ultimately define new communities. 'Just how the relations between social structure and human agency fall out is evidently different from place to place and depends crucially on the particular arena of encounter' (Livingstone 1992: 357). Certainly an important issue for people who migrate to the island is that they have not lived the experiences that create the 'nebulous threads' of island culture. It is not possible for them to fully understand the 'subterranean level of meaning' that would allow them to truly belong in island culture (Cohen 1982: 11). Young people, in particular, are affected by their family associations. For anxious teenagers, the expectations of particular behaviours can overwhelm personal goals and personalities in ways that are both intimidating and stultifying. In the interviews conducted on the island over a period of twelve years, it was not uncommon for families to be characterized in single phrases as 'they're trouble', 'good, hard-working folk', having 'no ambition' or 'snooty'. Referring to a community characteristic, one islander talked about the independence that is 'bred into them' and how hard it is 'to get them to pull together unless it is life-threatening'. On the one hand, Grand Mananers talked with pride about the importance of 'community' and on the other, they acknowledged the competitiveness of fishing and the need for privacy in a small community where everyone seems to know everyone else's affairs. Or, as one person said, 'they think they know'. For youth the changes that have impacted the island during the 1990s are particularly important because it is youth who are the community of tomorrow.

It has been suggested that as globalization begins to diminish the strength of community-based relationships, there is a concurrent possibility for the growing significance of structures such as schools, workplaces, and voluntary organizations as purveyors of identity and a new form of community solidarity (Dunn 1998: 61). On Grand Manan the school has a complex relationship with the community, arising partly out of the relative isolation from mainland institutions, and partly out of its ambivalent situation within a value framework that traditionally has not highly prioritized education.

Some of the forces associated with globalization that have been most important for the island through the 1990s include: (1) the new ferry (1990); (2) the degradation of the wild fishery and the introduction of salmon aquaculture by multinational corporations (1991); (3) new federal initiatives that brought aboriginal persons to the island as fishers (1999); (4) the restructuring within the agro-food industry that led to the closure of several Connors fish packing plants on the mainland and the creation of a second shift at the Grand Manan plant, requiring significant numbers of new workers (1999); (5) complete restructuring within the aquaculture industry (2003–05); and (6) the closure of the Seal Cove plant in 2004. At the peak of economic optimism in 2000–01, islanders were less concerned with the loss of the wild stocks of cod and haddock than they were with the growth of salmon aquaculture, which seemed to be privatizing large areas of the marine commons. At the same time many islanders were seizing opportunities to invest in this relatively young industry without adequate financial foundations to sustain them through price declines associated with increased global production and the onslaught of disease that forced the closure of sites several years later. The period from 1995 to 2006 was especially tumultuous, bringing change to virtually every aspect of island life. Institutions were part of this transformation, specifically the school, churches, and the Boys and Girls Club. Thus, inherent insecurities were beginning to permeate the Grand Manan community at the time of migration to the island by the Newfoundland families studied. Despite the growth of aquaculture and the opening of a second shift at the fish plant, islanders sensed that the changes they were experiencing were undermining their historic structures, and that they were losing control of their futures. The instabilities on Grand Manan were linked to broader issues of global restructuring that threatened to undermine the traditional strengths provided in the families' lineages, religious affiliations, and strong community norms. During times of significant change there are inevitable challenges to social cohesion, and on Grand Manan the restructuring occurring during this period had a devastating effect on many of the social relations that previously had sustained the community.

The Newfoundlanders

Arriving on the island at the beginning of this period of significant change, and unwitting participants in the dynamic of community change, were a group of families from Newfoundland. Both the process of adapting to the new community and the responses of islanders to the new arrivals created a reflexive relationship that changed the meaning of community on Grand Manan. This relationship was informed by: (a) the differences in historical experiences, as tied to the fishery and other resource-based activities; (b) the similarities of fishery experience, educational, and income levels; and (c) the strong sense of place and history rooted in family lineage that defines Grand Manan culture. Woven into the complex processes of establishing new ways of being were the social networks of the youth, both Newfoundland migrants and Grand Mananers. Insofar as all of the Newfoundland migrants shared common language and racial characteristics with Grand Mananers, some of the obvious sources of dissension were alleviated. Nonetheless, even the shared experience as fishers did not fully mitigate the deep sense of alienation experienced by many of the Newfoundland families as they tried to make new lives for themselves. As this chapter makes clear, family relationships are crucially affected by the geography of migration and the particular problems for youth of negotiating new social networks.

When the first Newfoundlanders arrived in the mid-90s, a new phenomenon was introduced to Grand Manan, that of entire families with young children migrating to the island. And in 1995, even the two families who arrived on a trial basis were the subject of conversations. Before this time, the few new arrivals had been new spouses of islanders, retired couples, or single women retirees, nurses, or teachers. The phenomenon of young families was new, and there was a subtle but significant distinction between how islanders perceived the culture of these new arrivals and how they viewed other people 'from away'. There was a grudging respect for them because it was known that they had shared many of the same experiences as fishers in their home communities and that their presence was a direct result of the collapse of fish stocks, a situation to which Grand Mananers could easily relate.

The arrival of the Newfoundlanders occurred in three phases, followed by their abrupt departure after 2004 due to the closure of the Connors fish plant. In the first phase (1992–97) only a few families were involved, having been encouraged by the recruitment efforts of Ron Benson, the local owner of a fish brokerage company and lobster pound. They were all related and all came from Comfort Cove or nearby Twillingate. In the second phase (1998–2002) there was a higher percentage of seasonal workers compared to those earlier arrivals who were considering the possibility of a permanent move. The second group had been recruited by Connors, through the efforts of the manager of the Seal Cove plant, David Green. Finally, in 2002–04, a group arrived from Fogo Island to work harvesting rockweed, recruited by Acadian Seaplants, a Nova Scotia-based company that had been operating on Grand Manan since 1996. By 2000–04 there were over 100 Newfoundlanders on Grand Manan, some as seasonal workers, others hoping to settle there permanently. Their long journey covered about 1,200 kilometres that involved an all-day trip, two ferry crossings and costs of over $1,500.

There was a clear and very significant distinction between those who were coming to the Island for seasonal work and those who were hoping to find a new permanent home. For those returning each year to their homes in Newfoundland, the availability of accommodation in a trailer park developed by the plant manager provided at least minimum requirements, at a minimal cost of $200 per month for a trailer site that had electricity and water. For those who intended to move permanently to Grand Manan, the choice of accommodation was limited to available rental houses or trailers. For these people it was important to find accommodation in the central village of Grand Harbour where recreational facilities and the school were located. As will be explained later, for these migrating families the effects of the spatial dislocation of moving to Grand Manan was exacerbated by the island residential location in a trailer park that was a significant distance from activity centres on the island. A second distinctive and very important characteristic of the migrating families was whether or not they included children and the ages of the children. In all cases, the mobility of couples as compared to families with children who had to be in school was an important factor in their experience of the island. Few of the migrants were single, and those few were usually related to other migrant families already living on Grand Manan.

The demographic characteristics of the migrating Newfoundlanders were important indicators of the nature of their struggles to integrate into the life of Grand Manan. The first few families from 1995 to 1998 had young children (ages 7 to 12), while the range of ages and family structures became more variable with the increasing numbers after 1998. Three children were enrolled in the school during this early period, a number that drew comment from islanders unused to having incoming families. Moreover, whereas Grand Mananers were curious about the original group, their interest became more focused when the larger group after 1999 threatened to take away jobs or encroach upon their territorial 'claims' to dulse grounds.[3] In the first large group of about 15 families who arrived in 1999 there were about eight children, ages 1 to 18 years old. Initially the families all viewed the island as an economic opportunity that would allow them to make enough money to bridge the period of unemployment through the winter months. They came to Grand Manan to work in the fish packing plant or on aquaculture sites. If there were children, one parent would work during the day while the other would work the night shift at the fish plant in Seal Cove. In their first year most returned to Newfoundland, leaving Grand Manan at the end of October when the fish plant closed for the year. For the few who stayed, there were particular problems associated with the school that I shall examine later. By 2000, there were 56 adults and 23 children (13 of whom were enrolled in school) from Newfoundland living on Grand Manan, representing 28 households, 13 of which were located in the trailer park, named by the Newfoundlanders with a large sign as the 'Comfort Cove Trailer Park'.

The specific location of the trailer park was a crucial factor affecting the experience of the Newfoundland families. Located two kilometres from

Seal Cove and the fish plant, it is in a relatively isolated site at the end of an unpaved road. Moreover, Seal Cove itself is the most southerly of the Grand Manan villages, located about 20 kilometres from the North Head ferry terminal (and the bustling tourist facilities associated with it) and about 10 kilometres from Grand Harbour, the most central of the villages where all of the key services, such as schools and recreation facilities, are located. This geography necessarily influenced the degree to which the Newfoundlanders could integrate into community life. As well, the relative isolation meant that Grand Mananers experienced the Newfoundland migrant families as separate; as visitors whose ties to the island were going to be temporary and shallow. One Newfoundlander woman said, with despair in her voice, 'We are the local tourist attraction' for Grand Mananers out on Sunday afternoon drives. She was describing the habit of islanders to drive around on Sundays, checking out new houses being built, the state of boats at the wharves or even just who was visiting whom. The trailer park became a destination that made the Newfoundlanders feel like they were on display like animals in a zoo.

Before exploring how the impact of seasonal and permanent migration decisions affected the Newfoundland families, I shall describe the Grand Manan community context in greater detail in order to provide an understanding of the challenges confronting Newfoundland families, and especially the challenges confronting children, for whom the school and social networks were crucial mediators of a sense of belonging.

Grand Manan Youth Culture: The Challenges for Newfoundlanders

The importance of youth networks developed and nurtured within schools and recreational activities must be viewed within any consideration of the probable futures for a small community. For young people, the variety of choices in friends, social groups, activities, and hobbies are significant underpinnings of the potential health and stability of their identities. The importance of this has been argued by Cindi Katz based on her studies of youth in America and Africa. 'The key construction of childhood and youth, in theory and practice, is that of a time or space "without walls" when and where all futures seem possible' (Katz 1998: 136). For the young people, who have arrived on Grand Manan with their parents, intending to secure their economic futures, the struggles are everyday realities of trying to be accepted while negotiating malleable identities. Childhoods are being constructed in different spaces with different rules than those of their parents, inevitably creating stresses for the entire family.

For Grand Manan the long history of dependence upon the sea and the relative isolation of the island from mainland institutions have been important factors constraining a sense of cultural connectivity with the mainland. A strong and distinct island culture has resisted social change in ways that have proved to be problematic in the context of youth whose choices for recreation are severely restricted, to sports for example. The small populations of each age group means there are fewer choices for a variety of activities. The result is a disproportionate attraction to the 'partiers' seen as the leaders and those who set the standards for acceptance. The normalization of high alcohol consumption amongst young teenagers and the hero status accorded to basketball stars are norms that have permeated the youth culture on Grand Manan to an extent that they have excluded possibilities for activities such as theatre clubs or debating societies as outlets for those not inclined to sports. The interviews conducted with youth during 1999 and 2000[4] illuminated some common attitudes and values that reflected a high level of frustration with their limited life choices. An interesting distinction emerged between the girls and the boys with respect to the importance of 'being my own boss' insofar as this category selected in a ranking exercise referred to the ability to control their lives. For the boys there was an assumption that they could do whatever they chose, whereas for the girls there was an inherent sense of having to consciously exercise decision-making in order to have the control they wanted over their lives. In a patriarchal society the gendered nature of decision-making, reflected in the use of spaces across the island, was clear in the interviews with youth. As potent conveyors of culture, gender relations that are played out in family relations and interactions with the community, in both spatial and social ways, define youth networks and the ways in which young people use the spaces of the island. Although spatial distinctions are gradually waning, the male dominance of the woods and trails that has characterized spatial relations on the island (Marshall 2009; Massey 1994, 1997) continues to be evident in youthful activities on their all terrain vehicles, at the camps in the woods and on the seawalls, and even in their small gatherings around the trucks on the wharves. The interview questions explored values, expectations, religion, sports, leadership, drugs and alcohol, village affiliation, and prospects for leaving the island. Interestingly, despite stated doubts about the value of going to church, 'religion' as a category of importance (along with family and friends) was always very high on their lists. The island youth continue to hold onto the strong values of family and religion despite their apparent resistance to parental control that would limit parties or late night gatherings.

One area that emerged from the interviews with Grand Manan teenagers as being especially significant because of its implications for community relations, now and in the future, was that of leadership. While leadership was never an explicit concern of the Newfoundlanders nor was it specifically discussed by Grand Mananers, all of the interviews with islanders and with institutional representatives (such as the school principal and the director of the Boys' and Girls' Club) referred to the problem of role models and the absence of leaders who would initiate and champion new projects or activities. For young people, the issue of who was seen as 'leaders' was a question that only seemed pertinent with respect to who was 'in' and who was being excluded. Asked about who the leaders are within their social groups, the young people invariably had a difficult time. At the same time as they acknowledged that it was girls who held the majority of school positions such as president of student council or editor of the yearbook, they also felt that it was the boys who were the 'role models', who would be seen as 'leaders'. In other words there was a contradictory sense of what was and what should be. One young twelfth-year captured this paradox perfectly in his responses. Having named all boys in his list of leaders I asked if there were any girls he might include. His response was, 'Well, maybe. They sort of are but really not'. Generally, however, it is the girls who complete high school in highest numbers and it is girls who will leave the island for higher education, often never returning. Between 1948 and 1996 females leaving the island accounted for over 50 per cent in 31 of the 48 years (Marshall 2009: 340). Within this dominant framework of apparent female determination for advancement but an acceptance of male prerogative, the roles of sports and family became decisive in affirming status within social groups. The same people tended to be on all the teams and those who participated in leadership roles tended to come from the same families year after year. All of these relationships had important implications for the Newfoundland youth who migrated to the island. For them the challenge was not only to understand a different social milieu but also to become accepted into a youth culture that affirmed the central importance of sports and family history.

The youth described a number of groups according to their status, with names such as 'Partiers', 'Animals', and 'Skunks' or 'Skanks'. The young people who were effectively marginalized by their lack of sports activities or family histories were sometimes grouped together as the 'troublemakers', a group who disproportionately was said to come from Seal Cove and to hang out at the Boys and Girls Club. Mainly male, they were the first to welcome the Newfoundlanders and, as a result, the Newfoundland youth became labelled in the same way. There was a suspicion that they were 'backward', they wore 'weird clothes' and were suspected of stealing. In fact, interviews with adults involved at the Boys and Girls Club and outside the school system did not corroborate this perception, suggesting that the affiliations of Newfoundlanders with the marginalized youth of Grand Manan may have been of necessity because of not being accepted into the higher status groups, but that invariably 'rough' or 'backward' were inappropriate descriptions that stuck by association rather than natural disposition. Another factor was the high costs associated with being part of a high school team that had to travel to tournaments, requiring both expensive sports shoes and travel allowances, expenses that the Newfoundland families certainly could not afford. When one Newfoundland mother described the relief she felt as her son began to take an interest in the computer and to coach some of his classmates, she was reflecting the relief of any mother who recognized the limited choices for youth who might be seen as 'nerds' and whose social networks would be irretrievably constrained as a result.

Exploring the Experience of Migration

The stories of the families, couples, and individuals who arrived on Grand Manan between 1996 and 2004 are most effectively illuminated in their own words. I begin with a discussion of the methodological challenges of meeting the new arrivals on Grand Manan and of gaining their trust as I sought to explore their experiences of migration and of their relationships with the Grand Manan community. As I make clear, their stories do not reflect a single reality. Instead they illuminate a variety of experiences according to family structures (couples or singles, children or not, ages). The nature of family and the choices that could be made for living accommodation and participation in Grand Manan activities were crucial factors determining the ways the Newfoundlanders experienced and perceived the community of Grand Manan.

Methodology

In the midst of a long-term ethnographic study of social and economic change on Grand Manan since 1995, I began to hear about the Newfoundlanders who had moved to the island. Both from the remarks people made and because of the long history without family migration, it seemed apparent that these new incomers would have a profound impact upon the community, and that their experiences as migrant families would be an important part of the changes occurring on the island. Their experiences would have to be examined and given a voice. The problem, however, was that despite islanders claiming to know 'who' they were and even where they lived, islanders did not in fact *know* them. As a result, initially it was difficult to arrange interviews. After a fruitless (and misguided!) attempt to knock on a door to request an interview without an introduction, eventually I was able to make contact with a family who had originally arrived on the island in 1991, moving to Grand Manan permanently in 1996. Through an islander involved in Scouting, it was learned that this Newfoundland mother had served as a volunteer Cub leader, and thereby become known to several Grand Mananers. Contacted by phone and using the islander as a reference, the Newfoundland mother agreed to meet me for an interview. 'Carly' (not her real name) was a Cub leader while her sons were registered members, and in agreeing to be interviewed she became a crucial entrée to the entire Newfoundland community on the island. Through 'Carly' it was possible to build up a network of interviewees, who represented the various patterns of migrating behaviours.

The specific methodology used for the interviews varied. There were 35 interviews, 10 of which were formal interviews in the homes of 5 families; 20 were informal interviews conducted both in trailers and outdoors on front porches (and one on a beach), and five were with representatives of the school, the Boys and Girls Club, and the fish packing plant. The interviews lasted from one to two hours each, focusing on the decision to migrate, experiences in moving, patterns of family ties both to their home villages and to others on Grand Manan, perceptions of the Grand Manan community, and expectations for the future. While the representatives of Grand Manan were individuals (such as the school principal) who impacted upon the lives of the Newfoundlanders directly, the interviews with the Newfoundlanders were usually as couples, occasionally with their children when the interview was in a trailer. Of the Newfoundlanders, only one family was not employed at the fish plant, and four individuals worked on aquaculture sites. Two workers worked cleaning cottages in the tourism sector, and two worked in a small restaurant.

As will become apparent in the discussion that follows, while the home origin in Newfoundland was common to all migrant families, and the stage of life cycle was fairly homogeneous, the research uncovered a complex variety of experiences and responses to the new 'home'. These various experiences are reflected in the variable patterns of movement that will be described. The different spatial patterns of movement reflected different family geographies insofar as family relationships and family structures were intricately and directly implicated. The interviews can be characterized as those that occurred at the trailer park and included families who returned each winter to Newfoundland, and those which occurred in small rental homes and trailers mainly in Grand Harbour. As discussed, the clear boundaries between those who migrated with the idea of a possible permanent move (and who sought to live in Grand Harbour) and those who were seasonal migrants meant that Carly did not know many of the post-1998 arrivals.

In 1999, Connors fish plant had actively encouraged even more families to migrate, resulting in the arrival of approximately 30 families between May and July that year alone. Based on school records of children registered for classes at the local school, approximately eight families elected to stay through the winter of 1999–2000. While some of these families considered themselves as 'permanent' and expressly described their settlement as finding a 'new home', others continued to vacillate. As one woman said, 'We're trying it this year, but if the school doesn't work out we'll leave in October next fall' (interview 2000). The sense of feeling 'between' communities, and of struggling with choices between economic security and social networks of belonging, was palpable in almost all the interviews.

Without exception, all interviewees reflected an indomitable spirit of determination even in the face of difficult challenges associated with the loss of their secure homes and Newfoundland futures and with the problems of integrating into the Grand Manan community. For those who were returning to Newfoundland seasonally, the challenges in relation to schooling were especially formidable. Children had to be taken out of their home schools in late May in order for the families to be able to arrive on Grand Manan in time for the beginning of the herring season and the opening of the fish plant. They not only had to leave friends before the end of the year, parents also had to negotiate special concessions with respect to marks and final evaluations that normally would be available only a month later. Furthermore, the timing of the return home to Newfoundland presented problems since the families had to stay on Grand Manan until the plant closed at the end of October. While the weather was a serious consideration (severe storms could affect ferry schedules and driving conditions) it was the stressful situation with school for the children that most concerned parents. Just as they were beginning to adjust to a new year and new friends, the children in seasonal migrant families would have to be taken out of school and then have to settle in at schools back in their home communities in Newfoundland.

Family Geographies and the Meaning of the Migration Experience

Carly

Even in the retelling of her story years later, 'Carly', whose family had been the first to migrate from Newfoundland, was tearful in recounting their experiences of arriving on Grand Manan. She described, getting off the ferry in 1991, feeling as though she had 'arrived in a Third World country'. 'What had we come to?' she recalls asking. 'It was like going back in time.' The arrival of her family (including her husband and two children) in 1991 was followed by several years of instability, including moving to Ontario and going back to Newfoundland for a time before returning to Grand Manan in 1996. In the end, the steady work was the deciding factor in the family staying on Grand Manan. Her husband 'was really happy to be working. Because at home it was hard; you couldn't make ends meet. As soon as you'd start to get ahead, you'd get laid off and you went on unemployment again. . . '. So here 'it was a full-time paycheque coming in every week; it was wonderful'. In the meantime, her husband's brother had migrated with his family in 1992, soon followed by two other single brothers in 1993. The support was important. 'It was exciting at first, but it was really lonely. We didn't know anybody. We were desperately lonely. When he [her husband] came home I'd be crying. . .'. When the second family arrived, she described her feelings, 'So now we had family, we're doing fine. . .and they fit in really well'. The spatiality expressed in these multiple moves was directly reflective of the social and economic challenges these families were facing in trying to find secure solutions to their long term family survival. In their choice to eventually find a permanent home in the centre of the island near all services, they were also establishing their geographic base for new lives within a context of island activity patterns. Unlike those who chose to become seasonal migrants over several years and who mainly lived in the isolated trailer park, their lives were gradually able to integrate into island relationships both in their physical interactions and in their social activity patterns.

However, the challenges facing their children soon became the major concern of both parents. 'Carly' was unequivocal in feeling that her son was ahead of the children on Grand Manan in his grade level. When they moved permanently in 1996, their son had been in grade one, and 'was doing science and geography in Newfoundland. But here, grade one is the same as kindergarten in Newfoundland!' 'It's lax here compared to at home. The kids don't take it seriously here; they're always misbehaving. I don't know how they get away with that.' She described the difficulties her sensitive son had in trying to integrate socially. 'The kids called him "Fudgie" and "Newfie". Didn't I pine for him! Grand Manan feels like they're the only people. They look at Newfoundland as being stupid. In Ontario it was a wonderful year 'cause people are from all over the world. You don't stand out.'

The Lambert Family

The experience of another family was quite different. When the Lambert's 22-year-old daughter migrated to Grand Manan in the spring of 1998, she found a job for her father whose choice in Newfoundland had been to 'live on welfare or move'. With one of their three children, they drove an old truck the more than 1,000 kilometres to Grand Manan. As a mysterious noise progressively got louder all the way from Stephenville, New Brunswick, the mother described getting off the ferry in North Head, 'glad that at least no one knew who we were', and arriving at their brother-in-law's, before the truck finally 'died'. The next day her husband started his new job in the aquaculture industry, feeding salmon on one of the multi-cage salmon sites. For their teenage daughter, leaving Newfoundland began with tears, but she has since decided on her own to stay. She found a job in the local bakery that continued through the winter when she was at school and in 2001 she was the second Newfoundlander to graduate from the school, and as her mother proudly told me, she graduated with honours. The mother described the pain of her daughter's early days on the island until she met a friend at school who introduced her into a social network. 'Now', the daughter said, 'I'm fine'.

The Wallace Family

For Newfoundlanders such as the Wallaces, who were able to find trailer sites or small homes that were more centrally located, in Grand Harbour in particular, there was a sense that a permanent move might be an attractive option. The Wallaces described their struggle to establish some security in their lives. They had both been 'in the fish' in Newfoundland, but when the cod moratorium was declared, Matthew was unable to change to crab since it would have been too expensive to refit his boat and gear. He sold back his licenses, which did not include any provision for a 'retirement package' even though he was 55 years old, resulting in their determination to find work elsewhere. Years earlier he had worked in a pulp mill, and before that in the mines at Buchans, Newfoundland, but both of those enterprises were also shutting down. Now, with a son of six years and a small baby of only six months, they were trying yet again to establish some sort of security for their family. While 'Matthew' worked the night shift at the plant, his wife 'Joanne' worked during the day. Neither had begun to consider the problems that their children would face as they entered the school system.

Two Teenage Friends (Jen and Mary)

Two teenage girls, 17 and 16 years old, had also come to the island for work upon finishing high school in Newfoundland. The ten-hour shifts, standing all day as they worked on the SAP line (semi-automated processing line), and having to lift heavy trays of fish caused severe back problems because they had to 'rock back and forth'. Being paid piecework meant that there was an extra incentive to move as quickly as possible. One of the girls said when she started she had been making $10 per hour, but that she hoped to be able to get up to $15 per hour by the end of the summer. Asked if she would have preferred the 'scissors line' (which was not automated) Mary shook her head, 'I'd be afraid I'd cut my fingers off, and there's the feed from the fish that burns the arms', describing the open sores caused by the acidity of 'feedy fish' from seiners which have taken in fish that have not emptied their stomachs.

The Miller Family

Another couple who arrived in 1999, the Millers, were somewhat older, with grown children of 33 and 27 years who stayed in Newfoundland. For this older couple the possibility of earning enough money to carry them through the winter seemed enticement enough to migrate to Grand Manan. Neither had worked in a fish plant before.

Helen was on the SAP line while Bart worked as a mechanic, both earning between $10 and $11.30 per hour. For them it was nice to have the work, but they wondered if they would recoup the expenses paid out for their new trailer and the travel to the island. There had not been jobs immediately upon arriving at the end of May, so they had left and gone to Nova Scotia to get temporary work for a couple of weeks until the fish packing plant opened. Indeed, they did not return the following year.

The Hodge Family

But for some Newfoundlanders the trips were an opportunity and the island situation was reminiscent of home. 'It's just like home in Comfort Cove, not so big of course'. Dave Hodge liked working nights, when time seemed to go faster and they could 'talk and carry on'. As one of the single men, he had arrived with two nieces and was living in a trailer with a friend, and was paid $12 per hour during his 42-hour week. They all had the numbers of hours they would need to qualify for Employment Insurance calculated carefully because their ability to survive the winter months back in Newfoundland was dependent on earning enough in Grand Manan. Dave estimated that between 10 and 20 of the new arrivals (having been on Grand Manan for less than five years) were related to him. Having worked for 18 years in a fish plant at home in Newfoundland he was comfortable with what he was doing. But the job on Grand Manan was, he said, much better, because there were medical benefits and it was not so hard on the hands as working in the crab plant in Newfoundland had been. Connors was 'definitely a better company' to work for he thought.

For some there was a feeling of real support and empathy from islanders. After the accidental death of a Newfoundlander caused by an allegedly drug-impaired Grand Mananer, islanders were especially contrite, expressing their concern with gifts of money and food. One of the Newfoundland women was talking to a clerk in a store and happened to mention that she was looking for a second-hand refrigerator. Hours later someone had retrieved one from the 'Island Waste' (dump) and had delivered it to her door. On another occasion, at the trailer park, an islander drove up and handed an envelope to one of the newly arrived couples standing beside the road. In it was money to help with their initial expenses. 'We've had boxes of food delivered to us; Grand Mananers have welcomed us'. For Grand Mananers there was widespread recognition of the sacrifices and struggles these Newfoundlanders were confronting. Nonetheless, it was the children who experienced the taunts at school, felt the boredom with no playmates or activities after school, and who had to deal with the low expectations of the teaching staff. It is these issues that created strife for the Newfoundland migrant families.

The Parsons Family

Variety and flexibility characterized the experiences of Newfoundlanders who migrated during this period after 1998, such as the Parsons, a couple who had arrived with a two-year-old and were willing to accept almost any job in order to earn the money they needed. They started living in a tourist cottage when they arrived in April, moving into a trailer when the cottage rates were increased for the tourist season beginning in mid-June. Months later they found a small house more centrally located. The husband, John, found some work with a small computer services company, while his wife Marg worked 30 to 32 hours per week in the kitchen of one of the inns in North Head. The wife's extended family on the island included relatives in five other families who were important for her, and her family's sense of security. Another couple, older and retired, had come hoping to supplement a meagre pension but were wondering whether they could sustain the late-night shift and the lack of sleep. When she arrived home at 3:30 a.m. it was only a few hours before others were getting up for the morning shift. 'There's not much sleep. I'm so tired all the time.' She wondered why there couldn't be an eight-hour shift that the 'women would like better'. They had worked in the Peace River area and in Aylmer, Quebec, so they were accustomed to the instabilities associated with migratory-labour living. Unlike most other families, they did not have relatives amongst those who had come to Grand Manan and as a result felt somewhat isolated from the sense of community many Newfoundlanders described. Their networks were restricted to those around them in the trailer park. They did not return for a second year.

Social Relations: Families and the School

Talking to a school representative in 1999 I heard about the problems of the seven or eight Newfoundland children who were in the school that year. She described their frequent moves and multiple family arrangements, from a tiny rental house to a slightly larger house in Grand Harbour, then a few months later to another house closer to the centre of the village. She felt that several of the children had 'got in with bad kids'. One boy, she said, simply did not want to be in school, and many of them are 'rough'. Many islanders acknowledged that the Newfoundland children were not being accepted by the island children and that it 'must be very hard for them'. When I first visited the trailer park in early June 1999, only three trailers had arrived and there was still no water or sewage servicing the sites. Within two weeks the laundry facility had been completed just as 10 trailers arrived and were newly installed on their sites. Fifteen-year-old Kimberly was there by herself walking her dog, Bear. She described the island as 'boring' and in the three weeks of their stay she had not met any island youth. Her parents were trying to find a site for their trailer closer to North Head where they felt there would be more activity and things to do. She and her 14-year-old brother felt very segregated from everything on the island. She was not even able to go to school because at her grade level it was considered too late in the year to enrol. In June that year estimates of how many youth were actually on the island varied from about 13 to 18 because some of them were not in school. One of the most disturbing perceptions that was described to me during this first year came from several teachers, island children and parents: the Newfoundland youngsters were 'almost illiterate', 'stupid', not up to the academic standards of the local school, and they were members of the group known as the 'Skunks', the troublemakers and marginalized kids. Moreover, the community was beginning to talk about the possibility that the Newfoundlanders might take jobs away from them; they were mentioned as the source of crime and there were complaints that they did not know the 'rules' about the dulse picking grounds. On the other hand, for the plant manager who had gone to Newfoundland to recruit them, they represented dependable, hard-working labour that he respected.

For Carly's family who had returned seasonally over a period of five years and who had experienced seasonal work in Ontario, Grand Manan provided secure incomes for both parents, albeit within a context of not knowing when it would start and when they could return to their home in Newfoundland, and as well the uncertain possibility of schooling for their children. They described the high costs associated with working in Ontario, such as rent and food, and the lack of freedom experienced by the young children compared to on Grand Manan. But the uncertainty as to when the plant would open (dependent upon the arrival of herring) was a perpetual cause of worry, especially when the delay extended into late June and even early July.

For 'Joanne', on the other hand, who had been a hairdresser in Newfoundland, there was the possibility of some informal work cutting hair before the shifts started in the plant, providing some measure of financial relief. For women in particular, learning the work was challenging given the hazardous nature of snipping heads and the possibility of cutting fingers or wrists in the process. (The men typically had different jobs, lifting trays for example.) Because it was piecework, as compared to an hourly wage, the women were encouraged to work quickly, knowing that the experienced workers could earn as much as $16 per hour. For most Newfoundland women the rate, even by the end of the season, was considerably lower. Asked about her dreams for the future, 'Joanne' replied that she would like to be in one place, especially 'when Julie starts school. But you gotta do what you gotta do'. The intrepid sense of doing what was right for their children was tempered by their own experience of having to move home and family every six months, leaving secure roots in Newfoundland and trying to establish some stability on Grand Manan. For one couple that had lived as far away as Yellowknife, the lure of Grand Manan was in part due to the relative proximity to Newfoundland. Having been twice called home from the Territories because of family emergencies, the geographic location of Grand Manan was a benefit that outweighed the higher earning potential in the north. As well, their living expenses were lower, not having to pay rents as high as $800 a month. Earnings on Grand Manan were considered adequate insofar as they were almost twice as high as what they had been able to earn in Newfoundland, assuming they could have found a job in 1999. Perhaps the most often mentioned problem was the long hours, which could be up to 16 or even 18 hours on a shift. But without exception, they were all focused on those hours as contributing to the employment insurance benefits (EI) that would sustain them through the winter.

Unhappily, underlying the perseverance and determination to assure an economic future for their families was the resignation and sadness related to their loss of social connections at home. Asked if they would consider moving permanently to the island, they laughed: 'No! It's too isolated; there's not even any malls'. For them the telephone calls home, several times each week, were a way to maintain contacts and feel they were not far away from their friends and family. The social relations for both seasonal migrants and for those who hoped to make permanent homes on Grand Manan were reflected in the choices for home locations, social activities, and even their ambitions for employment. Even as the phone calls were illuminating the ongoing sense of alienation in their newly adopted community, these Newfoundlanders struggled to balance their spatial ambiguities and instabilities with secure economic futures.

By 2000, less than two years after the initial influx to the trailer park, many migrants were beginning to participate in local church services partly as a way to integrate into the Grand Manan community. Others were making the move to permanent residency, moving out of the trailer park and beginning to look for jobs in the stores and restaurants.

In considering the spatial dimensions of family relationships, the story of one family is particularly poignant, highlighting the depths of meaning associated with family migration. The story of 'Carly's' family illuminates both the universal struggles of young teenagers and the specific dilemmas associated with the trauma of migration histories involving children and their experiences of school networks and the educational system. Having migrated on a 'trial basis' to Grand Manan in 1991, and moved away and then back again permanently in 1996, this family of four finally settled into their own house in 1998, feeling a sense of pride in ownership and a commitment to fully engage with their new community. By the fall of 2000, one son (now aged 14) who had started in grade one on Grand Manan was in grade eight and the other son in grade four, but each with very different associations with their classmates and teachers. Unlike nine-year-old 'Sam', 'Tyler' seemed to feel terrorized by children at school, according to his mother. He continued to talk about 'going home', and for the parents there was a feeling that 'we can't let go; it's always drawing you back'. Comparing his island experience to that in Ontario where he had been in grade two, she noted that there were so many newcomers in Ontario that it was 'normal' to be from away. Everyone seemed to fit in. On Grand Manan newcomers were not accepted; they did not belong. 'We're neither here nor there. We talk about it all the time; it makes you not settled'. As late as 2001 they were still thinking of returning to Newfoundland, and her husband had sent out resumés to no avail. She had decided that home schooling was too much of a commitment for her, but she recognized that 'Tyler' internalized his struggles and was not able to cope with the daily challenges of school. A new teacher from Newfoundland had given her some hope, someone 'sent from heaven' she thought. The whole class seemed to come together, and the teacher would not tolerate any negative attitude, according to 'Carly'. While she felt that their lives were better than they would have been in Newfoundland because both parents had been able to find secure jobs that enabled them to buy a house and a new car, there was constant worry about their sons. They blamed the school for lack of adequate oversight of rudeness, taunts, and teasing amongst the students. 'It's like anarchy', she said. There was also a feeling that teachers assumed 'there's a problem with Newfoundland children'. That her perception might have validity was affirmed in several of my conversations with school staff. She tried to encourage 'Tyler' to join activities outside school such as Scouts, and became a Cub leader so that she could participate with him. But he seemed more interested in the computer that they had acquired a year earlier. For her, one hope was that other youth were coming to 'Tyler' for help in learning computer skills, a possible way she thought for him to gain acceptance. The involvement of both parents in the Pentecostal Church had allowed them to feel more accepted themselves, but had not seemed to help 'Tyler' who continued to be isolated. By the time he graduated from high school six years later both parents felt he had acquired a measure of confidence. Going to a community college on the mainland and having summer jobs on Grand Manan seemed to further encourage his growth and they were feeling more relaxed about his resilience and adaptation to their move 10 years earlier. The tragedy of his suicide in the summer of 2008 reverberated across the island, devastating his family and raising questions about the role of community, school, and social networks in the migration experience.

For another family whose children were 13 and 9 years old, there was an explicit recognition of the special challenges for teenagers. Their decision to stay through the winter in 2000 was based on their concern about the effect on the children of leaving school in May in Newfoundland, integrating on Grand Manan for a few weeks, then again leaving at the end of October just when the children would be establishing new friendships in school. The problems associated with itinerant seasonal moves seemed to them to outweigh the possible disadvantages of staying through the winter on Grand Manan. The mother acknowledged that her 13-year-old son did not like school on Grand Manan, but she was not convinced that he would like it any better back home in Newfoundland. He did not like being called 'Newfie' but she felt he had been able to ignore it. Her daughter had a lot of friends, and the parents both felt that, although they lived in a trailer, their location in the middle of Grand Harbour (as compared to the trailer park) was a much better choice since they were closer to islanders. For the Newfoundlanders even the spatial choices on the island had significant implications for their possible integration into

Grand Manan networks. Interestingly, it was their son who expressed concern about the amount of drugs amongst his peers and told his parents he would prefer to be home in Newfoundland where they were not a problem. According to his parents, 'he wants to go back because there's so much dope here, and he's kinda afraid of peer pressure'. His mother felt that there were drugs in Newfoundland as well and that he would have to deal with the problem anywhere. She described her son's behaviour as being quite irascible during the school week but being much calmer on the weekends. On the other hand she acknowledged that it was hard being away from their community in Newfoundland and that ultimately it would be the ability of the children to adapt that would determine whether or not they would return the next year. 'We have to do what's right for the kids.'

After being laid off in Newfoundland, both parents felt that Grand Manan represented an opportunity for financial security. Having worked in a crab plant for 15 years the mother had developed a severe allergy, an asthmatic reaction that would not allow her to continue. Working 50-hour weeks at $6 to 7.00 per hour in Newfoundland the family had struggled when she had to leave her job to work in home care, while the father's job at a sawmill was terminated in 1998. They migrated to Grand Manan hoping for a better future for their children. She worked the night shift, on the SAP line, and was happy that her pay had gone up from about $7.00 per hour when she started to over $16.00 an hour two years later. The manager at the Connors plant had been very helpful in finding housing, and the people at the plant seemed friendly. But adapting to a new community and the school and youth networks was not easy for the family. 'It would be easier if we were just a young couple, with no kids.' Unquestionably the hardest thing for them was 'getting the kids adjusted and keeping them happy. For me, I miss home, but it's OK'. The next year brought about a different resolution to the socio-economic tension that they were describing. With the children completing the school year in 2000–01, grandparents from Newfoundland arrived at the end of the summer to take them home while the parents stayed on until the fish plant closed in early November. Then the parents too returned to Newfoundland, not coming back to Grand Manan again. Meantime, by 2001 the numbers arriving each summer were increasing, with 27 trailers that summer in what was now named 'Comfort Cove Trailer Park'.

For those who arrived on Grand Manan without knowing whether or not they wanted to stay permanently, a key deciding factor was often the children and their ability, or not, to integrate into the new community. For one family whose daughter was 15 when they arrived, the problems at first seemed insurmountable as she struggled to find friends. Eventually, after about four months she found a friend who was able to transform her attitude to island life, and encouraged her parents to stay over the winter. While the mother worked in a local restaurant, her husband worked on an aquaculture site, feeding salmon. Their daughter not only was the second Newfoundland graduate at the local high school, as her mother told me proudly, she was on the high school honours list and won prizes in three subjects. Within two and a half years the commitment to stay had been made and they sold their house in Newfoundland, thereby breaking another link with home.

But for another young family the decision to sell their home was a necessity in order to be able to sustain themselves on Grand Manan in their first year. 'It was hard, but we'll give it five years, or as long as there's work.' In the end the difficulties their 10-year-old daughter had in school eventually caused them to move back to Newfoundland late in the fall of 2001 after three years on the island. While they both had jobs in the Seal Cove fish plant and the father enjoyed the opportunity to work in the lobster fishery during the winter, their concerns about the youth networks as their daughter was about to enter the high school became a major factor in their decision to return. Their daughter was 'outgoing' and had lots of friends, they said, but a period of increasing drug problems on the island (1999–2002) culminated in several deaths of twenty-year-old youths within less than two years. Despite the lure of being able to make more money, even given the higher costs of living compared to Newfoundland, they felt their daughter's schooling and her social networks had to take priority. 'Our main goal is to give her a better life. Our parents couldn't afford for us to go to college. We were very poor growing up. We want to afford an education for her.' But the problem of family stability on Grand Manan was a concern. This same family pointed out that 'The most confusing day on Grand Manan is Father's Day'. Hearing stories at school of family problems, their daughter often talked about her friends who were moving between houses through the week. With their daughter entering high school and needing to begin accumulating credits, the very different school systems of the two provinces was another factor in their decision. Whereas in Newfoundland the students can accumulate credits towards graduation beginning in grade seven, in New Brunswick they do not start until years later with fewer needed at final graduation. As well, the parents felt that the 'racist comments' were too frequent, and they had even contemplated going to the media based on unwarranted (in their opinion) comments made by the teaching staff. Given an essay topic such as 'Life and Times on Grand Manan' their son had no context to begin writing, and was not permitted to write a comparable essay related to Newfoundland. For the parents, their son's education was being hampered by both the provincial system of rules and curriculum and by the ways in which the local teachers were applying them.

In the case of another family who had been coming for five years, they tried to stay into mid-autumn with their children but decided it was too difficult for them. In subsequent years, the final years of high school, the mother returned to Newfoundland in late August with the children, while the husband stayed until the first weeks of the lobster season in November.

While there was consensus about the problems for children, there was also general agreement that the adult Grand Mananers were friendly. The work was harder than in Newfoundland because it was 'piecework' rather than hourly wages, but they all agreed that the money was much better. Because Grand Mananers did not like the night shift, the incomers did not feel they were taking jobs away from locals, allowing for a more comfortable working environment. For the migrants from Newfoundland the sacrifices all related to the lack of family stability and the unhappy social environments for their children at school. Trying to reach a balance between these social challenges and the economic imperatives of survival was the challenge all the migrant families faced. The strategies to mitigate these challenges ranged from 'trying it for the winter' to sending children home in August with a parent or grandparent while one or both parents stayed on to work.

After five years of seasonal moves so that both parents could work, one couple described the strain of trying to ensure proper education for their daughter. As 'Austin' rushed around to winterize their camper, sealing windows, flushing toilets, putting in antifreeze, his wife admitted that the older their children got the less sure they were about returning. The long trip each day along a pot-holed dirt road to meet the school bus was only one of their frustrations. Another stress involved not knowing exactly when the plant would close. They were committed to staying until closing, and yet, during some weeks at the end of the season there would be only six hours of work. Only the young or retired couples without children were able to have a measure of stability within the context of the semi-annual move. When the fish plant was finally closed in December 2004 because of the sale and restructuring of Connors Brothers (originally owned by the Weston Corporation) the decision whether to 'stay or go' became much more clear. By the summer of 2005 fewer than eight Newfoundland families remained on the island, and of those some were there because of marriages to islanders. The trailer park was open for tourists that summer but was closed a year later. The school records for 2005–06 indicate only a few children originally from Newfoundland, some of whose parents had bought homes, had children graduate from the school (2001), married islanders (2004), and had children of their own (2005).

Conclusions: Geography and the Experience of Family Migration

In exploring the meaning of migration for families and their children, this discussion has drawn upon the experiences of two historic communities, both having long histories of dependence upon the sea, as they meet and must negotiate new ways of belonging. Incorporated in this discussion has been the notion of evolving identities and the stresses that necessarily affect family relationships in their transformation. Two cultures that might appear similar with respect to language, race, and occupation, nevertheless are shown to have different norms and understandings that are played out especially within the context of families and school networks. The young people who had migrated with their families to the island were essentially outcasts within the school and in the context of sports and other recreational activities such as the Boys and Girls Club. They did not have the camps in the woods that Grand Manan families had enjoyed for decades; they did not have the money to buy ATVs or sports equipment; they were, in fact, both socially and materially cut off from the mainstream social groups that might have provided some solace as they adapted to the new community culture. While Grand Manan youth themselves reflected many problems of substance abuse and suicide attempts, the new migrants had been drawn into a society where dysfunctional relationships could not be escaped. For the families who watched children struggling at school the stress often proved overwhelming, leading to decisions to return to Newfoundland. Had the fish plant not closed in late 2004 or had the problems of the aquaculture industry not caused the closing of almost half the sites in 2005, it is possible that more of the Newfoundlanders might have decided to stay. However, for most of those with young children, the prospect of many difficult years as their children grew into teenagers certainly would not have balanced out the possible economic benefits. The Newfoundland family who lost their son to suicide in 2008 will always question the decision to stay. No one will ever know for certain what role the decision to migrate might have had in this tragic outcome. For the parents there is certainly a cloud of doubt that no economic security can ever erase.

For all youth, the teenage years can be tortuous, but for those negotiating new social networks defined by long family histories and affirmed in large part by the school, there is an even greater challenge. Family geographies involve complex relationships of family structure and life cycles embedded within social and community cultures that have developed over generations. For youth these geographies are central to both their identities and their futures. The policy implications of these relationships need to be taken into account by all levels of government. Few choices allow for few options in social and emotional development. The ways that social networks are supported or can be opened to change, especially in the context of youth will be crucial to the relative success of migration policies.

Questions for Further Thought

1. To what extent should migration be considered a legitimate solution to regional development problems?
2. In what ways could one expect that family relationships would be impacted by the migratory movements of a large group such as the Newfoundlanders? How might these relationships be reflected in the wider community?
3. What are some of the particular issues that must be considered for both sending and receiving communities?

Further Reading

Boyle, P., K. Halfacree, and V. Robinson. 1998. *Exploring Contemporary Migration.* NY: Longman.

Carey, R.A. 1999. *Against the Tide: The Fate of the New England Fishermen.* NY: Houghton Mifflin.

Cohen, A., ed. 1982. *Belonging: Identity and Social Organisation in British Rural Cultures.* ISER, Paper no. 11, Memorial University, St. John's

Gurak, D.T., and F.E. Caces. 1992. 'Migration networks and the shaping of migration systems'. Ch. 9 in *International Migration Systems: A Global Approach.* Oxford: Clarendon. 150–76.

Harris, M. 1998. *Lament for an Ocean.* Toronto: McClelland & Stewart

King, R. 1998. 'Islands and migration', Ch. 6 in *Insularity and Development–International Perspectives on Islands.* E.Biagini and B.Hoyle, NY: Wellington. 93–115.

King, R., and J. Connell, eds. 1999. *Small Worlds, Global Lives: Islands and Migration.* London, NY: Pinter.

Lippard, L. 1997. *The Lure of the Local: Senses of Place in a Multicentered Society.* NY: W.W. Norton & Co.

Olwig, K.F. 1986. 'Children's attitudes to the island community: the aftermath of out-migration on Nevis'. Ch. 7 in *Land and Development in the Caribbean.* J. Besson and J. Momsen, eds. London: Macmillan. 153–70.

Richling, B. 1984. 'You'd never starve here: return migration to rural Newfoundland'. *Canadian Review of Sociology and Anthropology,* 22 (2), 236–49.

———. 1985. 'Stuck up on the rock: resettlement and community development in Hopedale, Labrador'. *Human Organisation.* 44 (4). 348–53

Sibey, D. 1995. *Geographies of Exclusion.* London: Routledge.

Sider, G. 2003. *Between History and Tomorrow: Making and Breaking Everyday Life in Rural Newfoundland.* Toronto: Broadview Press.

Skelton, T., and G. Valentine, eds. 1998. *Cool Places: Geographies of Youth Cultures.* London: Routledge.

Endnotes

1. The sardine industry on the east coast of Canada dates back over a hundred years. The original Grand Manan packing plant that opened in 1947 burned in 1985, putting over 100 people out of work. Rebuilt and reopened in 1988, it included both the traditional 'scissors' line and a new automated line that allowed the plant to increase its productivity significantly. By 1992 record amounts of sardines were being packed by the approximately 100 women who worked there, paid as pieceworkers. Regarded as desirable work because of the possibility to earn as much as $15 per hour, the eventual closure of the plant in December 2004 was devastating to the community.
2. The Statistics Canada data show a population of 2,610

in 2001, and 2,460 in 2006, a decrease of 5.7 per cent. Interviews with school officials, members of the Village Council, and youth themselves all suggest that young families are continuing to leave the island, and that young people who graduate from the high school are increasingly likely to leave for further education and not return. This is a significant change that has happened since 2005 but which began in 2002 following a series of tragic accidental deaths related to substance abuse.

3. Dulse is a seaweed that grows at the extreme edge of the intertidal zone and can be harvested only at extreme low tides. It is regarded as a delicacy by islanders but is also sold into the global market through two brokers on the island. Because it requires no capital investment or licenses it is an important niche activity that sustains many families through poor fish harvests or between the various fishing seasons.
4. See methodology discussion.

References

Boyd, M. 1989. 'Family and personal networks in international migration: recent developments and new agendas'. *International Migration Review* 23, 638–70.

Cohen, A. 1985. *The Symbolic Construction of Community.* London & New York: Tavistock Publications.

Cohen, A., ed. 1982. *Belonging: Identity and Social Organization in British Rural Cultures.* Social and Economic Papers No. 11, Institute of Social and Economic Research: Memorial University, Newfoundland.

Dunn, R.G. 1998. *Identity Crises: a Social Critique of Postmodernity.* Minneapolis, MN: University of Minnesota Press.

Fielding, A. 1992. 'Migration and culture'. Pp. 201–12 in A. Champion and A. Fielding, eds. *Migration Processes and Patterns. Population Redistribution in the United Kingdom.* Vol. 2. London: Bellhaven Press.

Foucault, M. 1980. *Power/Knowledge: selected interviews and other writings, 1972–1977. ed. & trans. Colin Gordon.* New York: Pantheon.

Geertz, C. 1973. *The Interpretation of Culture.* New York: Basic Books.

Halfacree, K.H., and P. Boyle. 1993. 'The challenge facing migration research: the case for a biographical approach'. *Progress in Human Geography* 17, 333–48.

Hall, S. 1987. *Identity: The Real Me.* ICA Documents No. 6. London: Institute of Contemporary Arts.

Katz, C. 1998. 'Disintegrating developments: global economic restructuring and the eroding of ecologies of youth'. Pp. 130–44 in T. Skelton and G. Valentine, eds. *Cool Places: Geographies of Youth Cultures.* London, New York: Routledge.

Livingstone, D.N. 1992. *The Geographical Tradition.* Oxford: Blackwell.

Marshall, J. 2009. *Tides of Change on Grand Manan Island: Culture and Belonging in a Fishing Community.* Montreal & Kingston: McGill-Queen's University Press.

Massey, D. 1998. 'The spatial construction of youth cultures'. Pp. 121–29 in T. Skelton and G. Valentine, eds. *Cool Places: Geographies of Youth Cultures.* New York & London: Routledge.

Massey, D. 1994. *Space, Place and Gender.* Minneapolis: University of Minnesota Press.

Mead, G.H. 1934. *Mind, Self and Society.* Chicago: University of Chicago Press.

Ogden, P.E., 2000. 'Migration'. Pp. 380–81 in R.J. Johnston, D. Gregory, G. Pratt, and M. Watts, eds. *Dictionary of Human Geography.* Blackwell: Oxford.

Pocius G. 1991. *A Place to Belong: Community Order and Everyday Spaces in Calvert, Newfoundland.* Montreal: McGill-Queen's University Press.

Pohjola, A. 1991. 'Social networks—help or hindrance to the migrant?' *International Migration* 29, 435–44.

Rogoff, I. 2000. *Terra Infirma: Geography's Visual Culture.* London: Routledge.

Smith, D.P. 2008 'The geographies of contemporary British families and family life' Unpublished paper presented at Family Life Conference: Interdisciplinary Perspectives; University of Newcastle, Newcastle-on-Tyne, 16 May 2008.

Chapter Seven

South Asian Transnationalism in Canada: Effects on Family Life

Lina Samuel

Introduction

This chapter provides a broad discussion on **transnational** families, with a focus on the South Asian[1] immigrant **family** and eldercare. There is growing interest in the immigrant household and the ways in which the experiences of migration and settlement impact the family structure and define immigrant identity construction. The research presented here is grounded in the 'lived experience' of female migrants to Canada from the state of Kerala, India. The study conducts an examination of the influence of migration on the roles of women within the family structure, the changing nature of patriarchal relations within the household, and the construction of migrant identities. The research brings attention to the gendered process of emigration, and the impact of the migration process on the construction of identities. Because social practices, customs, and traditions around the family are inherently spatial, geography plays an important role in the reproduction of those practices. As Doreen Massey (1985: 18) writes, 'social practices are constituted spatially'. In the present study the first generation migrants are responding to a global demand for professional nurses, an opportunity that alters the family and **gender roles** attached to the household.

The narratives of the women in the study support the conceptualization of migration as a gendered process which is mediated through patriarchal relations, and framed within a particular historical and economic context. While migration provides the women in the study with financial and professional opportunities, they are also exposed to a fragmented family structure as they lose the support of their **family of orientation** left behind in India. This loss of familial support is evident when women begin their own families, and is intensified as these same women enter their senior years. Gendered relations and ideologies (within the family) are negotiated, reaffirmed, and reconfigured within the transnational context (Malher and Pessar 2001).

The chapter begins with a review of how migration has been theorized, both in the past and in the contemporary literature. This leads to an examination of transnationalism and **diaspora**. The history of South Asian migration to Canada is then discussed, with particular focus on South Asian women and their families within the diaspora. Much of the literature dealing with South Asian immigration focuses on male migrants coming to Canada and setting up 'bachelor societies'. The research in this chapter, alternatively, sheds light on the stories of women as they left their homes and families to start new lives in Canada. The narratives presented in this chapter highlight the journey taken by women from India to Canada in the early 1970s. As professional women, the respondents in this study were armed with education, job skills, and fluency in English, which facilitated easy entry into employment. Their narratives around settlement and adaptation point to the struggles in raising teenagers and young adults and the challenges of constructing avenues of eldercare outside the context of traditional Indian family expectations.

Four interrelated questions are examined in the chapter. These are (1) How do women's roles as mothers, wives, and salaried professional workers influence the constructions of traditional gender roles within the family? (2) In what ways do women reconstruct the extended family structure which they left behind in Kerala? (3) How has the experience of aging altered the migration experience and migrant women's connections to the dominant society? (4) How do these experiences of immigrant settlement, adaptation, and integration impact the construction of migrant identities?

Literature Review

Theorizing Migration

There has been considerable change in the way migration has been theorized. Early economic explanations were rooted in urbanization processes, access to employment opportunities, and push-and-pull factors (Harris and Todaro 1970; Lee 1966; Ravenstein 1885). More contemporary writers on migration (Papastergiadias 2000, Van Hear 1998) point out that current flows of migrants are fundamentally different from earlier movements. Current flows of global migration are much more multidirectional in nature. Migrant labour is no longer moving to the urban centres of the North and West, but is now moving to the new industrial 'epicentres within the south and the east' (Papastergiasidas 2000: 6–7). Disparities in levels of human security between places of origin and destination, for instance, have been found to play a significant role in migration decisions (Weiner 1993).

Furthermore, a number of personal and socio-cultural factors have become increasingly important, including the desire for independence, the need to break away from traditional constraints of social organization, and community conflicts. Relocation to a different social context liberates people from many of the social, cultural, and mental constraints they face at home from parents and extended family (Kurien 2002: 6), and may improve one's status (Van Hear 1998: 14). Experiences of exclusion and marginalization may also force people to migrate to new regions (Nambiar 1995: 16). Life cycle choices play a factor in migration, such as marriage, divorce, graduation from school, starting a career, the birth of children, and retirement (Greenwood 1975).

Transnationalism and Transnational Families

Both Safran (1990) and Cohen (1997) refer to diaspora as a dispersion of a group of people from a centre (homeland) to peripheral regions. Diasporas tend to also include a collective memory (real or imagined) of the homeland. Cohen (1997), in particular, highlights the voluntary nature of contemporary diasporas as well as the strong links between dispersed groups. These connections between groups are a key characteristic of the modern diaspora. Transnationalism can be seen as a more inclusive term which embraces the general definitions of diaspora, but tends to focus upon populations that are 'contiguous rather than scattered' (Van Hear 1998: 6). Transnationalism specifically refers to immigrants who despite having left home, and often through the process of movement across borders, maintain social, economic, and cultural contact with both sending and receiving countries (Wong 2007). Similar to the formation of diasporas, these transnational communities may involve the voluntary or involuntary movement of groups back to their place of origin. Transnationalism specifically forces us to think about the unequal linkages between the more developed and less developed worlds, a relationship which is grounded in the history of colonization, decolonization, and the globalization of late capitalism (Das Gupta 1997).

Research confirms that immigrant communities maintain transnational relationships: financial, social, and cultural. Research on Chinese families, for example, illustrates how transnational migration, economic opportunities, and globalization have led to new family forms both in China, and in the host countries where families must adapt to changing work relationships which impact the family form (Wong 2007). Research on transnational cousin marriages between British Pakistani women and Pakistani men move beyond traditional studies which focus on marriage as 'strategies' for migration (Charsley 2007). Such marriages are understood as reducing elements of 'risk' for families as they choose suitable partners for their children. Children's marriages necessitate a complex set of negotiations for the families, and this is reflected in this research on South Asian immigrants from Kerala.

Transnational research on families also examines the relationship between adult migrant children and their aging parents who are left back home (Baldassar 2007). This research examines how care of the elderly is managed over long distances. The research illustrates how family relations and family identities are maintained across time and distance and how these relationships are not determined or limited by particular locations and state borders (Baldassar 2007: 276). Studies on transnational care, which is sustained through remittances, phone calls, and brief visits back home, are framed by the institution of the family, and as a result is sharply gendered through the differing social roles of men and women in families (King and Vullnetari 2009: 31). The research presented here on Keralite women migrants extends this examination.

Faist (2000: 202–10) writes that the construction of transnational social spaces necessitates specific types of ties between the migrant and his/her home community. One form of transnational connection is kinship groups based on ties of reciprocity. This can take the form of financial remittances back to the home country. Transnational circuits can be in the form of trade networks and movement of goods. Kinship solidarity can also arise from a shared conception of collective identity and belonging. A recurring theme is that communities that do not share proximity form other non-spatial linkages and connections through exchange, reciprocity, and solidarity to achieve a high degree of social cohesion and a common basket of symbolic and collective representations (Kivisto 2001: 569).

Emerging diasporic consciousness and transnational linkages provide a context to account for new immigrant identities and the growth of transnational family communities. The following sections detail the history of South Asian migration to Canada and the specific experiences of women migrants.

South Asian Migration to Canada

Immigrants from India have been coming to Canada since 1904. The first South Asians to land in Canada were Sikhs who arrived in British Columbia. Their arrival coincided with an already rising hostility toward Chinese and Japanese immigration. South Asians thus were identified as part of the larger 'Asian Menace' (Buchignani 1980: 122). The federal government responded by imposing a $500 head tax, and began to restrict an already rigid immigration policy. In 1907 South Asians were disenfranchised in the province of British Colombia (Buchignani et al. 1985).

Provincial disenfranchisement led to federal disenfranchisement and a denial of economic rights in terms of labour contracts, and access to improved employment opportunities (Ralston 1996). In 1908 the 'Continuous Passage' stipulation was passed which stated that immigrants may be prohibited from landing in Canada unless they came from the country of their birth or citizenship by a 'continuous journey' (Bhatti 1980). This legislation was directed specifically at South Asians as it was impossible for them to meet this condition. Then in 1909 the government banned the immigration of South Asians, including the wives and children of those who had already arrived. The Immigration Act of 1910 explicitly used racial terminology for the first time and restricted British Indian immigration by name (Ralston 1996).

The ban on South Asian immigration between the years 1909 and1947 was relaxed after the end of the Second World War. In 1967, regulations were implemented that disregarded race, ethnicity, and nationality in the selection of immigrants. In 1968 the 'merit point system' was introduced. Points were given for education, skills, occupational demand, and knowledge of one of the official languages (Isajiw 1999). As a result between 1967 and 1974 South Asians responded in large numbers to Canada's need for professionals (Das Gupta 1986: 68). During this immigration period women were largely classified as dependents even in cases where the wife held better qualifications than the husband. Besides the stigma of dependency, this status also has several legal implications, for example accessibility to government sponsored services such as subsidized language training programs (Boyd 1986; Ghosh 1983). After 1974, applicants entering into the labour force (excluding immediate family members) had to enter into an occupation for which there was demand in Canada (Canada, Department of Immigration, Annual Report 1973–74: xi). The 1978 regulations specified 'designated occupation' and 'designated area' to both meet necessary skill shortages as well as to encourage immigrants to settle away from the major metropolitan areas (Hawkins 1989). Encouraging immigrants to fulfill Canadian labour force needs in the more peripheral, rural regions both dispersed and concentrated South Asians within the country. Although South Asians were encouraged to settle away from the larger metropolitan areas, they nevertheless concentrated in geographic places in Canada.

A new emphasis on family reunification in 1987 led to greater numbers of female applicants (Ralston 1996: 35–6). Data collected by Citizenship and Immigration Canada show that in 1999, 2000, and 2001 the number of female immigrants, as principle applicants and dependents, outnumbered male applicants who were applying to enter into Canada (Citizenship and Immigration Canada, 2002). The 2001 Census shows that about 60 per cent of immigrants were economic immigrants applying based on specific qualifications and skills, whereas 25–30 per cent were family reunification immigrants who were sponsored by relatives. About 10 per cent entered as refugees (Momirov and Kilbride 2005: 99). Though explicit racist immigration regulations against non-traditional, non-white, source countries have been removed, and the focus has been on filling labour force needs in the Canadian economy, the racist history of Canadian immigration legislation informs the experiences of many migrants from Asian countries.

The research reported on in this chapter focuses on the experiences of one specific subgroup, the Syrian Christian community. Within this group attention is paid to different historical points of departure as well as differences emerging from gender and generation. As Anthias (1998: 564) writes, 'diaspora is constituted as much in difference and division as it is in commonality and solidarity'. The stories of the respondents speak to this difference, and reaffirm the fact that though

South Asians may share similar characteristics, they do not belong to a single conceptual category. While members of the Syrian Christian community originate from the subcontinent of India, their experiences, qualifications, and religious affiliation, distinguish them from other Indian communities. The research presented here points to these differences and elaborates on the specific life-course experiences of migrant women who, having migrated in the early 1970s, are now over the age of 60, and face specific challenges stemming from the need for eldercare. The chapter speaks to the challenges they experienced in setting up new lives and recreating families in a Canadian context. Having left close-knit extended family structures in Kerala to face isolated communities in Ontario, Canada, these early migrants were forced to recreate families and reconstruct intimate bonds to replace the families they had left behind.

Community History and Social Context

It is important to highlight, briefly, a few of the salient historical characteristics of this community, and their social/cultural context. An important characteristic of the state of Kerala is the strength of the Christian minority community. Christians in Kerala account for 20 per cent of the religious community, as compared to 2.4 per cent in the entire country (Percot 2006). The Syrian community claims that the origin of Christianity in Kerala stems from St. Thomas the Apostle who landed in north Kerala in 52 CE (Alexander Mar Thoma Metropolitan 1985). The Syrian Christians in Kerala are well placed in the caste hierarchy of the state as early conversions came from the Brahmin and other high caste Hindu families. There is evidence that these early Christians and upper caste Hindus intermarried until the end of the sixteenth century (Bayly 1989). Hinduism, with its all-embracing caste organization provides the framework for the entire society (both Christian and Muslim). Despite the very distinct regional languages and religions found in the Indian subcontinent, the behaviours and value systems of Indian people reflect the dominant Hindu cultural tradition (Ibrahim Ohnishi and Sandhu 1997). This influence and prevalence of Hindu ideology is particularly important when discussing issues around family, gender roles, and responsibilities. Family and extended kinship relations provide the basis of individual and community identity in Indian society (Choudhry 2001: 378).

The relationship between these three dominant religious groups (Hindus, Muslims, and Christians) changed drastically under colonialism. The arrival, in particular, of the British in the early eighteenth century was important for the Syrian Christian community as they flourished under British rule. The British forged alliances with the Syrian Christians in order to strengthen their political control. The Syrian Christians led the other groups in English language education, and a large majority went on to become teachers, lecturers, doctors, nurses, and administrators in colonial institutions (Kurien 2002). The Christian community continues to educate and send nurses around the world where countries actively recruit nursing professionals. In India, according to Percot (2006), nursing training has increased, particularly in the private sector as nursing labour markets continue to open in both the West and in the Persian Gulf countries. The majority of first generation respondents in the study were nursing professionals.

South Asian Women in Families

Discussions of South Asian immigrant families are dependent on the experiences of women and the dominant gender ideology in Indian society. Women's roles in the household and their responsibility for housework and childcare are rooted in the ideology of **patriarchy** and the sexual division of labour, both of which have been historically and socially situated (Mazumdar and Sharma 1990). Writers on migrant women point out that cultural background and social positioning in the home country must be taken into consideration when defining experiences in the receiving country (Buijs 1993; Simon and Brettlee 1986). Thus, this section begins with a discussion of some of the common themes which emerge from the literature on South Asian migrant women and their families. Contemporary writings on South Asian immigrants point to not only the gendered dimension of the migration experience, but as well experiences of **racialization**.

South Asian women who have moved from India to Canada often experience a change in their economic roles as well as their status as women within the family (Ghosh 1981). There is often a conflict between the traditional role of the housewife and the role they acquire as a working woman in the paid labour force. Religious ideology (which is informed by both Hinduism and Christianity) and the dominant cultural practice place Indian women in a secondary and subservient position relative to men. There is a structure of ambivalence in the family setting as these women are inherently displaced from one homeland to another. South Asian women, in making a space for themselves in Canada must negotiate an identity which satisfies the two contradictory poles in their lives on a daily basis: the dominant western tradition which mark their new home and their own traditions and values which are carried with them. Josephine Naidoo (1980, 1988) in her research of South Asian women, refers to this dialectic as the South Asian women's 'dualist' view of life.

Adults who emigrated from India have been socialized in a traditional value system. When raising their own children in Canada, parents tend to frame their parenting practices around those traditional family and cultural paradigms (Ghosh 2000: 287–88). Girls, in particular, experience these traditions, parental expectations, and limitations in ways that reflect underlying power differentials between men and women. Dasgupta (1998: 8) writes that a South Asian woman's identity has never been conceived of on individual terms; rather her identity is always fused with the identity of the men in her life: father, husband, son. In this way, a South Asian woman is defined by her familial roles of daughter, sister, daughter-in-law, wife, mother, and grandmother. While these roles provide her with some protection and stability, they also place limitations and confine her in her everyday life. These restrictions can impact life choices ranging from education and friendships, to choices in marriage partners.

The experiences of South Asian women and their families also need to be understood within the larger context of racialized discourse. As Aujla (2000: 41) writes, 'in the same spirit as colonial cartography, South Asians have been "mapped" and inscribed by the dominant culture through a racialized discourse and state practices since they began migrating to Canada in the late nineteenth century'. The writings of Brah (1996) which examine the relationship between ethnicity, culture, and identity among South Asians in Britain assert that any examination of culture must be understood within the context of power relations among different groups.

Handa (2003), in her study of second generation South Asian youth in Toronto, goes beyond the dominant 'culture-clash' model and points to how notions of race and gender position young South Asian women within the diasporic context of Canada. Radhakrishnan (2003), in his examination of Indian immigrants in the United States, writes that the Indian immigrant is reborn as an 'ethnic minority American citizen'. In this article, the author asks whether the remaking of the immigrant in American society points to empowerment or marginalization. This new American citizen 'must think of her Indian self as an ethnic self that defers to her nationalized American status' (Radhakrishnan 2003: 121). In this way the immigrants' ethnic identity can be seen as a *strategic* response to a shift in time and place. The migrants' response is one that takes into consideration marginalization in the host society and tensions between two divided loyalties. Writings on transnational Caribbean families assert that the transnational and cross-cultural family networks, rituals, and celebrations play a strong role in shaping Caribbean identity. These networks can be seen as a source of social capital which takes into account experiences of marginalization and racialization in the host country (Reynolds 2006).

Talbani and Hasanli (2000), in their study on adolescent females and female socialization point to strategies used by families in maintaining traditional roles. Both generations of migrants (mothers and daughters) are expected to adapt to the host society while at the same time act as guardians of group culture and tradition. There is additional pressure placed upon the second generation daughter as her behaviour reflects the level of assimilation of the family to the host country, and she is doubly placed with the burden of 'cultural continuity' (Joshi 2000: 5–7). The author points out that young women in particular, within the community and household, are forced to maintain and uphold the cultural traditions of the group, often specifically through marriage practices (as discussed later in this chapter). Migration, in general, causes tension within and outside the group as community members and their families attempt to redefine and renegotiate their roles and identities (Talbani and Hasanali 2000). The process of negotiating cultural change and making adjustments to family life is difficult and marked with new challenges despite the length of stay or the destination of migration.

There has been minimal research on the aging South Asian population in Canada. Studies (Choudhry U.K. 2001; Ng et al. 2007) point to some common issues of isolation, loneliness, rising family conflict, and economic dependence on adult children. Studies show that widowed South Asian women tend to live in extended family structures while men are more likely to live alone (Ng et al. 2007). Existing research fails to provide any understanding of the changing nature of identity among immigrants as they enter retirement. The study presented here, with a focus on the first generation cohort, reveals the complex nature of identity construction as immigrants enter the final stage in their journey as seniors in Canada. The relationships within the family changes as women enter into a phase of heightened dependency, and too often, rising family conflict.

Research Methods

In this study of immigrant families from Kerala, India, a semi-structured questionnaire was used. In the larger study[2] (of which only a portion is reported upon here) data was gathered by the author from 64 individuals all living in the Greater Toronto Area (GTA) of southern Ontario, Canada. For this paper, I draw on the responses of first generation immigrant women and how spatial change, and separation from their homeland and family of orientation impacts gender roles and responsibilities, and their sense of self in Canadian society. I divided the larger group into three cohorts. The first generation cohort (17 respondents) all arrived in Canada between 1965 and 1977. Except for two women who came with their husbands and young children, all the remaining members of this cohort were nurses who immigrated as independents, responding to a labour need for nurses. The average age of this first generation group at the time of the study was 65.9 years of age.

The data gathered from the second generation cohort (23 respondents) and the third cohort (24 respondents) were analyzed as part of the larger study, but are not directly reported upon here.

The second generation cohort consists of the adult children of the first generation migrants. The third cohort is made up of more recent migrants who arrived in Canada between 1993 and 2005 and came directly from India, as well as from Malaysia, the Persian Gulf, South Africa, and the United States.

Research Findings

From the Paddy Fields of Kerala to the Snow Banks of the GTA

The research findings presented here detail the journey of first generation respondents and the struggles they experienced raising families. The stories of these respondents speak to a number of themes: the pursuit of higher education, liberation from traditional structures, freedom from parents, and the need to help with the finances of the households they left behind. Many of these women left at the age of 17 or 18, entered nursing schools which were out of state, and then left for abroad, despite resistance by family members who wanted them to remain home and marry. I have chosen responses from women whose tales are typical of this flight from home.

Susan, a retired nurse, reflects on her parent's reaction to her entry into nursing,

> 'My parents reacted very bad . . . Those days girls don't go for nursing from any family. I was the first one in my complete family to go for nursing. They think it is bad, girls can go the bad way, any decent family does not send their girls for nursing. My family was very much against me going to nursing'.

Like most of my respondents, Susan had trouble getting permission from her parents to enter nursing school. Nursing was not seen as a desirable occupation for girls from 'good families' and attending nursing school was a source of conflict in many families.

Isabelle, 65 years of age and a retired nurse, remembers how difficult it was to get approval from her father to go to Canada.

> 'So I got my first appointment and I am happy . . . When I got the appointment I know that is a job, so I wrote to my father to send the money [for ticket] , but my father said "no you are not going anywhere". They were waiting after graduating to get me married and settled in India, and we were not that rich . . . just middle class. Then I thought my father won't send the money, he expected me to come back home. . . . I was determined to go overseas. They cannot do anything. There was a lot problems at that time anyway, but I got the ticket and then I resigned from the hospital . . . I told my dad just to let me go for one year and I will come back. My father was upset, my mother was upset, everyone was upset. There were a lot of problems. . . .'

Though Isabelle did not obtain the permission of her father, she secured her airline flight through her employer in Canada, and left India. Similar to other respondents, Isabelle had a strong desire to leave Kerala, go abroad, and be financially independent. Independence and pursuit of individual goals are not valued in the Indian family structure. Family needs and desires are perceived as more important than individual goals. The pursuit of individual ambitions is contrary to the collectivist nature of the Indian family structure (Choudhry 2001).

Migration for women and employment abroad allowed women greater independence and autonomy within the family. In the traditional Indian family, daughters tend to be in a state of dependence to their fathers and this dependence then transfers to their in-laws upon marriage. By securing employment, particularly professional employment, women were able to free themselves from this dependence on parents, and from pressures to marry young. In a society which values interdependence among family members these women were challenging firmly held traditional gender expectations by pursuing careers. The respondents in this study were cognizant of their future had they stayed at home with their families. Had they not left Kerala they would have married early without the chance for professional education, employment, and financial independence. Professional employment abroad increases their own status within the community as women, and improves their economic position as independent wage earners. Their resistance against family pressures to stay at home and marry after finishing high school speaks to their personal character as independent thinkers, and as women who were active in transforming their lives.

Raising Families Far From Home

Despite their success in gaining meaningful paid professional employment, these early immigrant women often faced expectations that compelled them to fulfill roles as mothers and housewives. While these women subscribe to the early second wave feminist movement values expressed by women like Betty Friedan (1963) who urged women to reject the myth of the 'happy housewife' and pursue professional careers and other 'meaningful work', the women in this respondent group still very much adhere to 'traditional' family responsibilities and duties. Despite their independence in leaving home and going abroad, when it came time for marriage, except for one respondent, all the women surveyed married in a traditional arranged marriage (which in Kerala is more loosely structured than in other parts of India) thus respecting parental wishes for suitable marriages.

Desai and Krishna Raj (1990), state that regardless of religious adherence, a women's survival is not socially conceivable without the extended family in South Asian culture. The social value which is placed on the familial role for women can also be seen as the cause for her subordination to men and her lack of access to important economic and political resources. This is even when she may be contributing an equal or greater amount of money to the family income. Dhruvarajan (1992) echoes this conclusion as she claims that patriarchal family norms provide the context for family behaviour. Housework and child care are defined in Indian society as women's work. These practices are legitimized in Indian society through a world view of male dominance and female subordination provided by and strengthened by traditional Hinduism (these sentiments are pervasive as well in the Syrian Christian community) (Dhurvarajan 1992). Though patriarchal relations are characteristic of Christian traditions, in Kerala, **familial ideology** is structured primarily through Hinduism. Women's unequal positioning in the family and the sexual division of labour have been heavily theorized by numerous feminists (Barrett 1988; Friedman 1963; Luxton 1989; Mies 1986). Among the respondents, although the family structure is very much grounded in patriarchal family norms there is considerable difference in the way the family roles are carried out. The changing social context and the changing work arrangements and access to paid wage labour allowed women greater power in negotiating domestic work. This is evident in discussions with 'Marnie', a retired nurse and mother of two girls,

> 'I had my first child in 1970, a girl. Those days there was no maternity leave, I had to resign from my job and I moved to Scarborough from Toronto. Before my husband came from Kerala I moved out here. A new hospital opened in Scarborough and I thought it would be nice to work in a new hospital. I had my first daughter during this time. I resigned from my job

for six weeks. Those days you have to go back to your job right away. You have to. [What did you do with your daughter?] My husband and I worked opposite times. I worked in the evenings and my husband worked days so we gave her to the babysitter for a few hours while we were changing shifts. My husband worked in a factory for a few years and then he went to night school for four years in the evening. We had to share all the housework somehow . . . I did all the cooking though'.

All the women in the first generation respondent group continued work after the birth of babies out of financial necessity and a lack of government support for paid maternity leave. For most who went back home to marry, there was the added issue of husbands whose educational qualifications were not recognized and thus they could not find appropriate employment. Often husbands had to retrain at a local college or university taking evening classes, or had to, more commonly, take up employment in manual labour.

For some families, the inability of men to find work in their fields of study led to tensions in their marriages. Rema, who immigrated to Canada in 1970 as an independent professional nurse, tells her story and points to the difficulty she had with a husband who did not find suitable work.

'I got married in 1974 and he came at the end, December 1974 [. . .] We got to know each other only through letters . . . [Was he able to find appropriate work?] No, he worked so many places. He was not so happy. He worked in factories here and there. He did not like it here, first of all. He was teaching there and he wanted to get a teaching job here, but he could not. . . He went back to India. He was here from 1974–1982. [In] 1978 we had our daughter. He left in 1982. He worked for a little bit and did not come back. We never had any contact. He just left. . . . he said he was leaving for vacation and he never came back . . . My daughter was only 4 years old. It was so hard.'

Rema's experiences of a failed marriage and her struggles to raise her daughter on her own highlight the stresses involved in the migration process within the family unit. As a woman who was working full time, she had to not only care for her daughter, but she felt compelled to ensure that her husband was well adjusted to the new environment. While most husbands were able to find employment, for women whose husbands did not find meaningful work, the tensions within the family increased. Pressures to maintain a 'good family' and a 'good home' fall on the shoulders of women. Though the women in this survey are all well educated and financially secure, they still grapple with the responsibility of maintaining a 'perfect' home for the family. The 'cult of domesticity' strengthened and solidified women's positioning and function in the home (Cott 2001: 129). Women are placed with the burden of maintaining the 'happy household' and servicing their husbands and children, despite their work in the public sphere. It is through the work of women that the household presents itself, its outer image, to both the dominant host community, and the Malayalee[3] community. Women in particular have to manage their home economically and emotionally, making sure all members are satisfied (echoing the findings of Luxton, 1980). Any distress or family problem is frowned upon and stigmatizes the family. For Rema, though the community knew of her family issues and were supportive, the stigma of her family problems and divorce fell heavily on her and her daughter.

Women's continued employment during these early years of having children and juggling work shifts with husbands allowed for a steady source of income (often the only source of income). There was a shift in gender roles from what is prescribed through traditional patriarchal family norms in Indian society. Women in all cases were responsible as main income earners and because of this they held considerable decision-making power when it came time to buy houses, cars, or taking vacations. One of the respondents stated that she went and bought the family house in an upscale Toronto suburb, doing so without even consulting her husband, as the purchase was based solely on her salary.

For many of these immigrant families the growing network of immigrants from Kerala provided an important source of support. They gathered on Sunday mornings initially in people's homes for prayer and the sharing of food. As the community grew in numbers and the church was built (now there are numerous Syrian churches in the Greater Toronto Area) these centres played a vital role in supporting community events and cultural development. The church and the priest were an important source of support as parents faced the challenges of raising adolescent children.

Raising Teenagers and Young Canadians

Parents from the Keralite community faced considerable difficulty in parenting their teenage and young adult children because of different familial and cultural expectations for children, children who were growing up *Canadians* and not Malayalee. One respondent, Isabelle, reflecting on that period of her life states,

> 'Since my children were teenagers psychologically I was terrified. I should not hide this. Even though I am living here, even though I am working here, even though I am making money . . . even though I have a good husband, I have everything, when your children turn into teenagers they are a different responsibility. The children are *not* Indians, they are Canadians, and everything is different!'

Adolescence, and the raising of young 'Canadians', posed new challenges around parenting practices and social/cultural identity. On the one hand, parents had high expectations for children in terms of educational and academic success. For the most part, their children fulfilled these dreams of attaining good university educations and securing professional employment. There is a community orientation toward high achievement in traditional areas of study (medicine, engineering, and teaching) and parents tend to encourage their children into these areas of study. However, the difficulties around dating and self-selection of marriage partners were situations for which the older, first generation respondents were not prepared. Responses varied from having very strict rules to attempting to set some marginal guidelines and hoping children would respect parental wishes.

This lack of parental control is something respondents were not expecting when they made their migration choices. For many parents it is around this time that they experience feelings of regret about leaving their home country where parents still have greater control over the marital choices of their children.

Edith laments her children's decisions around marriage. Her son married out of the community and her daughter (aged 31) remains single which is a source of unhappiness and regret for both her and her husband. She was very quiet and melancholy during the interview and felt let down and disappointed by her children.

> 'We had strict rules about dating. They did not like it very much but those were our rules. No dating. In high school we were very strict. They did not like this. They had friends, but . . . [for the most part they listened?] Yes, for the most part, but after 18 . . . it was a different story. My daughter was living in residence so I could not very well see what she was doing. My son stayed home though, he went to the

local campus. My son was almost 24 when he started bringing in his girlfriend. This was hard for us. [Was she Malayalee?] No, no . . . she was not. My son is married, he has two kids, she is Canadian, her parents are from West Indies. My daughter has a boyfriend who is also Canadian. My children were not interested in other Malayalees.

In India you could arrange a marriage and just get them married, but here you cannot do that. They say they do not like it, they do not want it. It is during these times that I regret coming here. Oh yes that is the only point where I wish I had stayed back in India. All my nieces and nephews back home are married to Malayalees and they were all arranged marriages . . . My daughter especially says "no, I do not want".'

The lack of parental involvement in marriage matters was difficult for parents who are used to arranging suitable partners for their children. This removal from choosing marriage partners and having minimal involvement is a key source of conflict in immigrant families. Marriage matters are family matters and the response of the second generation in choosing and self-selecting marriage partners challenged traditional family patterns. Neither the parents, nor the community at large, had an easy time dealing emotionally with this change and the perceived threat to community values and culture.

A dialogue with Marni over the marriages of her daughters, who both married out of the community, is quite interesting. Her husband joined the discussion as well. The conversation illustrates the disappointments parents feel when children neglect to fulfill parental hopes and desires.

'[Marni] We would have preferred them to marry Malayalees.

[Husband] We are not 100 per cent happy with that to tell you the truth. But we are okay with it now. We have to be really.

[Marni] We have no choice but to be happy about it. So we are okay with it, we are not happy about it. That is one thing, if you were in India this would not have happened. They would have married the right person, they would have gone by the rules. It is a traditional matter.

[Husband] Here we have a say but not really . . . basically they are not living in your household all the time they are outside, they are independent. Just like my daughter said, "it is *my* life".

[Marni] In India they will never say that. They will never say that. "I want to do what I want to do". This would never happen. This is a forbidden thing in Kerala. . . . If you are in India your children will fulfill your dreams too. Here they do not think of the parents, their dreams, their expectations. Here they will say "this is my life . . . I want to do what I want to do". They have their own expectations'.

The conversation here reflects numerous contradictions. On the one hand the respondent entered into a profession (nursing) that was not approved of by her parents and then left her family to seek employment overseas, and thus she 'lived her own life'. On the other hand, however, she is unable to make a link between her own ambitions to the desires and hopes of her daughters. Remarkably her girls have the same sense of adventure as their mother. Marni, however, sees her daughters' pursuit of happiness as a challenge to her own, because it challenges her cultural identity and her values as Malayalee.

Conflicts around marriage stem from parental pressure in maintaining family cultural backgrounds. As parents it was important for them to encourage their children to marry within the cultural group and maintain, to some extent, parental expectations. For second generation children, who are attempting to form their own cultural identities separate from their parents, these pressures to marry Malayalees from Kerala are not important to their settlement in dominant Canadian society and to their identity as South Asian Canadians. Their struggles for independence and 'living their own lives' did not resonate with their parents who expect more collectivist, family-oriented behaviours from their adult children.

The responses show that it is during this time of dating and marriage that parents are confronted with the dilemma of either enforcing strict guidelines to preserve their cultural traditions, or allowing children to pursue their own individual desires. Respondents from the first generation group made numerous return trips back to Kerala with their children in an attempt to continue linkages with the home community and foster traditional values and customs. Respondents also took grown children back to India in the hopes of finding suitable mates, but these attempts did not always succeed. Instead parents faced resistance, and at times outright rejection by their children of traditional marriage practices and expectations.

Migrant Identities and Belonging: Being 'Nowhere'

This section highlights the ambivalent position of women as they attempt to define themselves and their place in Canadian society. For women who have challenged traditional gender roles and left the comforts of their home and family to begin lives overseas, their current positioning and their reflection on the choices they made point to feelings of regret and indifference. The respondents in this group are all over the age of 60, and particularly for women who are widowed or divorced, they are worried and feeling increasingly anxious about being isolated in their old age.

Respondents were particularly reflective and thoughtful when discussing choices they had made in the course of their lives. With regard to 'gains' made as a result of immigration, respondents were quick to point to financial security, material assets such as houses and cars, and fulfilling dreams of educating children. However, when asked to think about things which they 'lost' as a result of migration respondents pointed to the loss of their families and the closeness they experienced with parents and siblings.

Susan, who immigrated in 1965, is indecisive about her life here and the gains she has made.

> 'I got a better education here. I learned a lot. Economically I think I am better off. I helped my family [in Kerala]. Not an awful lot, because I had expenses going to school here, I did save a bit. . . . It is hard to say. They [family back home in Kerala] are all doing pretty good [financially], they have boys. I don't know if I am better off than them. I sometimes feel that they are better off than I am.
>
> [What did you lose?]
>
> My family, my siblings, my culture to a point . . . ah . . . I have not, but my children have lost their culture. When I came to Canada at that time there were not many people from our community. There was not enough people and there was not enough time to begin with. I was busy working all the time, the children go to school and they come home, they had different activities. They were busy. We did go to church and the children came too.'

Though all the members in the respondent group were active in setting up community prayer meetings and cultural celebrations, this did not encourage children to marry within the community. As the second generation children reached adulthood, their participation in these community churches and cultural associations diminished.

Rina also expresses some mixed emotions over her choice to emigrate. Again, much of this regret stems from children who did not fulfill her wishes for marriage within the community.

> 'When I think about it, at my time, the reasons [to migrate] were fulfilled. The education of my children I am very happy about. When the children are grownup, I am used to that custom from home that they try and get the children married, the family will press on you to marry, here when it comes time to marry, they do not do anything and we do not have much voice in it. So that is a bit difficult. Sometimes I sit and think that if I had stayed back home they would be married by now . . . then again . . . this is something that has to come from them. . . . I should not feel unhappy about this . . . I have mixed feelings . . . Then I think, things will happen according to God's will. It is not I who plans these things. I can cry to God only.'

While she had led a successful career, these achievements were overshadowed by the behaviours of her children who refused to meet her expectations around marriage.

Eldercare: Growing Old in a Foreign Land

Respondents in the older cohort pondered their current positioning as elderly women who are not being looked after by their children. As the first cohort ages there is a growing recognition within the wider community that their needs for elder care are not being met by children as they see they ought to. According to traditional family structure, most elderly parents live with their children. Studies on elderly immigrants show that, while many South Asian elderly persons continue to live within an extended family setting, the 'traditional reverence for parents has diminished' (Choudhry 2001: 377). There is a rise in intergenerational conflict as parents retire and enter an age of expected dependence upon their children. Respondents in this study express disappointment as children neglect to take care of their aging parents according to the traditional customs in Kerala.

Marni, like many of the elderly respondents, states quite clearly that she is not expecting her adult children to look after her. However, she not only has regrets and misgivings over the marriage choice of her girls, but now as she ages she worries about her future and how she will manage alone.

> 'I am worried about my old age. That is the question. Who is going to look after us Malayalees? We do not have an old age home. Other cultural groups like the Italians, or the Portuguese all have homes and drop in centers. [Is there a stigma attached to old age homes?] The Indian culture does not allow you leave your parents in an old age home, you keep your parents with you and look after them. Which you cannot do it here. . . In the end you have to go to a nursing home. I do worry that my kids won't look after me, but how can they look after me? It is not their fault. They have their life, their children, their work . . . I have no idea what life is going to be like. When I think about the future, I wish sometimes that I had not come here and that I had stayed in India. I do feel this sometimes. Even by coming here, what did I gain? What did I achieve? I worked hard all my life, in the end what did I get? I do have some regret. When you are young you do not feel that way, you keep going, going, going . . . when it comes to an end you ask, what is it all for? In the end you are worrying, what is your future, who is going to look after you? What

> happens, it is big worry, when you are young you never think about it. Once you retire it is important. You work so hard all your life. In India, you might not have all these material things, but you are better off in some way.'

Rema who is living alone also worries about her future. While she is independent now in terms of movement, she is worried about a time when she will need more direct care. Her narrative is full of remorse and regret as she enters this stage of life alone.

> 'My concern is what is going to happen to me. An old age home . . . my generation everyone is retired. [Any regrets?] No, I do not know . . . when I think of my sisters they are all in India, they all have a good time, they have servants, they are not alone. They can see each other. Here I am all alone, I have no relatives. In that way I regret coming. Financially I am okay. Otherwise I feel . . . friends you cannot depend on them. If I have somebody here, a couple of sisters here then I would feel more at home here . . . there are lots of Malayalees nearby.
>
> [Do you feel like you belong here? Do you feel Canadian?]
>
> No . . . no. I feel Malayalee first. I am a citizen here, but still I do not think I belong to here. [Why not?] I do not belong here. [You have been here for so long.] I have been here 37 years. [Do you feel like you belong in Kerala?] No, not in Kerala either. Nowhere . . . I am in Nowhere. Do you understand that? Do you understand what I mean? That is why I feel sometimes why I came here? At this point in my life I feel like I belong nowhere.'

While the women in this community are financially secure, they share the same issues around isolation, loneliness, and rising family conflict as adult children are reluctant to provide care for their aging parents in ways parents see as culturally appropriate. While the church was a source of support for many of their years in Canada, the respondents in this study feel that both the church and the larger community are ignoring the needs of the seniors. Issues around isolation and safety are growing concerns especially for elderly women who are living alone.

When I asked Marni about *belonging* she laughed it off and stated quite adamantly that in her heart she was Indian.

> [Do you feel like you belong here? Do you feel Canadian?]
>
> No, I do not feel Canadian. You could live here for one hundred years and you will still not feel like you belong here, that you are Canadian. You never feel that belonging. No matter how many years you are here. Maybe your generation, like my children generation, they may feel connected here. They probably feel more Canadian, rather than the first generation. I am still Malayalee, I am still an Indian. My mind is still in India.'

Rema's earlier remark about belonging 'nowhere' is quite significant. For many of the women in this respondent group, particularly those who are widowed or divorced there is a profound sense of isolation and loneliness as their children, who were brought up in Canadian culture, do not have the same sense of responsibilities toward eldercare. For these respondents, this perceived lack of responsibility by children exacerbates their feelings of regret about their migration journey.

This lack of attachment to place or nation can be seen as the ambivalence of **post-colonial identity**. There is a reluctance in committing to a place, this feeling of belonging 'nowhere' is very significant for the elderly immigrants and reflects a larger sense of detachment from singular notions of nationhood and citizenship. The '. . . post-colonial identity is articulated in multiple modalities—the moment of experience, the mode of writing [the performance of everyday realities] or representation' (Ganguly 1992: 46). It is between the performance of their everyday lives and representation that their subjective identities appear (within lived space). In other words, their identities emerge from their lived reality, and this reality is becoming increasingly unstable as they age in a foreign land. Identity, as Bhabha (1990: 192) writes 'is always poised uncertainly, tenebrously, between shadow and substance'. From the responses of these women who at the end of their professional careers and perhaps at the final stage of their migration journeys, are in a place between 'shadow and substance', a place (metaphorically or geographically) filled with regret, remorse, and sorrow. Are the respondents' current positions of 'being nowhere' the end point of diasporic identity?

These women have challenged the dominant negative stereotypes of Indian women, which define them as passive and submissive, and resisted the gender roles within their families by pursing educational and career opportunities. Marni, who has lived here since 1965, however is determined to hold on to an identity based on her past homeland. This stems from a construction of a past which in her mind is more ideal than her present life, which she identifies as characterized by children who have defied her wishes and ignored her dreams and desires. Marni is like many of the first generation respondents who participated in this study. Her story is filled with unfulfilled hopes and mixed feelings over the choices she made over 30 years ago, and which she cannot now change.

As these women reflect on their migration journeys they are ambivalent and conflicted over the gains they have made. These mixed emotions stem not only from children who challenged them, but as well, from not feeling connected to the place in which they have settled for over 30 years. As they enter old age alone, the added dimension of eldercare (or lack of eldercare) further exacerbates their feelings of isolation and regret. Maryanne, another first generation respondent, states, 'Whatever comes I will deal with it, that is the way I feel. My children so far tell me you are going to be in a retirement home or nursing home, we will come and visit you. . . . What can we do. We have no choice'. There is often a lack of real choice for these women as they face these issues largely alone. As a community, either through the churches or the associations, there has been no real attempt to address the needs of this aging group of community members.

Conclusions

The process of **deterritorialization** of cultural identities among migrant groups refers to the way in which people feel they are still connected to communities despite the fact that they do not share a 'common territory with all the other members' (Papastergiadis 2000: 115–16). Thus, despite the lack of propinquity they still feel very much connected to, and a part of, the cultural group in Kerala. For the respondents in this study the strong family unit combined with the diasporic community helped to foster and develop these feelings of connection to families back home as well as to 'new families' in the Greater Toronto Area of southern Ontario. Strong community networks such as the church facilitated the connections to families back in Kerala, both for the first generation immigrants and their children.

Despite the distance from the homeland, migrant families from Kerala are very much connected and influenced by their culture and sense of community. Transnational migrants are defined by their economic, social, and cultural linkages to both their country of origin and their country of residence. The continued connections to networks and organizations both in Canada and in Kerala attest to the nature of transnationalism and the spatial construction of diasporic identities. Identity construction among immigrant families involves a relationship to the 'homeland', although the degrees and specificities of each group's and individual's situations vary. Customs around marriage and responsibilities toward the family, however, can set them in opposition to the dominant Canadian culture. Thus, not only are identities created within, but they are also formed in relation to, the 'other'. As Hall (1996) writes, the individual's inner core is not autonomous and self-sufficient, but is instead dependent on others. Identity for this sociological subject is formed in 'interaction' between self and society.

Transnational migrant identities are created within the context of families and the migration experience. First generation women who arrived in the mid 1960s and 1970s formed their identities within the context of migration and settlement, and faced the challenges of raising children in a new environment. These women faced new struggles as they were forced to redefine, and reconstruct their identities as the result of children who did not subscribe to the same cultural values and traditions. This is despite the fact that these women defied traditional gender expectations by furthering their careers and pursuing professional employment, breaking with traditional gender roles which placed women within the confines of domestic work. While many girls of their cohort were getting married, these women not only left their villages, but left their country in search of a new life. These early migrants did however maintain contact through letters and remittances, and eventually most women returned to marry a man of their family's choosing. Their households remained intact, 'stretched through time and space' (Baldassar 2007: 287). For these women, having left home at a young age for school and then travelling abroad, the responsibility toward the family remained very real in their minds. The transnational family structure is one where the duties of family responsibility remain, although they are influenced by the context of place and space. The experiences of these early immigrants support research on transnational families which states that 'much like . . . diasporic identities and relationships, family identities and kin relations can be maintained across time and distance and not necessarily or completely be determined by particular localities or by state borders' (Baldassar 2007: 276). The respondents in this first generation group show a strong linkage to the homeland, and maintained their relationships with parents, siblings, and friends through letters, phone calls, and numerous visits that included their second generation children.

Any examination of identity must be rooted within the context of the family, as the family is the key site for the socialization of children (Malinowski 1913; Murdock 1949; Parsons 1955). For migrants, the family is the setting where conflict emerges, but as well the family is the site where many find strength and support. Research on migrant families points to the ways in which family and kinship networks and relationships act as social capital for young people and assist them in building ethnic solidarity and unity with groups in the host country as well as in their home country. In the same way the local family presence, as well as transnational family linkages back home in Kerala, provide sources of support for the migrant family. Transnational family networks which reaffirm and reproduce family values and traditions, though at times contested, play a central role in shaping migrant identities and family geographies.

Questions for Further Thought

1. Raising teenagers and young adults is a difficult stage of parenting in all cultures. In what ways does the immigration process heighten these difficulties?
2. How has the experience of aging for immigrant women altered how they identify with Canada? How does aging influence the migration experience?
3. In terms of social policy, what are the ways in which governments could give support to aging immigrant populations?
4. How would you approach the situation as a young adult with an aging parent? What are some of the ways in which you could give support to aging parents? What discussions need to take place within the family unit to prepare for this situation?

Further Reading

Arber, S., and J. Ginn, eds. 1995. *Connecting Gender and Ageing: Sociological Reflections.* Buckingham: Open University Press.

Baldassar, L. 2007. 'Transnational families and aged care: the mobility of care and the migrancy of ageing'. *Journal of Ethnic and Migration Studies* 33, 2: 275–97.

Chappell, N. 1989. 'Aging in the family'. Pp. 185–206 in G.N. Ramu, ed. *Marriage and the Family in Canada Today*. Toronto: Prentice Hall.

Choudhry, U.K. 2001. 'Uprooting and resettlement experiences of South Asian immigrant women'. *Western Journal of Nursing Research* 23, 4: 376–93.

Faist, T. 2000. *The Volume and Dynamics of International Migration and Transnational Social Spaces.* Oxford: Clarendon Press.

Endnotes

1. As Ghosh (2007: 224) points out, by using 'South Asian' as an analytical category we dismiss the numerous differences which exist within and between various groups from the Indian subcontinent. South Asian refers in a broad sense to individuals who trace their origins to the South Asian subcontinent: India, Pakistan, Republic of Myanmar, Nepal, Sri Lanka, Bangladesh as well as those who have migrated from South Asia to other parts of the world.
2. For this study a combination of purposive sampling and snowball sampling techniques were used. I met with women who were originally sourced from key figures in the community. With an understanding of the general population I phoned individuals and informed them of the study. Meeting with women often led to further contacts, and referrals to other members of the community who were interested in participating. In the larger study data was gathered from 64 individuals both first and second generation immigrants from Kerala.
3. The language in the state of Kerala is Malayalam, and the inhabitants are referred to as Malayalees.

References

Alexander Mar Thoma Metropolitan. 1985. *The Mar Thoma Church Heritage and Mission*. Houston: T & C Copy and Printing.

Anthias, F. 1998. 'Evaluating "diaspora": beyond ethnicity?'. *Sociology* 32, 3: 557–80.

Aujla, A. 2000.'Others in their own land: second generation South Asian Canadian women, racism, and the persistence of colonial discourse'. *Canadian Women Studies* 20, 2: 41–7.

Baldassar, L. 2007. 'Transnational families and aged care:

the mobility of care and the migrancy of ageing'. *Journal of Ethnic and Migration Studies* 33, 2: 275–97.

Barrett, M. 1988. *Women's Oppression Today: the Marxist/ Feminist Encounter.* London, New York: Verso.

Bayly, S. 1989. *Saints, Goddesses and Kings.* Cambridge, UK: Cambridge University Press.

Bhabha, H.K. 1990. 'Interrogating identity: the postcolonial prerogative'. D.T. Goldberg. *The Anatomy of Racism.* Minneapolis: University Of Minnesota Press.

Bhatti, F.M. 1980. 'A comparative study of British and Canadian experience with South Asian immigration'. Pp. 43–61 in V.K. Ujimoto and G. Hirabayashi, eds. *Visible Minorities and Multiculturalism: Asians in Canada.* Toronto: Butterworth and Co.

Boyd, M. 1986. 'Immigrant women in Canada'. Pp. 45–61 in R.J. Simon and C.B. Brettell, eds. *International Migration.* New Jersey: Rowman and Allanheld.

Brah, A. 1996. *Cartographies of Diaspora: Contesting Identities.* London: Routledge.

Buchignani, N.L. 1980. 'Accommodation, adaptation, and policy: dimensions of the South Asian experiences in Canada'. Pp. 121–50 in V.K. Ujimoto and G. Hirabayashi, eds. *Visible Minorities And Multiculturalism: Asians in Canada.* Toronto: Butter worth and Co.

Buchignani, N., D.M. Indra, and R. Srivastava. 1985. *Continuous Journey: A Social History of South Asians in Canada.* Toronto: McClelland and Stewart.

Buijs, G. 1993. 'Introduction'. Pp. 1–19 in G. Buijs, ed. *Migrant Women: Crossing Boundaries and Changing Identities.* Oxford: Berg Publishers.

Canada. Department of Immigration. 1973–74. Annual Report.

———. Department of Citizenship and Immigration Canada. Annual Report, (2002).

Charsely, K. 2007. 'Risk, trust, gender and transnational cousin marriage among British Pakistanis'. *Ethnic and Racial Studies* 30, 6: 1117–31.

Choudhry, U.K. 2001.'Uprooting and resettlement experiences of South Asian immigrant women'. *Western Journal of Nursing Research* 23, 4: 376–93.

Cohen, R. 1997. *Global Diasporas: An Introduction.* Seattle: University of Washington Press.

Das DasGupta, S., ed. 1997. *A Patchwork Shawl: Chronicles of South Asian Women in America.* New Jersey and London: Rutgers University Press.

Das Gupta, T. 1986. 'Looking under the mosaic: South Asian immigrant women'. *Women and Ethnicity: Bulletin of the Multicultural History Society of Ontario* 8, 2: 67–9.

Desai, N., and M.K. Raj. 1990. *Women and Society in India.* New Delhi: Ajanta.

Dhruvarajan, V. 1992. 'Conjugal power among first generation Hindu Asian Indians in Canada'. *International Journal of Sociology of the Family* 22: 1–34.

Faist, T. 2000. *The Volume and Dynamics of International Migration and Transnational Social Spaces.* Oxford: Clarendon Press.

Friedan, B. 1963. *The Feminine Mystique.* New York: Dell.

Ganguly, K. 1992.'Migrant identities: personal memory and the construction of Selfhood'. *Cultural Studies* 6,1: 27–50.

Ghosh, R. 1981. 'Minority within a minority—on being South Asian and female in Canada'. Pp. 413–26 in *Women in the Family and Economy: An International And Comparative Survey.* London, England: Greenwood Press.

———. 1983.'Sarees and the maple leaf: Indian women in Canada'. Pp. 90–9 in G. Kurien and R.P. Srivastava, eds. *Overseas Indians: A Study in Adaptation.* New Delhi: Vikas Publishing House.

———. 2000. 'Identity and social integration: girls from a minority ethno-cultural group in Canada'. *McGill Journal of Education* 35, 3: 279–96.

Ghosh, S. 2007. 'Transnational ties and intra-immigrant groups settlement experiences: a case study of Indian Bengalis and Bangladeshis in Toronto'. *GeoJournal* 68: 223–42.

Greenwood, M.J. 1975. 'Research on international migration in U.S.: a survey'. *Journal of Economic Literature*, 13.

Hall, S. 1996. 'Introduction: who needs "identity" '. Pp. 1–17 in S. Hall and P. du Gay, eds. *Questions of Cultural Identity.* London: Sage.

Handa, A. 2003. *Of silk saris and mini-skirts: South Asian girls walk the tightrope of culture.* Toronto: Women's Press.

Harris, J., and M. Todaro. 1970. 'Migration, unemployment and development: a two sector analysis'. *American Economic Review* 60: 126–42.

Hawkins, F. 1989. *Critical Years in Immigration: Canada and Australia Compared.* Kingston and Montreal: McGill-Queens University Press.

Ibrahim, F., H. Ohnishi, and D.S. Sandhu. 1997. 'Asian American identity development: a culturally specific model for South Asian Americans'. *Journal of Multicultural Counselling and Development* 25, 1: 34–46.

Isajiw, W.W. 1999. *Understanding Diversity: Ethnicity and Race in the Canadian Context.* Toronto: Thompson Educational Publishing, Inc.

Joshi, V. 2000. *Indian Daughters Abroad: Growing up in Australia.* New Delhi:Sterling Publishers Pvt. Ltd.

King, R., and J. Vullnetari. 2009. 'The intersections of gender and generation in Albanian migration, remittances and transnational care'. *Swedish Society for Anthropology and Geography* 91: 19–38.

Kivisto, P. 2001. 'Theorizing transnational immigration: a critical review of current efforts'. *Ethnic and Racial Studies* 24, 4: 549–77.

Kurien, P.A. 2002. *Kaleidoscopic Ethnicity: International Migration and the Reconstruction of Community Identities in India*. New Brunswick, NJ: Rutgers University Press.
Lee, E.G. 1966. 'A theory of migration', *Demography* 3, 1.
Luxton, M. 1980. *More Than a Labour of Love: Three Generations of Women's Work In the Home*. Toronto: Women's Press.
Mahler, S.J., and P.R. Pessar. 2001. 'Gendered geographies of power: analyzing gender across transnational spaces'. *Identities* 7, 4: 441–59.
Malinoski, B. 1913. *The Family Among the Australian Aborigines*. London: University Of London Press.
Massey, D. 1985. 'New directions in space'. Pp. 9–19 in D. Gregory and J. Urry, eds., *Social Relations and Spatial Structure*. London, Macmillan.
Mazumdar, V., and K. Sharma. 1990. 'Sexual division of labour and the subordination of women: a reappraisal from India'. Pp. 185–97 in I. Tinker, ed., *Persistent Inequalities: Women and World Development*. New York: Oxford University Press.
Mies, M. 1986. *Patriarchy and Accumulation on a World Scale: Women in the International Division of Labour*. London, Zed Books Ltd.
Momirov, J., and K. Murphy Kilbride. 2005. 'Family lives of native peoples, immigrants, and visible minorities'. Pp. 87–113 in N. Mandell and A. Duffy, eds. *Canadian Families: Diversity, Conflict and Change 3rd ed.* Toronto: Thomson and Nelson.
Murdock, G.P. 1949. *Social Structure*. New York: The Free Press.
Naidoo, J.C. 1980. 'East Indian women in the Canadian context: a study in social psychology'. In V.K. Ujimoto and G. Hirabayshi, eds. *Visible Minorities and Multiculturalism: Asians in Canada*. Toronto: Butterworth and Co.
Naidoo, J.C. 1988. ' Canadian South Asian women in transition: a dualistic view of life'. *Journal of Comparative Family Studies* 19, 2: 311–27.
Ng, C.F., H.C. Northcott, and S. McIrvin Abu-Laban. 2007. 'Housing and living arrangements of South Asian immigrant seniors in Edmonton, Alberta'. *Canadian Journal of Aging* 26, 3: 185–94.
Papastergiadis, N. 2000. *The Turbulence of Migration*. Cambridge: Polity Press.
Parsons, T. 1955. 'The American family: its relations to personality and to the social structure'. Pp. 3–33 in T. Parsons and R.F. Bales, ed. *Family: Socialization and the Interaction Process*. Glencoe, Illinois: The Free Press.
Percot, M. 2006. 'Indian nurses in the Gulf: two generations of female migration'. *South Asia Research* 26, 1: 41–62.
Radhakrishnan, R. 2003. 'Ethnicity in an age of diaspora'. Pp. 119–31 in J. Evans Braziel and A. Mannur, eds. *Theorizing Diaspora*. Oxford: Blackwell Publishing.
Ralston, H. 1996. *The Lived Experience of South Asian Immigrant Women In Atlantic Canada*. Lewiston: The Edwin Mellen Press .
Ravenstein, E.G. 1885. 'The laws of migration'. *Journal of the Royal Statistical Society* June.
Reynolds, T. 2006. 'Caribbean families, social capital and young people's diasporic identities'. *Ethnic and Racial Studies* 29, 6: 1087–103.
Safran, W. 1990. 'Diasporas in modern society: myths of homeland and return', *Diaspora* 1 (1990 trans 1991): 83–99.
Simon, R.J., and C.B. Brettell, eds. 1986. *International Migration: The Female Experience*. Towtowa, NJ: Rowman and Allenheid.
Talbani, A., and P. Hasanali. 2000. 'Adolescent females between tradition and modernity: gender role socialization in South Asian immigrant culture'. *Journal of Adolescence* 23: 615–27.
Van Hear, N. 1998. *New Diasporas: The Mass Exodus, Dispersal and Regrouping of Migrant Communities*. London: UCL Press.
Weiner, M. 1993. 'Security, stability and international migration'. Pp. 1–35 in M. Weiner, ed. *International Migration and Security*. Boulder: Westview Press.
Wong, B.P. 2007. 'Immigration, globalization, and the Chinese American family'. Pp. 212–28, J.E. Lansford, K. Deater-Deckard, and M.H. Bornstein, eds. *Immigrant Families in Contemporary Society*. New York: The Guilford Press.

Chapter Eight

'Doing it for the children': Family Factors in Skilled Migration from South Africa to Canada

Suzanne Huot and Belinda Dodson

Introduction

The research on which this chapter is based was not intended to focus on families. Rather, it was designed to investigate the phenomenon of **skilled migration**, specifically the migration of skilled professionals from South Africa to Canada. This has been a small but significant flow of people over several decades, with South Africans in Canada now numbering in the tens of thousands (Grant 2006). We were interested in the motives for people leaving South Africa and for choosing Canada as a destination and whether they perceived themselves to be part of a South African '**brain drain**'. In particular, we wanted to investigate how skilled migrants' motives and experiences differed between people who had left during the **apartheid** era (1948–90) and those who had left since the end of apartheid. But in what had seemed like a straightforward examination of political push factors from South Africa and skilled-labour market pull factors in Canada, children emerged as perhaps the single most important factor in their parents' migration motives, decisions, and experiences.

In this chapter, we draw on family geographies as an analytical category to explore how adults who emigrated from South Africa to Canada in these two distinct political phases were influenced by their position as parents. The findings presented are based on a case study of skilled South Africans living in London, Ontario, Canada. London is a mid-sized Canadian city located in southwestern Ontario, with a 2006 population of approximately 352,395, of which 326,105 are Canadian citizens. The majority of London's residents in the labour force are employed in sales and service occupations (N=47,040); followed by business, finance, and administration occupations (N=34,135). As many as 13,560 London residents are employed in health occupations, the main focus of our study (Statistics Canada 2007).

The study employed a mixed-method, sequential approach. The first stage was a questionnaire survey of a sample of South African-origin skilled professionals based in London. The second stage, from which this chapter largely draws, was a series of nine interviews with a sub-sample of medical professionals from the surveyed group. This group is of particular interest given the 'brain drain' of South African health professionals, as discussed further below. Owing to the small and dispersed nature of South African immigrants living in London, participants were recruited using a number of strategies. Initial contacts were made through the Protea Club, a social organization for South Africans in London. Personal contacts of the authors were also approached with an invitation to participate in the study. Advertisements, as well as reputational and snowball sampling were used to identify additional participants who fit the selection criteria of 'skilled South African immigrant currently practicing or residing in London'.

In addition to completing a questionnaire, participants took part in a single, semi-structured, in-depth interview, conducted either in their homes or places of work. While the questionnaire elicited data on the objective dimensions of participants' migration, the interviews explored the more subjective dimensions of the migration experience. The interview guide was divided into four themes: the context in which migration had occurred; family and personal factors; transnational and diasporic attitudes and behaviours; and education and occupation. Verbatim interview transcripts were entered into NVivo for analysis. Transcripts were read line-by-line and coded according to key themes. The themes across transcripts were then compiled to highlight similarities and differences amongst participants' responses. It was in the interview stage, when respondents were free to express their personal reflections on their migration experience, that family factors in general and children in particular came to the fore. At the time of the interview, participants had lived in Canada from as long as thirty years to less than six months, spanning the apartheid and post-apartheid eras. All respondents were white South Africans. While this does not reflect the racial variation of the South African population, it does reflect the disproportionate concentration of skills, and especially medical training, among white South Africans relative to those of other races. For the purposes of this chapter, only those respondents with children are included in the analysis. The findings presented are derived from the experiences of six men and two women, including two married couples.

The chapter begins with an overview of the place of children in the literature on international migration, identifying a case of 'missing children' in literature on skilled migration and calling for attention to family geographies within skilled migration studies. Following that is a section outlining the patterns in emigration from South Africa and immigration of South Africans to Canada, relating these to political developments in South Africa and to immigration policy shifts in Canada. We then present the stories of the participants, each describing their own particular migration trajectory. This provides a framing and foundation for the main body of the chapter, which elucidates the dominant themes that emerged in relation to the role and influence of children in their parents' migration process, from the initial decision to move to their integration into Canadian society. What emerges is that skilled migration is very much a family affair, and that when skilled professionals decide to leave their country of origin to settle in a new country, they are often doing so not only, or even primarily, for themselves, or for narrow career-related motives, but 'for the sake of the children'.

Children and Families in the Literature on Skilled Migration

Much of the literature on international *skilled* migration is largely blind to the role of children and other family relationships in the migration process. This is not the case in the work on immigration and **transnationalism** more broadly, in which children and family feature prominently. There is a considerable literature on children left behind by transnational labour migrants, especially women migrating to other countries to work as domestic workers or in the global care economy (e.g., Parrenas 2005; Silvey 2006). Migration scholars have also paid attention to the so-called 'astronaut families' moving between places like Hong Kong and Canada. These transnational families are split between countries, strategically employing a migration strategy to enhance familial educational or professional opportunities (e.g., Ho 2002; Kobayashi and Preston 2007; Waters 2002 and 2005). Various other transnational family forms and practices have also been extensively studied by migration scholars in diverse geographical contexts (e.g., Baldassar 2007; Falicov 2007; Hardill 2004; Levitt 2001; McGregor 2008; Sorensen 2005; Svasek and Skrbis 2007). There is an even longer tradition of enquiry into immigrant families themselves. Children and family dynamics seem to have attracted particular attention where 'ethnic' immigrant flows are the focus: in studies of generational differences in acculturation and the acquisition of language skills (Falicov 2007; Hwang 2006); or the involvement of children in immigrant family businesses (Couton and Gaudet 2008; Kim 2006). Children in refugee and other forced migration flows have likewise attracted scholarly attention, as has child trafficking and international adoption (Jaye 2008; Kaime 2004; Lovelock 2000; Manzo 2005; Selman 2002 and 2006; Xu 2007). Literature on immigration policy is also full of references to children and family, in part reflecting immigration policies themselves, which determine who qualifies for admission and right of residence, and under what conditions and restrictions, on the basis of marriage or kinship status (Dodson 2002; Dodson and Crush 2004; Hyndman 1999).

Family-related migration, that is, when family ties and reunification are the actual basis for legal admission to a country, has attracted varying levels of attention in different geographical contexts. Kofman (2004) demonstrates how family migration has been neglected in European migration studies—certainly relative to either settler societies such as Australia and Canada or areas with strong transnational networks, such as the Asia-Pacific region. Particularly in Europe, but also in other Northern contexts, both the academic and the policy focus have instead been on labour and skilled migration, with family migration relegated to secondary status in both policy and scholarship. Furthermore, and for a number of reasons, skilled migration tends to be written about in terms that are curiously people-less, and certainly largely childless. Skilled migration is commonly portrayed in terms of an economic logic, with skills, rather than people, moving: disembodied, impersonal, quantified flows, responding either to global forces of skilled labour supply and demand, or to the corporate needs of multinational companies in an increasingly globalized economy (Scott 2006). Skilled migrants are represented either as obeying the forces of labour supply and demand (Tani 2008) or, in a latter-day version of 'rational economic man' (or woman), assumed to be calculating economic optimizers, seeking to maximize economic gain by 'selling' their skills on the global labour market (Findlay and Li 1998; Findlay, Li, Jowett, and Skeldon 1996). National immigration policies are also debated and constructed in these terms, with countries—and, within them, different jurisdictions—competing to 'attract skills', only secondarily acknowledged as skilled *people* (Citizenship and Immigration Canada 2008; City of London 2008).

Recent years have seen some advancement in understanding and 'peopling' skilled migrant flows. This is especially true for work on gender and skilled migration, which is more likely to address family geographies, albeit often indirectly. Work by migration scholars such as Boyle, Cooke, Halfacree and Smith (2003), Cooke (2008), Iredale (2005), Kofman (2000 and 2004), Mahler and Pessar (2006) and Raghuram (2004) draws attention to the fact that skills are in reality embodied, gendered people, connected to other people by bonds of emotional, marital, kinship, and other social ties. There is, for example, a significant body of work on the so-called 'wife's sacrifice' in international migration, where the wife (most commonly, although it can be the husband) of a skilled migrant experiences a downward shift in their own employment or earnings status after accompanying a spouse on a career-related move to a new country. The sacrifice is found to be even greater in families with children (Boyle et al. 2003; Salaff and Greve 2004). Not only does the mother often become even more important as a caregiver following a move away from existing family and other social support networks, but in some cases there may be legal restrictions or prohibitions on a spouse's employment. A special issue of *Women's Studies International Forum* (Raghuram and Kofman 2004) addressed the skilling, re-skilling, and deskilling of female migrants from Asia, challenging the over-emphasis on their portrayal as caregivers and sex workers, and highlighting instead the presence of women in skilled migration streams. Papers drew attention to the wider social milieu in which skilled migration occurs, thus offering a more holistic perspective, incorporating family and household dynamics. Children, generally perceived as being mainly women's responsibility, have largely been erased from view in the predominantly masculinist discourse on highly skilled migration; but included, even foregrounded, in narratives of female, lower skilled migration (e.g., Piper and Roces 2003; Silvey 2004). Contributions such as Raghuram and Kofman's thus mark an important development, admitting not only women but also children into the academic discourse on skilled migration.

Such work on gender and migration has taken place within, and contributed to, broader theoretical and conceptual advances in both human geography and migration studies. These have created an intellectual climate more amenable to the admission of personal, emotional, and family factors in explanation and analysis. Insights from social and cultural theory, relating to embodiment, subjectivity, identity, and the like, have significantly enriched contemporary understandings of migration's personal and emotional geographies. Espin, for example, in her study of the role of gender and emotion in women's experiences of migration, notes how

> When immigrants cross borders, they also cross emotional and behavioural boundaries. Becoming a member of a new society stretches the boundaries of what is possible because one's life and roles change, and with them, identities change as well. Boundaries are crossed when new identities and roles are incorporated into life. Most immigrants who, either eagerly or reluctantly cross geographical borders, do not fully suspect all the emotional and behavioural boundaries they are about to cross (1997: 445).

We argue that enquiry needs to be open not only to these broader, non-economic realms, but also to the emotional experiences and familial contexts of the human agents involved: to skilled *migrants* and not merely skilled *migration*. As our case study demonstrates, neglecting family geographies would have provided only a partial understanding of the participants' migration experiences.

South African Pushes and Canadian Pulls: Politics, Policies, and Patterns of Migration

South Africa's turbulent political history—most notably the 'four lost decades' of apartheid (O'Meara 1996)—has presented a host of powerful motives for people to leave the country. During apartheid, numerous South Africans of all races emigrated to escape the morally repugnant system of racial classification and discrimination on which the country's entire social, political, and economic organization, indeed its very geography, was based. Anti-apartheid activists, both black and white, risked being detained and imprisoned, even assassinated, by the apartheid state. Several chose to go into exile, or fled to escape arrest, with many continuing their activism in the international anti-apartheid movement. White male South Africans were conscripted into the army, deployed both to fight liberation forces in neighbouring countries and to crack down on internal political opposition. Unsurprisingly, many left the country to avoid doing mandatory national service. For white South Africans, Australia, Britain, and Canada were the favoured destinations, while many black South Africans went into exile in southern African countries such as Zambia. The end of apartheid—marked by the release of Nelson Mandela from prison in 1990 and the first non-racial democratic elections in 1994—saw several of these political exiles return (Peberdy 1999). Others chose to stay abroad, having established roots, raised families, and developed successful careers in their countries of immigration.

Ironically, the end of apartheid also brought about increased emigration from South Africa as the dawn of democracy raised new fears and brought new realities. There was an upsurge in violent crime, including murder, car hijacking, and armed robbery (Kynoch 2005; Louw 1997). Policies of race-based affirmative action led to disillusionment among young white South Africans with regard to their employment and career prospects (Crush 2000). These changes, combined with the opening up of the rest of the world due to fewer travel and visa restrictions on people holding South African passports, led many to seek 'pastures new' abroad. Although not restricted to white South Africans, this has been largely a white efflux, in part because it is mainly members of this population who have the skills or ancestry that give them rights of residence in countries abroad. Similar patterns of emigration have occurred in other contexts of rapid political change and uncertainty about the future, such as Hong Kong in the lead up to the handover to China in 1997 (Findlay and Li 1998). While most South African emigrants are classified as skilled immigrants, and may have been granted admission to their new countries of residence as such, migrants' motives for leaving are clearly grounded in larger social and political forces. Somewhat unfairly, these post-apartheid emigrants have been accused of fleeing democracy, suggesting a racist reactionary stance, whereas for many it appears that it is the fear of crime and, in particular, wanting a better and safer future for their children, that is driving post-apartheid emigration (Crush 2000).

Policy changes at the receiving end also affect levels of immigration. The late 1980s and 1990s were characterized by shifts in the immigration policies of many countries towards more skills-based criteria for admission, aimed at more effectively balancing worker and family class immigration (Akbari 1999; Appleyard 2001). These shifts were rooted in a number of international and global developments, including falling birth rates in Western countries and resulting perceived demographic shortfall in the number of future working adults (Gould and Findlay 1994; Passaris 1998). Canadian policies were no exception to this general trend (Li 2003). These shifts occurred over the same time period as the politically turbulent end of apartheid in South Africa. The advent of more skills-based immigration policies in Canada, combined with incentives to leave South Africa due to the prevailing political uncertainty, created a situation where those with the skills and capital to leave could more readily do so.

Our sample included medical professionals from both pre- and post-1990 phases of emigration from South Africa. Immigrant doctors continue to be in demand in certain Canadian provinces and hence make up a significant share of the South African–Canadian migrant stream (Grant and Oertel 1997; Labonte, Packer, and Klassen 2006). The associated 'brain drain' has attracted considerable academic and policy attention. Here too, however, the children and families of medical professionals have been largely left out of the picture, with the focus justifiably on skills loss, the impact on source countries' health care systems (Chikanda 2004), and possibilities for reparation (Labonte and Parker 2006). Yet a study of medical doctors living in South Africa by Bhorat, Meyer, and Mlatsheni (2002), which asked questions about their emigration potential, found that fully 60 per cent felt that the 'future of their children' would be better in their preferred country of destination—given as Australia, the UK, the US, and Canada. Thus it would seem that as with skilled migrants more generally, the migration decisions of medical professionals are often made for personal or family reasons, albeit located within wider frameworks of professional opportunities, socio-political considerations, and policy constraints. This is certainly borne out by our own study findings.

Skilled South Africans' Migrant Stories

In this section, we introduce our respondents by presenting a series of vignettes telling the stories of their migration from South Africa to Canada. Pseudonyms are used to protect their identities. We intersperse summarized and paraphrased accounts of their migration histories with descriptions using the respondents' own words, taken from the interview transcripts.

During Apartheid (Up to 1990)

Sarah and Benjamin

Sarah and Benjamin came to Canada in 1972 with their two youngest children. They migrated because he was offered a job in Canada and South Africa 'was under apartheid, so it wasn't really nice' (Sarah). As Benjamin put it, 'We had talked about leaving South Africa before I was offered a job, everybody did in those days, but we weren't making any definite plans'. Benjamin and Sarah, now both in their 80s, continue to live in London, where they feel they have made their home. They are an example of a professional couple who moved for a combination of political and career reasons. As a family, they settled and integrated into Canadian society, while continuing to take pride in their South African origins and family connections.

Douglas

Douglas came to Canada in 1974, when he was 32 years old. The political climate in South Africa was the main incentive for his emigration, 'It had reached a sort of plateau stage . . . It was a very heavily policed society; dissent was not tolerated, and it looked like things were never going to change, but funny enough, they started to change after we left'. He described the racial inequalities and escalating violence, and while he admits his career as a psychologist facilitated his immigration to Canada, he emphasized that his motives were purely political. He did not want to raise his children in the prevailing societal context. Overall, Douglas and his family's migration and settlement experience in Canada and London has been a positive one and they have made good friends in the city. No one else from his family left South Africa, and one of his sons as well as his extended family still live there, maintaining transnational family ties.

Carl

Carl was 35 years old when he arrived in Canada in 1988. He and his wife made a joint decision to immigrate, but he admits that it may have been more his decision than hers. Carl was also dissatisfied with South Africa's political context and was opposed to apartheid. Carl thinks London is a nice place to raise a family, but said so with reservation because he admitted to having some trouble bringing up his children. While their experiences in Canada have been mainly positive, he does still have some anxieties about his children and worries that they have had things 'too easily' as a result of their successes in Canada.

Political Transition and Post-Apartheid (Post-1990)

Marc and Karen

Marc and Karen immigrated to Canada in 2001. Upon an initial visit, they had been impressed by the vast differences that they noticed between the two countries, but still had no intentions of moving. When they did eventually emigrate, they moved directly to London, where Marc had been recruited for his research. Their family is Christian and their faith was central to their migration decision and experience. Marc also admits that they were primarily considering their children's future. Being a white, Afrikaner family in post-apartheid South Africa led to concerns about what kind of life their children would have there. The family enjoys the safety and sense of security in London. Karen made sacrifices for her family, as she never re-qualified as a physician after moving to Canada. Guided by their faith, they expressed little doubt that their move to Canada was the right thing for them as a family.

Andrew

Andrew was approximately 40 years old when he came to Canada in 1992. He first got a job in Timmins (in northern Ontario) and lived there with his family before moving to London in 2003. He and his spouse were unsure whether they would leave South Africa, but worried about political extremists, especially considering they were raising four children. While he does not regret immigrating, he acknowledges the challenges faced by his children, who had to leave school and friends behind. They already had family (Andrew's wife's cousin) in Ontario and that helped with the transition. Like the others, Andrew left family members behind in South Africa. They keep in close contact, have had family members visit, and have returned to South Africa twice since emigrating.

Jeff

Jeff first moved to the UK, and then to the US before settling in London, Ontario. He and his family arrived in 2004 and had only been in Canada for five months at the time of the interview. At the time of his family's emigration, the South African political context was turbulent due to political change and violence. He listed two primary reasons for leaving. The first was that their child had just been born and they worried about the future. Second, he did not want to serve in the country's national defence force. Their initial impressions of Canada and London are largely positive. They prefer living in Canada relative to Britain, particularly enjoying the range of activities and opportunities available to their children.

Children and Family in the Migration Experience

Having introduced our respondents, we now go on to explore the role of family in their processes of emigration, immigration, and settlement, from the original decision to leave South Africa and their choice of Canada and of London as a destination, to their later reflections on both their lives in Canada and on the places and people they left behind. We do this by paraphrasing and quoting from our in-depth interviews. Bear in mind that these themes emerged in interviews designed primarily to enquire not about children, but rather about people's professional experiences as skilled migrants. This makes the respondents' emphasis on children and family all the more noteworthy.

Family in Migration Decision-making

Whether the participants migrated during or following South Africa's apartheid era, the decision to leave was taken neither lightly nor unilaterally. Certainly the weight of the decision was evident in the way that respondents described and explained their choice to emigrate. Respondents were all too conscious that they were making a life-altering choice that would affect their spouses and children who came with them as well as extended family members left behind. Family considerations were paramount, and family members were clearly involved in making the decision to emigrate.

Douglas, for example, while claiming primary responsibility for the family decision to emigrate, notes how family factored into the decision.

> 'Certainly leaving South Africa was my choice but I felt I had no choice. If I was going to stay there, I would probably have ended up in prison. I had two small children and my wife and I thought I wouldn't have too much of a choice and so while I chose to leave, I also sort of had to . . . My wife was involved in the anti-apartheid movement, but it was a joint decision to leave. We all came to Canada at the same time. Looking ahead for our kids, I didn't see any hope.'

When Marc was offered a job in London, he too saw it as an opportunity to secure his children's futures. Yet he was uncertain of how to broach the subject with them and referred to their emigration as a 'radical situation', given that his children were in school and university and had established friends and social networks in South Africa. He and his wife decided that if their children approved the decision, then they would emigrate. Upon their return from a visit to Canada, they had a family meeting at which his youngest daughter said, 'Don't you worry, Dad, we know what you're about to tell us, we're going to Canada, we can even tell you when you decided. That very night when all of us came together and we prayed for you'. Marc went on, 'It was difficult for the children, and I knew, and they knew, and they told us when we came back from Winnipeg, do not fear, we refuse to fear. That's when everything was cleared for them'. In this case, the move was truly a family decision. Marc drew on his faith to reinforce the choice his family made to come to Canada,

> 'I remember when we came to the decision to leave—you will see in my office there hangs a picture titled 'New Horizons'. We are Christian and we very much believe in happenings, we did not leave South Africa for anything else other than the happenings as we were directed'.

Emigrants also had to consider the opinions and wishes of their extended families. For instance, Benjamin's sister stayed in South Africa and thought his and Sarah's choice to leave was a bad idea. Douglas's nuclear family was the only one in his or his wife's extended families to leave the country, but both families understood and supported their decision. He stated that 'practically, without exception, every young white South African at that time wanted to go'. Carl explained that 'there are people on both sides, the people who say they will never leave, like my sister, she's very anti-emigration, [. . .] but most people at least have thought about it'. Andrew thinks that his extended family was disappointed by their decision to leave but also thinks they 'recognized why we were doing it, there was never any anger or recrimination. They wished the best for us, and obviously they were missing grandchildren and so on, but they accepted our decision and supported us all the way through'. Some respondents had family members who had already left South Africa, and were thus able to make decisions informed by their experiences.

In several ways, then, nuclear and extended family were part of the migration decision-making process. Were we to do the research again, and with a more explicit focus on family, we would perhaps interview people's spouses and children separately in order to gauge whether they too felt that they had had a say in the decision to emigrate, or whether they felt the decision had been imposed on them by a spouse or parent. But to the respondents themselves, involvement of their partners and other family members appears to have been central. Respondents were mindful of and consulted with their spouses and children, albeit to varying degrees, when weighing the potential costs and benefits of uprooting and moving to an entirely different continent.

Reasons for Leaving South Africa

While participants had all considered and involved their families in the decision-making, their stated reasons for emigrating differed depending on whether they had left during or after apartheid. For many of the respondents who emigrated during apartheid, political considerations were a primary incentive. Sarah and Benjamin have been in Canada the longest of all the participants. They talked about the negative impacts of the apartheid government. As socio-political conditions deteriorated over the 1970s and 1980s, others were also motivated to leave. Douglas, another apartheid-era emigrant, had been arrested on false charges under the Suppression of Communism Act and risked losing his university teaching privileges. Luckily he was ultimately found not guilty. He was a highly skilled health professional, which enabled his immigration to Canada. 'I think my career facilitated the move, but it wasn't the reason we left. . . . I certainly didn't move for career purposes and I was prepared to give up my career to move, it was really political.' He related that at the time they left, people were being shot through their home windows at night. His family was relatively safe because they were white and living in a wealthy community, but he and his wife did not want to raise their children in such a context. Similarly, Carl's family emigrated after the government 'lurched to the right and not toward any significant change'. He also did not foresee a positive future and thought 'the country was dead. That's the main reason we left. I'd always been unhappy about the future, a future in South Africa, for political reasons'.

When apartheid ended, the country experienced quite radical transformation, both positive and negative. Thus, the incentives to leave changed rather than disappeared. A rise in crime and violence in the years following the end of apartheid led to increased levels of fear and insecurity. For the respondents who came to Canada in the post-apartheid era, this fear replaced the political incentives of the earlier emigrants as one of the primary motivations for emigration. Marc explains how concerns for one's own and one's family's safety have become a part of everyday life in some areas of South Africa.

> 'The people there are adapting to that, they walk around with their arms (i.e., guns) at their sides and that's about it; they are confined within their own cages and to be safe, and that's it, that's how it is . . . If you just drive in a vehicle there for just a half hour, you would go in your house, lock it up, and just fall down on your sofa and say: God thanks I am home without injury. It's just another way of living, and you live tense and there is none of that here.'

Marc contrasts this with the sense of safety he and his family feel in Canada.

> 'You can just go anywhere you want with ease here and for that reason the quality of life cannot be equalized . . . We enjoy walking, driving and everything here that we do, constantly cycling and everything. You couldn't do that; you couldn't dare that in South Africa. Like

> what we're doing now, my daughter walking to a job opportunity at a restaurant back home . . . there's no such thing, that's what's worth paying for by moving . . . a sense of security, especially for your children.'

Karen agrees that crime is the reason that they left. She exclaimed that 'There is no other country like South Africa, we miss that, but we don't miss the trouble, the constant feeling of being on guard and hand on your purse and locking everything'.

While Jeff arrived in Canada during the post-apartheid era, he and his family had first moved to the UK during South Africa's political transition, and he described how seeing South Africa now sometimes makes him wish he had not left, 'Our reasons for emigrating changed with the change in regime. Once the apartheid government was deposed, from an ideological and political point of view, we would have liked to have been there'. Yet Jeff went on to observe that 'the violence that was bad when we left has gotten worse' and still supports his rationale for having emigrated, primarily because of the concern he and his wife had for their children, 'You know, we don't fear for our lives every time we go to a shopping mall'. Carl, an apartheid-era emigrant, also described how different Canada was from South Africa, where his house and car had been broken into on a number of occasions. Such crime had become a part of his family's life and they had grown used to having burglar bars and an alarm system. He was initially very surprised by how much lower crime rates were in Canada, although he emphasizes that this was not a conscious reason for emigrating.

Although there is a noticeable shift in emigration motives between the two time periods—from broader socio-political factors during apartheid to safety and security after apartheid—there are striking commonalities in the way that these were refracted through the lenses of children and family. People expressed not wanting to bring up their children in apartheid's 'sick society', or fearing for their family's safety. The role of parents as protectors of their children, saving them from psychological or physical harm, came through very strongly in almost all the interviews, no matter when the particular respondent had emigrated.

Children's Future Opportunities

As suggested in the paragraphs above, looking toward the future, and especially their children's futures, figured heavily as an incentive to migrate. Marc was especially vocal about the lack of opportunities for white South Africans in the post-apartheid context,

> 'The work opportunity for the young people was going nowhere due to the so-called affirmative action in South Africa, which is now even worse than then, it's worse than ever, but you will get a highly skilled person just turned down to a lower level or asked to leave, no matter how many years, just to be replaced by some incompetent, I state that again, incompetent, unqualified person to make up the so-called wrongs of the apartheid regime . . . as if it were to be corrected overnight'.

As a result of this perspective, Marc's father supported the family's decision to leave. He said his father told him that if he had had to raise children at that time, he would likely also have emigrated. Karen echoes the family's concerns about the future. She explains that while she had become used to the changes that had occurred over time in the country, not only did the rising crime rates concern her, but she also considered her children's stage in life, 'They were all nearly done their education, not completely, but a big part of it was done, but for them to raise a family and to make a living, the circumstances there, there is a lot of uncertainty'. Regardless of when and why they left, several respondents justified their decisions on the basis of their children's interests, and minimized any regret they may have felt. For instance, Andrew explains, 'We're not sorry we left, there's still a lot of trouble. We wanted opportunities for the children'. He goes on to say, 'We left for reasons that we thought were valid and I think they're still there in many ways, but there's no hatred of the country'.

While the post-apartheid emigrants' responses highlighted concerns over what children might do after completing their education, some participants, especially those that had left during apartheid, were unhappy with the state of their children's schooling in South Africa. Douglas's children were young when his family moved to Canada. They were reaching school age and he did not want them to attend school in South Africa, 'My wife and I had two young children who were four and five years old, just reaching the age of going into the school system and quite frankly I didn't want my children to go through the school system there. It was a sick society, it really was, so it was a chance to get them out of there'. Just the fact of having children served as an incentive for some to migrate. Both Jeff and Carl described how having a child changed their priorities. As Jeff put it,

> 'My wife and I came with our daughter, who was only 14 months old at the time. But it was really our daughter being born . . . [My wife] was really not at all keen, she had a really big family who she didn't really want to leave in South Africa, but once our daughter was born, then she changed and was thinking, you know, our daughter had to have a future and it didn't look there was going to be a future in South Africa'.

Carl expressed similar sentiments, 'We left for two basic reasons, the first is that we had a baby that had just been born and the future was pretty unpleasant right then, nothing specific. We left for our children's future'.

Wives' Sacrifices

Although the majority of respondents were male, the theme of wives' sacrifices arose in several interviews, both by men as they discussed their spouses and by female participants as they described their own experiences. Sarah sacrificed her career prospects by emigrating. Herself a trained physician, just like her husband, she was unsuccessful in obtaining suitable employment in Canada. She sees this as partly due to the power of local professional networks, but also to her role as wife and mother, responsible for helping her family adjust to their new life in Canada, 'You know, one gets jobs when you get known. I was offered jobs in South Africa, I worked in Soweto, in Alexandra, in townships, I worked all over, in general hospitals. I'm a physician too, and here I tried a bit, but I had to adjust my family first'. Carl's wife also had to change careers after moving to Canada. She had a diploma in dress design and had successfully worked for a large company in South Africa, travelling to Europe and being able to take her daughter to work with her. Moving to Canada with a young child and having another soon after arriving, made things difficult for her, 'She spent a lot of time at home and I was away from home about a third of the time, travelling across Canada for work. It was a really tough time for her actually, a really tough time, I think she had a much tougher time than me'. Douglas too thinks it was more difficult for his wife to leave South Africa than it was for him, although he emphasized her political regrets as well as her professional sacrifices. She had been a social worker and activist who had taken part in protest marches at Pollsmoor Prison, where a number of apartheid-era political prisoners had been held. While Douglas did not want to return to the apartheid society in South Africa, 'she thought she should go back and do something for the country'.

Karen, who left after apartheid, described how her life changed both after having children and again after emigrating. In South Africa, she had worked part-time while she was a student and had married before finishing university. After becoming a mother, she only worked once or twice a week and did not start working full-time until her youngest child was nearly ready to begin school. Marc explained the challenges Karen faced when trying to work in Canada.

> 'People told my wife that it would be easy to qualify, you just go and it's two hearings and then you qualify, blah blah blah, but it's just not that easy, when you just go into that situation, it's a heck of an introduction and commitment that you have to make, because you have to make new relations and at that time there was number one, the age involved, number two accommodating children in your own house, trying to survive.'

Yet these sacrifices to Karen's career were worth it to her. As she explained, 'I have a very emotional personality and for me a relationship is much more important than anything else if my family is there, I can live anywhere, if my family is happy, I don't care where we live'. Marc and Karen maintain dual citizenship so that they can practice medicine and dentistry when they return to South Africa with missions, but Marc clearly remains the family breadwinner.

What is especially noteworthy about the experiences of these female migrants, all of whom were themselves skilled individuals, is the gendering of reproductive labour, including women assuming responsibility for 'settling the children' and integrating their families into their new social milieu. Many noted how this was made additionally difficult by the loss of extended family and other social networks. The few women we spoke to seemed to regard this as inevitable, even natural. Again, had family been our primary focus instead of something that emerged spontaneously during the interviews, we might have pursued this more systematically; but even our preliminary findings lend strong support to the 'wife's sacrifice' that has been widely documented in the migration literature (Boyle et al. 2003; Salaff and Greve 2004).

Reasons for Choosing Canada and London, Ontario

Not only did these South African men and women have to choose to leave their country, but they also had to select a destination in which to begin their new lives. The reasons participants cited for coming to Canada rather than any other country, and then specifically to London, varied considerably. For some, it was the availability of professional opportunities; others already had family members living in the province; and some were drawn by Canadian society. Sarah and Benjamin came to London primarily because Benjamin had been offered a job. Marc and Karen also moved directly to London because he had been recruited for a position. They had both enjoyed their previous experiences in Canada. On a prior visit to Vancouver, Karen had told her husband, 'If we would ever have to leave South Africa for whatever reason, I would like to come to Canada'. They enjoyed the differences they noticed between the two countries, especially in regard to safety, 'I was impressed at the attitude of the people, they seemed to have respect for other people; they treat people with more respect . . . there is no respect for life anymore in South Africa, with the high crime rate'.

Before moving permanently, Carl and his wife had come to Canada in order for Carl to write exams for his professional accreditation and they too were impressed by their experiences. For Andrew, not only did his family figure in his decision to emigrate, but it also influenced the family's city of destination. Although he had family in London, he first moved to Timmins. As he explains,

> 'I toured Oshawa and I went to Sarnia, and there were opportunities but a lot of them said, well there's a maternity leave opening, or that type of thing, so perhaps it was get to know the system first and then get in more easily, but in Timmins there was an opportunity to get a full time job, [which] I had to look for, with a family to support'.

Securing employment was not an end in itself, but a means to an end: to support and provide a brighter future for his children.

Douglas, who came to Canada in the 1970s, expressed wider social and political motives for choosing Canada. He had been impressed by the image of the country that was being disseminated internationally while Prime Minister Trudeau was in office. He and his family were pleased that their first impressions of Canada matched their expectations of a more just, tolerant, and multicultural society. 'I worked in a cinema and one of the women in this group was talking and her sister had just adopted a Korean baby, and she was talking about this new little person coming into this family, and I just couldn't imagine it, the whole concept in South Africa, just little things like that.' Whether their stated motives for choosing London and Canada were framed in political or professional terms, children almost always formed part of respondents' rationale.

Children and the Settlement Experience

Given that the future of their children was a primary incentive for emigrating, their children's adaptation and settlement in Canada was important to the respondents, regardless of when they had emigrated. For both apartheid-era and post-apartheid immigrants, their family experiences have generally been positive. Many of the participants, whether or not children had been a primary factor in their choice of destination, felt that London had proved to be a good city in which to raise a family. For instance, Douglas stated that 'London is a great place to raise kids, it's manageable, comfortable. [. . .] London has huge benefits. The city is quite manageable, what's available here, it's a got a theatre, it's got a symphony, it's got really everything you need'. Carl also enjoyed living in a smaller city like London and made good friends in his neighbourhood. 'Our best friends are a Canadian couple who have a young girl around our daughter's age, and another immigrant couple that's been in Canada for a long time and they had a son who was at school with our daughter.' Many have made good family friends in their neighbourhoods, as expressed here by Douglas.

> 'We used to live on the other side of the park in one of those old houses and most of the people who lived in the neighbourhood were from all over the world and many were academics and professionals and so on. Because none of us had extended families here, we sort of became each other's families. When your kids are young it's good to have neighbours that will be compatible and probably that remains, as qualities remain.'

He feels that his children have done 'quite well' in Canada. His daughter is a lawyer and his son is a medical researcher.

Carl expressed some reservations, for while his family has had a good life in Canada, he sometimes wonders if the problems his children did experience here may have been avoided elsewhere.

> 'I think London is a nice place to raise a family. I say that with reservation because we've had a lot of trouble bringing up children, such as we may have had anywhere. For instance, our son ran into problems with marijuana at school and stuff like that; drug use is much more prevalent here. Not because it's London, but because it's a wealthy, sophisticated society

where I think these things tend to become problems.'

He went on:

'I just worry sometimes, because I guess I'm a worrier, I worry that our kids have had things too easily. They've travelled to Europe and South America and Central America and South Africa and have been on vacation and they've been able to do a heck of a lot. I just worry that maybe things have come so easily, that this is life, and maybe things come easy and don't get the work side of it, you know.'

Andrew's daughter too had a difficult time adapting, 'It wasn't easy for the older ones; our oldest daughter was thirteen so it was difficult for her to blend in and that's when you're really needing a group of people to be with'. Getting their children involved in the community was a priority for many of the participants. Marc, Karen, and their family joined a church in London and feel that this made their integration much easier. Andrew enrolled his children in sports, music, and other activities. Jeff's kids were similarly involved in a number of things, many of which they could participate in as a family. Spending time with their children was clearly a priority for most respondents, both as a motive for choosing a destination and as a means of integrating themselves into a new society.

Transnational Families

Family concerns were not restricted to children and nuclear family members. Transnational and extended family relationships featured in numerous ways in these migrants' lives and stories and thus should not be excluded from considerations of family geographies. For instance, they figured as a factor attracting participants to a particular location if they already had extended family members living there; as something they themselves had created in the process of emigration; and as an important, actively maintained link back 'home' to South Africa. As mentioned above, choosing Canada was an easier decision for those that already had family or friends living in the country. Sarah felt that having friends in Canada also made it easier to adapt, as did Carl, whose sister already lived in London. Knowing someone else in the country especially reassured Andrew about the decision to migrate to Canada.

'My wife's cousin had come to Ontario years before we did. We had been in contact with them and they had settled in and were quite happy, so really I think that was the main reason. I don't think we would have just gone to a country where there was no one. I spoke with them when I came out in '91 to write the exams, there's certainly an advantage to having some base that you could work from. It would have been very much more difficult just to arrive and try to find out how things work. We stayed with them for the first week, the whole family, and we got the cold weather gear and that type of thing, they took us sledding down the hill, eased us into it. It was a big help having a family member.'

Some of the participants also helped friends and family members that left South Africa after they did. Carl's family sponsored the families of two of his brothers-in-law for immigration, one of whom later also brought his wife's parents to Canada.

'My brother-in-law was an engineer and he and his wife decided that all of their children were going to high school and they decided that if they were going to move they would need to move right away. They were concerned, again, about the future and violence because of their kids, and so they emigrated.

We actually sponsored them and they came, but went to Vancouver. My wife's other brother, he and his wife had a little girl late in life, when his wife was about 40 or so, and they decided that things didn't look good and they were going to come. They emigrated and brought her parents with them.'

Emigration creates various transnational family forms. Sarah and Benjamin, for example, experienced the temporary separation of members of their nuclear family at the time of migration. Benjamin moved a month prior to Sarah and the children. One of their daughters did not immigrate to Canada until years after they had. She is currently living in Toronto, while their other daughter has returned to Cape Town. Their son joined them in London a while after they had moved. Marc and Karen also left a son behind. He was older and is now married and working in Cape Town. Their second son initially stayed behind to complete his schooling before coming to Canada.

Leaving extended family members behind in South Africa was also raised by some respondents, especially when reflecting on their decision to emigrate and the process of leave-taking. Douglas admitted not regretting the emigration itself, but the way they had undertaken it. He feels that they caused considerable emotional pain to his own and his wife's parents:

'If I had my time over again, would I do the same? It's a very difficult question and I'll tell you why. We started applying to Canada when we were in our late twenties and in your late twenties you're full of adventure, you're concerned with making your own way in the world, and looking back now, now that I'm in my sixties, you realize how much hurt you caused to the people around you. We told our parents, and took our kids and just left, you know. So the answer is yes, I would do it again, but I would do it differently. I would do it differently, hopefully to minimize the kind of hurt. We continued to see them until they all died. We went back every year, but just the hurt we must have caused them. It's amazing, we could have done much better'.

The emotional costs of leaving loved ones behind obviously weighed heavily on the minds of many respondents. All of them go to considerable lengths to remain in contact with family members in South Africa and many have returned to visit. As Douglas observed, 'We contact them regularly, almost every night at the moment, and we visit because we have family there, we wouldn't go otherwise'. Andrew has 'been back to South Africa twice I think, family reasons not financial or business or anything'. Like many of the participants, Andrew has also had family members come to visit in Canada, which helps to ease the anxiety some expressed at having abandoned family members.

'Also with them visiting, we've kept in contact, so there were no bad feelings. We keep close contact with all of our family there, they come and visit us here and take back experiences. I think they were disappointed by our decision to leave but I think they recognized why we were doing it; there was never any anger or recrimination. They wished the best for us, and obviously they were missing grandchildren and so on, but they accepted our decision [and] supported us all the way through.'

While missing their family members and expressing nostalgia about the lives and places they had left behind, almost all of our respondents felt that they had made the right decision.

'I feel that Canada has been very good to us, and we've always been grateful for the fact that we were allowed to come and settle in Canada because, whatever has happened, we've made a good life here and I think we've done well.' (Carl)

'It was exciting to move, a bit of mixed feelings but it was something new, and what we'd wanted to do so, if you approach it in that way, any sort of discomfort or downside, we'd always try to see it in a more positive light. We sort of had to make it work.' (Andrew)

Conclusions and Reflections

Having not set out to uncover children and family relationships in skilled migrants' experiences, we were surprised by the breadth and depth with which children came up in interviews. 'Our children's future' was expressed as a primary motive both for the decision to leave South Africa and for choosing Canada and the city of London as destinations, whether respondents had left during or after apartheid. All of these emigrants felt that their children's futures were threatened in South Africa, socially and psychologically, by apartheid's racist socio-political environment or, for those who left post-1990, by affirmative action and high levels of crime. Children also featured prominently in the discussions of settlement and integration into Canadian society, as well as in their reflections on whether they had 'done the right thing' by moving. Wives justified their career sacrifices in terms of having helped their children through the upheavals and disruptions of migration, and respondents of either gender countered regret at family members left behind with the belief that their children had 'done well' in Canada.

What these findings suggest, with relevance beyond our particular case study, is that even where employment opportunities and skill needs in the destination country are the means by which immigration is made possible, employment and economic concerns are often not the primary motive for emigration. Securing a job in a foreign country may be the means, but the desired end of skilled migration, from the migrants' point of view, is often to improve their family's circumstances, not only economically, but also socially and politically. From this perspective, families and children are every bit as important as skills gaps, labour market need, and the skilled immigration policies of recipient countries in understanding migrants' motives, behaviour, and experiences. This has implications for both skilled migration research and immigration policy. Each stands to benefit by more fully considering such emotional geographies (Espin 1997), not just in parallel but in intersection and interaction with the more conventional economic geographies of skilled migration.

Our findings also speak more broadly to the emerging scholarship on geographies of family and family life. If, as we argue, the skilled migration literature has prioritized economic factors at the expense of the personal, emotional, and familial aspects of migration, then so too there is a risk that research on family geographies may emphasize material, physically shared, *local* sites and spaces, such as homes, schools, communities, leisure spaces, and care facilities, at the expense of the long-distance, often virtual spaces of *transnational* family life. For immigrants, professional success and social integration in the country of destination can never, it seems, fully assuage their sense of loss and guilt at leaving family members behind, while the fulfilment of family obligations and maintenance of family relations at a distance present considerable emotional, financial, and logistical challenges. As shown by our own and some of the other chapters in this volume, these multi-local, stretched spatialities of family life warrant further scholarly attention, especially in the 'age of migration' (Castles and Miller 1993) in which families now live.

Questions for Further Thought

1. How might reframing the debate in terms of the movement of 'people' rather than of 'skills' enable considerations of family and family life in the analysis and understanding of skilled migration?
2. In what ways do countries' immigration policies create or reinforce particular immigrant flows in terms of education, skills, occupation, class, gender, and family status?
3. What does the persistence of 'children's future' as a motive for emigration from South Africa from the apartheid era into the post-apartheid era suggest about the importance of family in migration decision-making?
4. How important are children and family in processes of immigrant settlement and integration, and are they any less important for skilled migrants?
5. In what ways are family geographies rescaled and reconfigured as a result of the international migration of family members, and what are the material, social, and emotional implications?

Further Reading

Dodson, B.J. 2002. 'Women in the brain drain: gender and skilled migration from South Africa'. Pp. 47–72 in D. McDonald and J. Crush, eds. *Destinations Unknown: Skilled Migration in Southern Africa*. Pretoria: Africa Institute and Kingston, Ontario: Southern African Migration Project.

Grant, H. 2006. 'From the Transvaal to the Prairies: the migration of South African physicians to Canada'. *Journal of Ethnic and Migration Studies* 32, 4: 681–95.

Kobayashi, A., and V. Preston. 2007. 'Transnationalism through the life course: Hong Kong immigrants in Canada'. *Asia and Pacific Viewpoint* 48, 2: 151–67.

Kofman, E. 2004. 'Family-related migration: a critical review of European studies'. *Journal of Ethnic and Migration Studies* 30, 4: 243–62.

Silvey, R. 2004. 'Power, difference, and mobility: feminist advances in migration studies'. *Progress in Human Geography* 28, 4: 1–17.

References

Akbari, A.H. 1999. 'Immigrant "quality" in canada: more direct evidence of human capital content, 1956–1994'. *International Migration Review* 33, 1: 156–75.

Appleyard, R. 2001. 'International migration policies: 1950–2000'. *International Migration* 39, 6: 7–20.

Baldassar, L. 2007. 'Transnational families and the provision of moral and emotional support: the relationship between truth and distance'. *Identities: Global Studies in Culture and Power* 14, 4: 385–409.

Bhorat, H., J. Meyer, and C. Mlatsheni. 2002. 'Skilled labour migration from developing countries: study on South and Southern Africa'. *International Migration Papers* 52: i–40. International Labour Office: Geneva.

Boyle, P., T. Cooke, K. Halfacree, and D. Smith. 2003. 'The effect of long-distance family migration and motherhood on partnered women's labour-market activity rates in Great Britain and the USA'. *Environment and Planning A* 35, 12: 2097–114.

Castles, S., and M. Miller. 1993. *The Age of Migration*. Basingstoke: Macmillan.

Chikanda, A. 2004. *Skilled Health Professionals' Migration and its Impact on Health Delivery in Zimbabwe*. Centre on Migration, Policy and Society Working Paper No. 4. University of Oxford.

Citizenship and Immigration Canada (CIC). *Skilled Workers and Professionals*. 2 March 2009. www.cic.gc.ca/english/immigrate/skilled/index.asp

City of London. *A Great Place to Work*. 2 March 2009. www.welcome.london.ca/working/

Cooke, T.J. 2008. 'Gender role beliefs and family migration'. *Population Space and Place* 14, 3: 163–75.

Couton, P., and S. Gaudet. 2008. 'Rethinking social participation: the case of immigrants in Canada'. *Journal of International Migration and Integration* 9, 1: 21–44.

Crush, J., ed. 2000. *Losing Our Minds: Skills Migration and the South African Brain Drain*. Southern African Migration Project Migration Policy Series No 18.

Dodson, B.J. 2002. 'Women in the brain drain: gender and skilled migration from South Africa'. Pp. 47–72 in D. McDonald and J. Crush, eds. *Destinations Unknown: Skilled Migration in Southern Africa*. Pretoria: Africa Institute and Kingston, Ontario: Southern African Migration Project.

Dodson, B., and J. Crush. 2004. 'A report on gender discrimination in South Africa's 2002 Immigration Act: masculinising the migrant'. *Feminist Review* 77: 96–119.

Espin, O.M. 1997. 'The role of gender and emotion in women's experience of migration'. *European Journal of Social Sciences* 10, 4: 445–55.

Falicov, C.J. 2007. 'Working with transnational immigrants: expanding meanings of family, community, and culture'. *Family Process* 46, 2: 157–71.

Findlay, A.M., and F.L.N. Li. 1998. 'A migration channels approach to the study of professionals moving to and from Hong Kong'. *International Migration Review* 32, 3: 682–703.

Findlay, A., F.L.N. Li, J. Jowett, and R. Skeldon. 1996. 'Skilled international migration and the global city: a study of expatriates in Hong Kong'. *Transactions of the Institute of British Geographers* 21: 49–61.

Gould, W.T.S., and A.M. Findlay. 1994. *Population Migration and the Changing World Order*. Toronto: John Wiley and Sons.

Grant, H. 2006. 'From the Transvaal to the Prairies: the migration of South African physicians to Canada'. *Journal of Ethnic and Migration Studies* 32, 4: 681–95.

Grant, H., and R. Oertel. 1997. 'The supply and migration of Canadian physicians, 1970–1995: why we should learn to love an immigrant doctor'. *Canadian Journal of Regional Sciences* 20, 1–2: 14 pages.

Hardill, I. 2004. 'Transnational living and moving experiences: intensified mobility and dual-career households'. *Population, Space and Place* 10: 375–89.

Ho, E.S. 2002. 'Multi-local residence, transnational networks: Chinese 'astronaut' families in New Zealand'. *Asian and Pacific Migration Journal* 11, 1: 145–64.

Hwang, W.C. 2006. 'Acculturative family distancing: theory, research, and clinical practice'. *Psychotherapy* 43, 4: 397–409.

Hyndman, J. 1999. 'Gender and Canadian immigration policy: a current snapshot'. *Canadian Women's Studies* 19, 3: 6–10.

Iredale, R. 2005. 'Gender, immigration policies and accreditation: valuing the skills of professional migrant women'. *Geoforum* 36, 2: 155–66.

Jaye, T. 2008. 'The security culture of the ECOWAS: origins, development and the challenges of child trafficking'. *Journal of Contemporary African Studies* 26, 2: 151–68.

Kaime, T. 2004. 'From lofty jargon to durable solutions: unaccompanied refugee children and the African Charter on the Rights and Welfare of the Child'. *International Journal of Refugee Law* 16, 3: 336–48.

Kim, D.Y. 2006. 'Stepping-stone to intergenerational mobility? The springboard, safety net, or mobility trap functions of Korean immigrant entrepreneurship for the second generation'. *International Migration Review* 40, 4: 927–62.

Kobayashi, A., and V. Preston. 2007. 'Transnationalism through the life course: Hong Kong immigrants in Canada'. *Asia and Pacific Viewpoint* 48, 2: 151–67.

Kofman, E. 2000. 'The invisibility of skilled female migrants and gender relations in studies of skilled migration in Europe'. *International Journal of Population Geography* 6: 45–59.

———. 2004. 'Family-related migration: a critical review of European studies'. *Journal of Ethnic and Migration Studies* 30, 4: 243–62.

Kynoch, G. 2005. 'Crime, conflict and politics in transition-era South Africa'. *African Affairs* 104, 416: 493–514.

Labonte, R., and C. Packer. 2006. 'The brain drain of physicians from developing countries to Canada: a matter of human rights'. www.globalhealthequity.ca/electronic library/The Brain Drain of Health Professionals from SSA—AMD Series 2 2006.pdf. *Human Rights Tribune* 12: 5 pages.

Labonte, R., C. Packer, and N. Klassen. 2006. 'Managing health professional migration from Sub-Saharan Africa to Canada: a stakeholder inquiry into policy options'. *Human Resources for Health* 4, 22: 1–15.

Levitt, P. 2001. 'Transnational migration: taking stock and future directions'. *Global Networks* 1, 3: 195–216.

Li, P.S. 2003. *Destinations Canada: Immigration Debates and Issues*. Toronto: Oxford University Press.

Louw, A. 1997. 'Surviving the transition: trends and perceptions of crime in South Africa'. *Social Indicators Research* 41, 1–3: 137–68.

Lovelock, K. 2000. 'Intercountry adoption as a migratory practice: a comparative analysis of intercountry adoption and immigration policy and practice in the United States, Canada and New Zealand in the post W.W.II period'. *International Migration Review* 34, 3: 907–49.

Mahler, S.J., and P. Pessar. 2006. 'Gender matters: ethnographers bring gender from the periphery toward the core. gender and migration revisited'. *International Migration Review* 40, 1: 27–63.

Manzo, K. 2005. 'Exploiting West Africa's children: trafficking, slavery and uneven development'. *Area* 37, 4: 393–401.

McGregor, J. 2008. 'Children and "African values": Zimbabwean professionals in Britain reconfiguring family life'. *Environment and Planning A* 40, 3: 596–614.

O'Meara, D. 1996. *Forty Lost Years: The Apartheid State and the Politics of the National Party, 1948–1994*. Athens: Ohio University Press.

Parrenas, R. 2005. 'Long distance intimacy: class, gender and intergenerational relations between mothers and children in Filipino transnational families'. *Global Networks* 5, 4: 317–36.

Passaris, C. 1998. 'The role of immigration in Canada's demographic outlook'. *International Migration* 36, 1: 93–105.

Peberdy, S. 1999. *Selecting immigrants: nationalism and national identity in South Africa's immigration policies, 1910–1998*. Ph. thesis, Queen's University, Canada.

Piper, N., and M. Roces. 2003. *Wife or Worker? Asian Women and Migration*. New York: Rowan and Littlefield.

Raghuram, P. 2004. 'Migration, gender and the IT sector: intersecting debate'. *Women's Studies International Forum* 38, 1: 89–107.

Raghuram, P., and E. Kofman. 2004. 'Out of Asia: skilling, re-skilling and deskilling of female migrants.' *Women's Studies International Forum* 27, 2: 95–100.

Ray, M. 2007. *What is the Scale of the Talent Shortage in South Africa? What Might its Impact be*? Johannesburg: Oval Office. http://test.ovaloffice.co.za/content.php?gId=3&aId=62

Robinson, V., and M. Carey. 2000. 'Peopling skilled international migration: Indian doctors in the UK'. *International Migration* 38, 1: 89–107.

Salaff, J.W., and A. Greve. 2004. 'Can women's social networks migrate?' *Women's Studies International Forum* 27, 2: 149–62.

Scott, S. 2006. 'The social morphology of skilled migration: the case of the British middle class in Paris'. *Journal of Ethnic and Migration Studies* 32, 7: 1105–29.

Selman, P. 2002. 'Intercountry adoption in the new millennium; the "quiet migration" revisited'. *Population Research and Policy Review* 21, 3: 205–25.

Selman, P.E. 2006. 'Trends in intercountry adoption: analysis of data from 20 receiving countries, 1998–2004'. *Journal of Population Research* 23, 2: 183–204.

Silvey, R. 2004. 'Power, difference, and mobility: feminist advances in migration studies'. *Progress in Human Geography* 28, 4: 1–17.

Silvey, R. 2006. 'Consuming the transnational family: Indonesian migrant domestic workers to Saudi Arabia'. *Global Networks* 6, 1: 23–40.

Sorensen, N.N. 2005. 'The global family: disintegration or transnationalization of the family?' *Dansk Sociologi* 16, 1: 71–89.

Svasek, M., and Z. Skrbis. 2007. 'Special issue: emotions and globalisation'. *Identities* 14, 4: 367–1026.

Tani, M. 2008. 'Short-term skilled labour movements and economic growth'. *International Migration* 46, 3: 161–87.

Waters, J.L. 2002. 'Flexible families? 'Astronaut' households and the experiences of lone mothers in Vancouver, British Columbia'. *Social and Cultural Geography* 3, 2: 117–34.

Waters, J.L. 2005. 'Transnational family strategies and education in the contemporary Chinese diaspora'. *Global Networks* 5, 4: 359–77.

Xu, Q. 2007. 'A child-centered refugee resettlement program in the United States'. *Journal of Immigrant and Refugee Studies* 5, 3: 37–59.

Chapter Nine

The Family and Fieldwork: Intimate Geographies and Countertopographies

Margaret Walton-Roberts

Introduction

In this chapter, I reflect on three sets of field experiences I've had over the last 10 years through a discussion of **family**, **mobility**, and **social reproduction**. I use Katz's (1994) notion of 'the field' as a flexible space where the political and personal merge and this is clearly illustrated when we think about family within the context of fieldwork. I employ Katz's (1994) critical lens on what constitutes the field, and how we, as scholars, can and should always and everywhere be in the field. Aligned with this, I follow Pratt and Rosner's (2006: 21) call to 'understand the intimate as a politicized sphere of over determined meaning that feminists can explore and use as a basis for thinking about globalization'. The 'intimate' clearly relates to the kind of social contacts that exist within families, and yet Pratt and Rosner's call is to push the spatial referent of the intimate. In this paper I use these themes of the field and the intimate to construct a trajectory linking family, global mobility, neo-liberal migrant labour market regimes, and the institutional context within which academic knowledge about these phenomena is produced. While this may appear to be merely intellectual gymnastics, the point of the exercise is to challenge the idea of distance, both social and spatial, that informs much of our thinking about overseas fieldwork, and instead highlights the intimate personal connectivity that shapes academic labour and the **production of knowledge**.

My approach to this work is framed by the intersections between contemporary human mobility and social reproduction. Labour migration circuits and globalization are mutually constituted, and the research networks academics create and utilize in order to pursue their studies of these phenomena are also expressive of some of the same processes (though admittedly accompanied by very different degrees of privation and subjugation for the actors involved). Global labour migration presents a glaring example of how 'social reproduction gets unhinged from production' (Katz 2001: 710), and in this chapter I forge connections between the experiences of the female researcher doing fieldwork and the experiences of female labour migrants in order to imagine and create 'geographies of care beyond the national community' (Pratt and Rosner 2006: 19). I take the liberty of using highly personal reflections drawn from my field experiences not as a solipsistic form of self therapy, but rather as a way to communicate with some empathy the gravity of migratory systems that cause *de facto* or *de jure* rupture to families throughout the international migration process. I also focus on the messy tapestry of research rather than the 'gods' eye view' of the production of knowledge, which is often scrubbed clean of the emotional back-story, ambivalence and contradictions of academic work (Blunt et al. 2006; Bondi 2004; Butterwick and Dawson 2005; Stacey 1988). I trace three sets of field experiences and offer what Katz calls a '**countertopography**' that draws contour lines that connect globalization and social reproduction across different places and contexts in order to reveal similarities in the way certain processes operate (Katz 2001: 721; 1992; Silvey 2005). As Katz suggests,

> My intent in invoking them [contour lines] is to imagine a politics that simultaneously retains the distinctness of the characteristics of a particular place and builds on its analytic connections to other places along 'contour lines' marking, not elevation, but rather a particular relation to a process—for example, the deskilling of workers or the retreat from social welfare. (2001:720).

By identifying commonalities across different landscapes and subject positions through the lens of family geographies, I hope to expose and direct attention to how similar labour regulatory processes create disjuncture between labour migration circuits and social reproduction. This attempt to insert the intimate into the seemingly disjointed framework of the field and the family strives to expose and challenge migratory and labour regimes that are shaped by processes of labour reification that deny people the right to simultaneously be employed and with their family (whatever structure it takes). I also critically reflect on the interest geographers often display in family and home, whilst paradoxically wrenching themselves from those very same anchors in a job market that has globalized and professionalized in ways that demand successful academics be footloose global agents. Such labour practices signal 'the interpellation of subjects as *life workers*—the rendering of permanently mobilized bodies in new kinds of technologies of power'. (Mitchell et al. 2003: 417). Though these demands apply regardless of gender, the pressures women face in balancing social and professional norms personalize and intensify the demands of 'making it' in the academy in very particular and uneven ways. Challenging this process demands that we create alternatives to how we construct the position of life workers and create 'an elsewhere within' (Bondi 2004 calling upon the work of Desbiens 1999) that constructs a place for family within the world of work that does not necessarily undermine the value of the work or the working subject.

My research has primarily been focused on Indian emigrants (temporary and permanent) who inhabit transnational migrant networks across sometimes spatially extensive ranges. Reflecting this expansive geography of migration, my research has also been spatially extensive, including fieldwork in British Columbia and Ontario in Canada, and parts of India and Britain. Much of my work has revealed the challenges immigrants face in maintaining family relations, and the strategies used to either reconstitute family in one location or maintain social reproduction during periods of spatial separation. Ironically, in pursuing this agenda my research activities took me away from my own family networks. This began with international graduate training and separation from my parents and partner. Later when I had my own family, the usual demands of being an academic led to separations. To explore these issues and the politically meaningful connections Katz (2001) calls for with her idea of contour lines, I offer three sets of reflexive field notes categorized as 'the daughter', 'the mother', and 'the academic'. Each of these frames is time and place specific, and refers to my own complex subject position as contextualized by both personal and structural dimensions. This recounting of positionality in relation to family and the field highlights the type of critical topographies Katz is advocating that we imagine, as well as enabling a connection between the global and the intimate as Pratt and Rosner (2006) encourage us to make.

A Note on Fieldwork

Geography boasts a strong tradition of fieldwork, but several feminist geographers have drawn attention to this tradition as a masculinist exercise where the researcher dominates an often feminized field, a distinctly contained space 'out there', separate from the researcher which is then visualized and consumed through distinctly sexualized interpretations (Rose 1993; Sparke 1996). However, many other feminist geographers have argued that fieldwork can provide important opportunities for challenging dominant norms and attitudes that create and maintain spatial and social binaries and that in our activities, as scholars, teachers, and community activists, the field can be understood as present in everything we do (Katz 1994). Although I understand Sparke's contention that 'fieldwork remains imbricated within masculinist modalities of power' (1996: 216), I want to complicate this construction of the field as masculinist. Throughout my work I have never felt in complete control of the research field, I often felt dictated to whether I was alone or accompanied. Also, doing transnational multiple site research, especially research that traces migrant networks, means you never really leave the field. I contend that the spatialization of my field is not that of a separate and distinct space and time 'over there' or 'back then', even though I might have sought and enforced such separation at times. Instead the field became a recursive, expansive, and continuous space that continually collides and butts up against the researcher's subject position. In this regard one's personal location is always invocated within such field experiences, and therefore the subject's social embeddedness plays an important role in the production of knowledge emanating from such fieldwork. Thus a more expansive sense of the field is required to permit the inclusion of this relevant 'back-story'.

The Field 1: The Daughter

My doctoral research was conducted across two main sites, Greater Vancouver, British Columbia and the Indian Punjab between March 1998 and March 2001 (Walton-Roberts 2001; 2003). One of the main objectives of my work was to understand how Punjabi Sikh families sought to spatially reconstitute and reproduce their family and community in Canada via the working of the federal immigration system. Ironically, as I pursued this research, I was increasingly drawn away from my own family and personal support networks.

My research involved the utilization of social networks maintained by Punjabi Canadian immigrants with relatives and friends back in Punjab, who then became my hosts during my overseas fieldwork. My interactions and everyday experiences of living with families in Punjab proved to be fundamental in helping me understand how social and spatial relations are developed and maintained between the two regions. I was fortunate to stay with families of different social status and transnational positionality: for example, one family were themselves transnational, moving between Punjab and Vancouver on an annual basis (Walton-Roberts and Pratt 2005), while another were Indian citizens and residents with multiple links to several overseas Indians connected to their village (Walton-Roberts 2003), but who themselves did not frequently travel. I developed a deeper rapport with these respondents due to multiple interviews, visits to people's homes, and other forms of recurrent contact over much of the research period. One family in particular was central to the success of my research. The Singhs are a middle-class Indian family. Mr. Singh and his wife are both educational professionals and they have two children, Manjit and Joyti (pseudonyms), who were teenagers at the time of this fieldwork. I stayed with the Singh family for seven weeks and witnessed their closeness, sharing sleeping quarters and preferring to be close when relaxing, reading, and doing any other work. Initially I withdrew from this interaction, but after several weeks, I become more comfortable and began to seek out greater intimacy; sitting with everyone on a small cot in the midday sun, moving my laptop into the bedroom with Manjit and Joyti in order to write field notes while they completed homework rather than sit alone in my bedroom. My experiences became paradoxical, at one level I enjoyed the closeness with the family, but at another I resented the lack of personal space, freedom, and independence that I was experiencing in northern India, and the lack of closeness I was able to experience with my own family. The lack of open space, coupled with the manner in which I was often accompanied, became increasingly frustrating for me. Recreation deemed appropriate for me as a woman always involved socializing with other women, chatting about jewellery, clothes, and food while eating plenty of it. There seemed little in the way of physical activity for women. Paradoxically, I felt isolated amidst such intimacy.

Gender, of course, was a factor in the way I was guided and chaperoned during my research. I always had a companion sent with me even if it was to visit a neighbour's house. After a few weeks I was resenting the fact that I had lost my independence and the freedom to structure my own research, so I decided to visit a research institute in Chandigarh where I could develop my own schedule and move freely about the city. My first appointment in Chandigarh was with an Indian trade official employed at the Canadian Consulate. Upon meeting him he expressed shock and disbelief that I was travelling around the city alone, and demanded to know where my chaperone was. His consternation upset me, and drew my attention to the fact that at that time I appeared to be the only single white woman in the city. As I became conscious of being out of place in Chandigarh, I began to resent the way public space in the city was so gendered, and my obviously incongruent placement within it. My stay in Chandigarh was difficult; the independence I had sought from the comfortable family setting with the Singhs came at a price, and I missed my partner and longed to be at 'home' regardless of whether that home was in Punjab or Canada. While I resented myself for not being stronger and more resilient, the imperative of planning research, data collection, interviews, and meetings overtook me, and I 'pulled myself together' and ventured back out onto the streets. Still, I only stayed in Chandigarh for a week, and returned to Phagwara for a few more days before heading back to Delhi, this time to meet my husband. His arrival was a great relief, and his company dramatically changed the way many people interpreted my presence in India. I was no longer a woman violating social norms, a legitimate target for male visual consumption. I was now in my proper place, with my husband. Though I objected to this patriarchal norm, I appreciated the comfort it offered me. It was 'easy' to be in the right place in India as a married woman, and not as a solo female academic. Family in this sense became my acceptable identity.

For me the expansive nature of the field during my doctoral work was particularly evident with respect to the presence of my father. A year before I left for India my father passed away suddenly, forcing me to reflect more critically on my absences from his life as I pursued my transatlantic academic training. In India his loss took on a resonance for me that was both saddening and comforting. The moments when I felt my father's presence most intensely—and therefore his death most emotionally—were often during my fieldwork with Sikh men both in Vancouver and Punjab. My father had been posted in India during

World War II, and I remembered how he fondly recalled his experiences from that time. On numerous occasions in Punjab I found myself comforted by the similarities between the Sikh belief in the value of physical labour and working the land, and my father's life-long connection to farming. The Sikh tradition of status being accorded with gender and age, and the honourable position of veterans were amply displayed in the field sites I visited, and this provided a sense of comfort and familiarity for me as I reflected on my father's life, even as I queried the intense patriarchy and hierarchy embedded in these social relations. The structure and organization of family relations was sometimes comforting, but also a jarring reminder of my own isolation from my family.

As I reflected on the loss of my father and the cumulative years of separation I had enforced on him as a result of my graduate training, I developed a personal awareness of the emotional toll of the separation parents feel when children go overseas (Miltiades 2002). One particular elderly Sikh man with glassy cataracts in his eyes, sun-weathered skin, and a mild disposition instantly reminded me of my father. In this instance, family became an expansive term that forged a comforting connection for me in the exile I was experiencing from my own family. I recognized the pain and desire that immigrants in Canada feel when they seek out channels to bring their parents to join them. I understood how valuable that connection across the generations is deemed to be and, as a corollary, how painful immigration regulations that circumscribe the family as nuclear can be when they leave no room to negotiate the inclusion of parents or grandparents. The designation and reconstruction of 'family' into a particular nuclear variant that weighs heavily in immigration policies of the **Global North** is a challenge to such social networks (McLaren 2006). These restrictive readings of family can result in an increase in social duties that are borne most heavily by immigrant women (Salaff and Greve 2004). Academically we conceptualize and debate these issues as policy or social justice concerns, but we do not often engage or reflect on them as personal ones, and it is perhaps the latter that offers a more powerful framework within which to imagine ways to reform immigration policy. How do we construct contour lines that connect across diverse places unless we have some sense of empathy with people who live lives so seemingly different from our own, in places distant from our own, yet at the same time are subjected to the same economic and social processes and connections? How do we understand and share the great concern Indian immigrants express about leaving their parents behind when they migrate, in a health and welfare system that is heavily dependent upon family members being advocates and providers for the elderly? The immigration system often works to prevent these adult children from fulfilling their filial duty, and thus their consternation becomes directed at the legislative restrictions of international immigration systems, as well as at the lack of state support for the elderly in India.

The Field 2: The Mother

Eight years on from my Ph.D. fieldwork, my family identity is now primarily shaped by being a mother. In September 2008 I returned to India, this time to Kerala in south India. After initially planning to take the whole family for part of a sabbatical leave, my husband decided that instead he and my three young sons should stay with his parents while I left for a condensed three month field season alone. Ambivalence regarding this opportunity set in long before I set foot in India. Thoughts of declining the funding received negative reactions from female colleagues; while others supported my rationalizing that it 'would provide a great opportunity for the kids to bond with their grandparents'. Practically every woman I told I would have to leave my kids for up to three months gasped, then said 'you're brave'. My first few days were distressing. Dealing with the challenges of India and the visceral horror of having left my husband and children for an extended period were heightened with any stressful or low blood sugar event. I turned to the ether for support. Suddenly my computer became the most powerful tool I possessed; a wireless connection became the most valuable asset I could pilfer; voice over internet protocol tools like Skype™ became a life vest as I video-conferenced with my family, friends, and close colleagues. The sorrow and privations I felt at being away from my family in a country that can exact an emotional toll on a good day led to a fundamental questioning of the real purpose of this exercise. To cope, I diverted my energy to research, and as I read about the experiences of migrant women and men faced with leaving their families for extended periods (Gallo 2006; George 2005; Oishi 2005) I discovered immense empathy and understanding. I also recognized the vital link between these seemingly 'personal' relations and the structural capitalist labour processes we operate under today, as Katz observes,

> when reproduction is highly mobile but social reproduction necessarily remains largely place-bound, all sorts of disjunctures occur across space, across boundaries, and across scale, which are as likely to draw upon sedimented inequalities in social relations as to provoke new ones. (2001: 716)

Katz specifically refers this to the rescaling of childhood, and in the case of migrant workers the impact on children and mothers is vivid (George 2005; Hoschschild 2001; Waters 2006). In my heightened state of emotion due to my separation from my children, I cried as I read Sheba George's accounts of Kerala nurses who left their families to work abroad and how the emotional scars of such separations were vividly recalled years after the event. In one case George relates how a woman had left her own son with her sister at age two and how, when he later joined her in the United States, he took to calling any new woman he met 'Mummy' (George 2005: 167). My own situation gave me a clearer insight into the sacrifices these migrant women were making to advance their family's fortunes, and I shared in their sense of disappointment and loss when such sacrifices did not result in optimal returns. To suffer such privation and then to see your children face downward mobility (Pratt and the Philippine Women Centre 2008), or to become trapped in an endless revolving door of contract labour that creates irreparable emotional distance between migrant mothers and their children (Oishi 2005), all began to mean something more intense for me as I coped with my own insecurities and doubts. Indeed, I began to resent the power of Skype™ to seemingly overcome, but at the same time emphasize, the distance I was from my family. Such technology becomes an important form of connection for immigrants as Lusis and Bauder (2008) have discussed, but it also becomes an enabler, in the most pejorative sense, of labour mobility for millions to work away from their families. Rather than just a convenience, such technology permits and supports the development of migration regimes that rest upon severing the migrant from their social embeddedness. This offers just one example of the 'myriad ways in which capitalist production and its entailments have pushed people to drastic limits of their own resilience, and how willing capitalists have been to draw on that resilience for their own ends' (Katz 2001: 718).

Of course my case was never comparable to that of many migrant women, and I was the first to dismiss my universalistic 'your pain is my pain' position as that of a spoilt self-centred academic; and yet, as I explored the reality of Kerala's international labour migration system I noted something rather distinct. In Punjab I had explored the efforts (both official and unofficial) directed at using family migratory routes to reconstitute the extended family in the destination site. But in Kerala, where nearly 90 per cent of labour migrants go to the Persian Gulf, there is no 'permanent' immigration. Rather, contract labour becomes circular or perpetual long term migration and, depending on the skill level and religion of the migrant, it may or may not include family members. For Muslims there is a greater propensity for men to travel for work in the Gulf and for women to remain behind in Kerala (Zachariah and Rajan 2007). In some of these cases men can be gone for years, and the economic justification for such protracted separation becomes less clear as time goes on. In this context the migrant becomes complicit in perpetuating the system. One labour agent I interviewed talked about how some men work in the Gulf long after they need to, and rather than spend time with their family they become slaves to the *idea* of the money, but never actually enjoy its use.

> 'Even if they have money they don't know how to spend . . . Even if they are earning money, they are just existing, that is not life; life should be filled with colours. Some people . . . go to Gulf; that is not living. He never had a good dress he never have a good foot (shoes) he is just earning for his family, he is like a slave'. (Interview with labour recruiting agent, Varkala, Kerala September 15, 2008)

Migrants respond to the structures of inequality they are positioned within, but can also become complicit agents within these structures of capitalist inequality, and the social costs in terms of family relations are often high (Oishi 2005). In a study of domestic violence in one area of Kerala over one-third of the nine case studies highlighted involved male perpetrators who had been temporary migrants in Gulf countries (Sobha 2008). Indeed the rise in violence against women in Kerala has raised great concern among activists and researchers questioning the rhetoric of gender empowerment at the centre of Kerala's development discourse (Rajan and Sreerupa 2007). Such extended separations clearly enhance the potential for family conflict, and add another dimension to the harm this global migratory system can inflict on women, men, and children.

While I reflect on these contour lines, I question whether equating my experiences with those of the migrants I research is disingenuous, or worse some form of epistemological violence. Nevertheless, I do recognize that neither of us has been *forced* to engage in this mobility, rather we are *compelled* by the pressures of economic context and, as a result become complicit with the process to varying degrees. My case was far more privileged and the rewards for my separation from my family potentially greater and more sustainable than those received by most migrant workers, who are often separated for much longer than I was. Sanctions for refusing the research fellowship would have been minimal, and my situation was certainly more spoilt than that of the migrants I meet in Kerala. Yet, I felt some recognition of the ambivalence that defines both the migrant experience (Osella and Osella 2008) and the academic process (Butterwick and Dawson 2005). Despite these differences, I still channelled the emotional stress I endured in the first few weeks of fieldwork to build an intimacy, a sense of visceral sympathy with what labour migration means for so many families.

On my second day in Kerala, during the special festival of Onam when families gather to eat together, my research assistant took me to meet his friend's mother, whose two sons were away in Saudi Arabia. As we enjoyed a typical Kerala thali lunch spread out on a banana leaf and eaten with the right hand, she spoke about her sons. Tears were always close to the surface for me as I listened to her and thought about my own family. Actually inhabiting that physical space of family separation I sensed the pain that people endure when families are separated for months and years at a time. In this case the elder son's wife and infant daughter were living with the mother, and while I thought of my own sadness, I was in awe of their 'no-nonsense' acceptance of this arrangement and the resilience of these women who seemed to cope admirably alone. I realized that the Canada-based research I had conducted in the past had never felt this personal, this visceral, even when I heard similar stories of family separation. Inhabiting and experiencing the pain of family separation, rather than distract me, seemed to forcefully clarify my fieldwork for me.

To cope with my personal turmoil I turned to family, friends, and colleagues. I spoke often to my sister on my cell phone (pre-paid international in rupees, relatively cheap, the enabler of international communications technology), and she told me to be strong. She told me that knowing that I did not *have* to stay actually made it easy to say 'yes, I will see it through'. 'Imagine those poor women', she told me, 'who have to work away from their kids and don't really have a choice', and she told me of a woman from Thailand she met in the UK who had to leave her baby at home for a year while she came abroad to work. As she told me about this I could *feel* how cruel it is when a system is constructed that results in mothers being separated from tiny children; how mean spirited, how inhumane such labour market systems are that want only the monochrome labour of these women, but not the colours of their lives.

At one point during my research in Kerala, to overcome the isolation I was feeling I travelled north to visit the Singh family who I had stayed with in Punjab during my Ph fieldwork. Eight years on and the Singhs were healthy and still busy. Their children were now adults and their daughter Joyti was a pharmacist, now married and had applied with her husband to move to Canada. Mr. Singh told me that their applications would soon be processed, and I noted a sense of sad inevitability in this turn of events, that despite the fact that both she and her husband had good jobs in India, they felt the need to emigrate. Manjit, their son, had left only the week before to study computing in the US (he later moved on to Australia). My visit to the Singhs was a last minute decision, and at first I was worried by Mr. Singh's seeming distraction and lack of interest in my stay, but after a day or two he explained to me that he had learnt that his son Manjit was very homesick and wanted to come home. Mr. Singh was very concerned and upset about his son. I recognized instantly how sad this situation was since I had also been experiencing this pain of separation on an intimate personal level. I suddenly understood how deeply embroiled we have all become in the modernity of mobility, as subjects with the power of agents, yet also powerfully compelled to inhabit certain subject positions that we battle against, resent and yet continue to engage with. We are fully interpolated into these global labour migration circuits.

The Field 3: The Academic

In a collection of narratives from 12 female Indian academics regarding their mothers of an earlier generation, Gulati and Bagchi (2005) reveal stories of sacrifice, dedication, and immense loss. Though the book is called *A Space of Her Own* many of the narratives reveal the lack of personal space these women have experienced in their lives. In the introduction to the narratives Elliot (2005: 19) speaks of how 'women's vulnerabilities to the purposes, tempers and fates of their husbands and families, both natal and affinal, stands out'. The narratives reveal 'how much women valued their relationships in families . . . oppressed though many were by family relations, they show a deep commitment to human solidarity'. The book's focus on the placement of academic women within their family relationships is a welcome addition to an often limited exploration of the links between family and academic work, and how academic institutions react to domestic demands. There are still too few discussions about how the socially embedded researcher conducts his or her research, and how this informs the research process and product. As Frohlick comments,

> it still strikes me as somewhat taboo to acknowledge the presence of our families, in other words to blur and even violate the boundaries of our field sites with visible traces of our personal lives and relationships, however important these relationships and biographies are in enabling us to understand the phenomenon we are studying. (2002: 52)

During my fieldwork in 2008 I felt like I had been exiled from my family, ensconced in India with the luxury to work for as long as I pleased, yet I felt immense dissatisfaction. Living with the dissonance between what I was supposed to embody—the self-possessed researcher—and the reality of my 'outlaw emotions' (Jagger 1992) of insecurity and doubt, was exhausting.

To 'make it' as an academic demands that you occupy circuits that are masculinist, in that the more one becomes the ideal institutional subject—bereft of any other responsibilities than those to the academy, the conference circuit, the 'sexy' research agenda—the more successful you can potentially become. Add to this the increasing audit culture of the academic sector and the rise in fiscal retreats linked to neo-liberal educational mantras (Butterwick and Dawson 2005; Castree and Sparke 2000; Demeritt 2000), and I begin to see how my self-imposed punishment is actually driven by my internalization of those expectations and my willingness to be subjugated to them. As with other aspects of neo-liberal society 'this devolution is not imposed; rather, it becomes the accepted norm through time as it infiltrates and articulates with other common-sense understandings in society' (Mitchell et al. 2003: 418). I have internalized these understandings and expectations. I also contemplate what my research will really achieve in the same vein as Butterwick and Dawson (2005) do. And, as if to add insult to my self-inflicted injury, during this period I received updates on my faculty association's collective bargaining and was informed that the administration wanted to remove employment equity data, and replace automatic annual career development increments with greater merit based awards (WLUFA 2008). This all struck me as only adding to my ongoing interpolation as an institutional subject who, in order to succeed, needed to keep outlaw emotions (and my family) in check. All of these trends do not bode well for any kind of serious work-life balance. In this long saga of the overworked academic, we have become our own worst enemies, and I think we now need to explore how to make the system work in more humane ways. Rebick (2005) charts the changes she has witnessed in the feminist movement and comments, 'Women were cruel to each other, there's no other way of putting it. I saw it and I was part of it. We don't want to be that way anymore. We have to create organizations that deal with conflict in a creative and supportive way' (cited in Goar 2005). I echo her self-critical awareness of how *we* have to become more activist in creating humane workplace structures and relationships, and that we have to hold our employers, ourselves, our colleagues, and our students to higher standards of accountability and awareness when it comes to building a more inclusive and just society.

Building such support structures has immense potential in the long run, but we can begin within the academy. Solem and Foote (2004) focus on the challenges facing early career geography faculty and contribute to a wider literature that emphasizes the importance of building supportive structures to assist in advancing graduate students and junior faculty. One of their findings correlates opportunities for professional networking with enhanced self-esteem. Throughout my separation from my family and my intense field experiences, it has been my self esteem that has been most battered. The challenges of fieldwork have left me elated at times, but at others severely tested. Any sense of failure due to my lack of family commitment fought my sense of failure in my research. Slights and negative comments from other academics or researchers reverberated in my mind for days, but when colleagues offered support, no matter how small, my perspective on my situation dramatically improved. All of this suggests collaborative frameworks are most valuable in creating a balanced supportive work environment. The current audit/merit culture, however, goes against that and inculcates a deeply unpleasant competitiveness that creates division and further marginalization within the academic community that can certainly extend to how we treat research subjects and colleagues. What is clear is that forging greater collective and collaborative structures within the academy demands changes in both the ethics of our research practice, and also in the nature of our 'office politics'. As Staeheli and Nagar (2002: 171) discuss, this 'suggest[s] the importance of talking *within* our worlds as a means of strengthening our praxis *across* worlds' (Staeheli and Nagar 2002: 171).

Academics present a particular type of highly privileged migrant worker. The academic field is global, and academics have the opportunity of extreme mobility. Such mobility, however, as with all migrant workers, unhinges the reproductive and productive spheres. Most academics are aware of the emotional carnage the hypermobile academic circuit can leave in its wake; broken relationships, isolated parents, and the child that is never born, though desired. For example, in Canada nearly one half of female academics between 35 and 39 have no children under the age of 12 living at home, this contrasts to less than a third of female physicians (Czernis 2004). There are any number of personal and professional geographical networks academics inhabit, and, in the case of dual academic couples, the geographical distances couples operate across can push the limits of intimate bonds, often to the breaking point. We all know colleagues who have struggled to find satisfactory employment in the same city, and those who do not may end up constructing elaborate commuting schedules that can be national or international in scope. Certainly the lengths taken to develop family friendly work spaces varies between countries, even departments (Hyndman 2009), but it is clear that the imposition of this style of work does result in alterations to the typical family form. The bifurcation in the academy to greater research intensity for those in tenured and tenure track positions, and greater casualization and insecurity for non-tenure positions highlights the ongoing positioning of the academic as a neo-liberal subject where certain forms of research output becomes one (perhaps the only) matrix or measure of importance. This serves to institutionalize competitive modes of working, and engenders the same compulsion that migrant workers face when they submit to the demands of mobility and family sacrifice for economic gain.

As academics, we do have some power to change our work culture. Rather than the externally imposed system of labour atomization supported by legislative and capitalist frameworks that isolate and fragment migrant labour, for the privileged academic subject the process is more nuanced and more complex, but still subjected to structural processes of fragmentation. It is also more self-generated, but this is not to say international labour migration is not also reproduced through the willing involvement of migrant subjects who advance the privileged image of mobility through the positive demonstration effect (remittances, conspicuous consumption, and excessive philanthropic activity or salary, research publications and international invitations). Despite these differences between the labour and academic migrant, they both operate in ways that unhinge the reproductive and productive practices of society, and the cost of maintaining both spheres falls disproportionately on certain groups.

My effort in this chapter has been to find ways to combine my professional and personal sides. I have come to see my actions not in the typical faceoff between working and stay-at-home mothers, but as an effort to incorporate, in sympathetic and constructive ways, my family and work needs. I see this as a political action that is not directed at others per se, but at ideas of institutional separation. I do not wither, as much, now when I attend conferences as a parent. At times I do wonder if my professionalism is undermined, but it appears I am not necessarily alone in this effort. For example, at the 2007 Association of American Geographers annual meeting in San Francisco and the 2008 Institute of British Geographers annual meeting in London, on-site childcare was provided for delegates and costs were subsidized. Suddenly children were being seen, heard, and cared for within the professional sphere. Beyond the 'feel good' debate about the need to recognize family, what this insertion of social reproduction into the workings of the academy does is recognize and authorize the socially embedded nature of human relations. Again as Katz argues,

> Redistributing responsibility for social reproduction back to capitalists and the state, transnationally and at all scales, would begin to recalibrate the costs and benefits of globalization in ways that would pinpoint its widely distributed costs and promulgate increased social justice and equality across classes, nations, localities and gender. (2001: 719)

Positive things can flow from such border transgressions as we can begin to talk about social reproduction as intellectuals, yet at the same time be given space to practice it more effectively as part of our everyday life. Focusing on bringing the practice of parenting into public and professional spaces of the academy also speaks to recent calls to focus on the conference as an important site of academic reproduction and academic capitalism (Lindley 2009). Rather than focus on the direct issues of critical questioning of speakers though, the insertion of the family into the space of the conference works to make our social and gendered responsibilities part of the agenda. From this position, we can actively work to extend such rights to others, especially those who are given little option but to uproot themselves to become mobile labourers bereft of their family and other support structures.

Conclusions

In this chapter, I have recounted field experiences that brought me emotionally closer to the communities I was researching, but also to my own conflicting position as a socially and economically embedded agent. The moments of intense reflexivity I experienced when my outlaw emotions were unchecked withdrew in an instant the lens of ethnicity and difference, revealing commonalities and connections I had suppressed under the guise of being the 'detached' authoritative researcher. The experiences of two Indian field periods, one as a daughter, and one as a mother both allowed for reflections on family that brought to the fore deeply empathetic ways of understanding the challenges faced by migratory subjects from the Global South. By allowing the personal to filter through to the professional, I created a space for these fleeting connections to act as transformative moments with political and emotional potential where I could begin to understand and appreciate the pain and joy of those who become subjected to the ever expanding world of modernity and mobility. I was also able to recognize how those same processes are infiltrating my location as an employee of an academic institution. In this way, I have attempted to draw a connection, or contour line as Katz suggests, between my academic position as researcher and the positions of migrant labourers I researched in order to suggest that we are both interpolated into the ever-widening circuits of global capital and labour mobility. We have to recognize that our position in these flows is not necessarily forced (as it is for millions of others), but that we are somewhat complicit in the mobility process, and do have the option to work at the margins to improve the situation, or to create as Desbiens (1999) terms it, 'an elsewhere within'. Such active reconstitution of these counter topographies requires a greater awareness of not only the similar trajectories we face, but also how we can change the political situation, at the margins perhaps, but only by advocating from a position that builds upon greater collaborative and co-operative frameworks.

Although the migratory processes and labour market exploitation of the academic and the migrant labourer are not the same, they are part of a continuum, a contour line that traces the degree to which the modern migrant has become interpolated into a system that depends, in varying degrees, upon wrenching people from their social support systems. In both contexts there is potential to work towards the reunification of families over time and space, but in both cases there are costs that are borne unequally along various axes of difference, be it class, gender, ethnicity, or sexuality. Striving to overcome this detachment of the reproductive and productive spheres comes through political action that impedes the ability of the state and capital to isolate the labour subject. Alas, the prevailing economic system is cast in such a way that it continues to re-inscribe difference and inequality as one of the baselines for its success. My effort here has been to offer a reading of how we might construct contour lines that can connect across various differences to expose the similar meta processes that separate the productive and reproductive realms. I seek to locate the family as a central dimension of both the academic work I do, and as a category of significance in understanding how global labour migration is managed and differentiated along the lines of race, class, and gender. The geography of the family is central to the workings of the productive economy and the academy, and keeping its tantrums and complexities out of sight only works to protect the productive economy from the true costs it inflicts. As such, bringing the reality of family geographies to the fore through intimate global connections or contour lines exposes the embodied costs of the capitalist market system, and more importantly, who bears them.

Acknowledgements

The research discussed in this chapter was made possible with various funding from the Shastri Indo-Canadian Institute, the Social Sciences and Humanities Research Council of Canada and Metropolis British Columbia. I also want to acknowledge the support of the staff and students at the Centre for Development Studies in Trivandrum, the research assistance of Jithin Raj, and the many respondents in Kerala and Punjab who kindly shared their time and thoughts with me. I am also immensely grateful to the reviewers and colleagues who offered suggestions and support during the writing and revision of this chapter.

Questions for Further Thought

1. How do you think the gender of the researcher might affect the type of qualitative research gathered? Consider this for research contexts in the Global North and South, and for research with elite and non-elite groups.
2. Think of two or three processes of change you are experiencing today that are truly global, and explore how they affect other communities around the world.
3. How do different national immigration policies deal with family reunification? How are such regulations differentiated by the class of migrant worker?

Further Reading

Katz, C. 2001. Vagabond capitalism and the necessity of social reproduction'. *Antipode*, 33, 4: 709–28.

Mitchell, K., S. Marston, and C. Katz. 2003. 'Life's work: an introduction, review and critique'. *Antipode*, 35, 3: 415–42.

Moss, P. 2001. *Placing Autobiography in Geography*. Syracuse, NY: Syracuse University Press.

Pratt, G. 2004. *Working Feminism*. Temple University Press: Philadelphia.

References

Blunt, A., P. Gruffudd, J. May, M. Ogborn, and D. Pinder, eds. 2003. *Cultural Geography in Practice*. Edward Arnold: New York.

Butterwick, S., and J. Dawson. 2005. 'Undone business: examining the production of academic labour'. *Women's Studies International Forum*, 28, 51:65.

Bondi, L. 2004. 'For a feminist geography of ambivalence'. *Gender, Place & Culture*, 11, 1: 3–15.

Castree, N., and M. Sparke. 2000. 'Professional geography and the corporatization of the university: experiences, evaluations and engagements'. *Antipode* 32: 222–29.

Czernis, L. 2004. 'Envisioning a family friendly campus'. CAUT Bulletin, December A3.

Demeritt, D. 2000. 'The new social contract for science: accountability, relevance and value in US and UK science and research policy'. *Antipode* 32: 308–29.

Desbiens, C. 1999. Feminism 'in' geography: elsewhere, beyond and the politics of paradoxical space. *Gender, Place and Culture*, 6: 179–85.

Elliott, C. 2005. 'Introduction'. Pp. 9-19 in L. Gulati and J. Bagchi, eds. *A Space of Her Own: Personal Narratives of Twelve Women*.

Frohlick. S. 2002. '"You brought your baby to base camp?" Families and Field Sites'. *Great Lakes Geographer*, 9, 1: 50–8

Gallo, E. 2006. 'Italy is not a good place for men: narratives of places, marriage and masculinity among Malayali migrants'. *Global Networks* 6, 4:357–72

George, S. 2005. *When Women Come First: Gender and Class in Transnational Migration*. University of California Press: Berkeley

Goar, C. 2005. 'The changing face of feminism'. *Toronto Star* Jan 12.

Gulati, L., and J. Bagchi. 2005. A *Space of Her Own: Personal Narratives of Twelve Women*. Sage: New Delhi.

Haraway, D. 1991. 'A cyborg manifesto: science, technology, and socialist-feminism in the late twentieth century'. *Simians, Cyborgs and Women: The Reinvention of Nature*. New York: Routledge, 149–81.

Hoschschild, A.R. 2001. 'Global care chains and emotional surplus value'. Pp. 130–46 in W. Hutton and A. Giddens, eds., *On The Edge: Living with Global Capitalism*. London: Vintage.

Hyndman, J. 2009. 'Balancing work and life: a geography of parental leave'. *Geoforum*, 40,1: 2–4.

Jaggar, A. 1992. 'Love and knowledge: emotion in feminist epistemology'. Pp. 145–71 in A. Jaggar and S. Bordo, eds. *Feminist reconstructions of being and knowing*. New Brunswick: Rutgers University Press.

Katz, C. 2001. 'Vagabond capitalism and the necessity of social reproduction'. *Antipode*, 33, 4: 709–28.

Katz, C. 1994. 'Playing the field: questions of fieldwork in geography'. *Professional Geographer*, 46(1): 67–72.

Katz, C. 1992. 'All the world is staged—intellectuals and the projects of ethnography'. *Environment and Planning D-Society & Space* 10(5): 495–510.

Lindley, T. 2009. 'Academic capitalism and professional reproduction at the conference'. *Acme: An International E-Journal for Critical Geographies*, 8, 2: 416–28

Lusis, T., and H. Bauder. 2008. '"Provincial" immigrants: the social, economic and transnational experiences of the Filipino Canadian community in three Ontario second tier cities'. CERIS working paper 62.

McLaren, A.T. 2006. 'Parental sponsorship—whose problematic? A consideration of South Asian women's immigration experiences in Vancouver'. Vancouver Centre of Excellence, Research on Immigration and Integration in the Metropolis (RIIM). Working Paper Series. No. 06–08.

Miltiades, H.B. 2002. 'The social and psychological effect of an adult child's emigration on non-immigrant Asian Indian elderly parents'. *Journal of Cross-Cultural Gerontology* 17: 33–55

Mitchell, K., S. Marston, and C. Katz. 2003. 'Life's work: an introduction, review and critique'. *Antipode*, 35, 3: 415–42.

Oishi, N. 2005. *Women in Motion: Globalization, State Policies, and Labour Migration in Asia*. Stanford, California: Stanford University Press.

Osella , C., and F. Osella. 2008. 'Nuancing the migrant experience: perspectives from Kerala, South India'. Pp. 146–78 in S. Koshy and R. Radhakrishnan, eds. *Transnational South Asians: the making of a neo-diaspora*. India: Oxford University Press.

Pratt, G., and The Philippine Women Centre of BC. 2008. '*Deskilling across the Generations: Reunification among Transnational Filipino Families in Vancouver*'. Metropolis British Columbia Working Paper 08–06.

Pratt G., and V. Rosner. 2006. 'Introduction: the global and the intimate.' *Women's Studies Quarterly*, 34: 13–24.

Rajan, S.I., and Sreerupa. 2007. 'Gender disparity in Kerala: A critical reinterpretation'. *The Enigma of the Kerala Women: A failed promise of Literacy*. S. Mukhopadhyay, ed. New Delhi: Social Science Press, 32:70.

Rebick, J. 2005. *Ten Thousand Roses: The Making of a Feminist Revolution*. Canada: Penguin.

Rose, G. 1993. *Feminism and Geography: The Limits of Geographical Knowledge*. Minneapolis: University of Minnesota Press.

Salaff, J.W., and A. Greve. 2004. 'Can women's social networks migrate?' *Women's Studies International Forum* 27(2): 149–62.

Silvey, R. 2005. 'Consuming the transnational family: Indonesian migrant domestic workers to Saudi Arabia'. *Global Networks-a Journal of Transnational Affairs* 6(1): 23–40.

Sobha, P.V. 2008. 'Dowry related violence in Kerala: case study analysis from police station area'. *Confronting Violence Against Women*. V. Menon and K.N. Nair, eds. Daanish Books CDS: Delhi, 243–58.

Solem, M., and K.E. Foote. 2004. 'Concerns, attitudes, and abilities of early career geography faculty'. *Annals of the Association of the American Geographers*, 94(4): 889–912.

Sparke, M. 1996. 'Displacing the field in fieldwork: masculinity, metaphor and space'. *BodySpace: Destabilizing Geographies of Gender and Sexuality*. N. Duncan, ed. New York: Routledge: 212–33.

Stacey, J. 1988. 'Can there be a feminist ethnography?' *Women's Studies International Forum*, 11, pp. 21–7.

Staeheli, L., and R. Nagar. 2002. 'Feminists talking across worlds'. *Gender Place and Culture*, 9, 2: 167–72.

Walton-Roberts, M. 2001. 'Embodied global flows: immigrant, capital and trading networks between Punjab, India and British Columbia, Canada'. Unpublished Ph Thesis University of British Columbia, October .

Walton-Roberts, M. 2003. 'Transnational geographies: Indian immigration to Canada'. *The Canadian Geographer*, 47, 3: 235–50.

Walton-Roberts, M., and G. Pratt. 2005. 'Mobile modernities: a South Asian family negotiates immigration, gender and class in Canada'. *Gender Place and Culture*, 12, 2. 173–96.

Waters, J.L. 2006. 'Geographies of cultural capital: education, international migration and family strategies between Hong Kong and Canada'. *Transactions of the Institute of British Geographers* 31(2): 179–92.

Wilfrid Laurier University Faculty Association (WLUFA) bargaining update 18 September 2008.

Zachariah, K.C., and S.I. Rajan. 2007. 'Migration, remittances and employment short-term trends and long-term implications'. Centre for Development Studies, Trivandrum Working paper series #395.

Chapter Ten

The View From Here: Family Images in Retirement Community Brochures

Susan Lucas and Rickie Sanders

Introduction

The study of **family** and the representations of family across the life course have recently emerged as important topics in aging and family studies. Much of the work focuses on the family as an analytical category distinct in its own right and examines the social, demographic, and economic trends that influence the definition and configuration of family. Of this literature Bedford and Blieszner (1997) complain that the study of older families has been relegated to an inconsequential position and that older families are typically treated as invariant monolithic entities. Only recently has this begun to change as researchers grapple with the realization that older families are not only a significant component of the pluralism that now exists in family structure but that the aging of the populations of most economically developed countries will cause a dramatic increase in the number of such families. Just as attention has begun to focus on older families, increased consideration of the spaces, places, and cultural landscapes that families occupy and the contexts in which they age has also occurred. Along with the other chapters in this volume, the research presented here contributes to this growing literature.

This chapter examines how family is being (re) visioned and (re)presented in the highly commodified and distinct spaces of retirement communities. To accomplish this we undertake a content analysis of the images embedded in the advertising brochures produced by retirement communities. While typically these images are analyzed with regard to their marketing potential, here we are concerned with how 'family' is (re)visioned, (re) thought, and (re)presented. The images used in **retirement community** promotional materials are seen as moments in a particular set of ideologies surrounding aging and family. Ideologies are broad but indispensable shared sets of values and beliefs through which individuals live out their relationships with a range of social structures, practices, and discourses. They are the means by which specific ideas about and images of aging and family in later life are made to seem natural and inevitable.

The most important of these ideologies is the general (re)conceptualization of retirement and old age as times of self-fulfilment and the more specific promotion of successful aging (Atchley 1980; Rowe and Kahn 1997, 1998) and the **ageless self** (Andrews 1999; Featherstone and Hepworth 1991; Kastenbaum 1993; Kaufman 1986). Like the (re)conceptualization of retirement and old age, the modern retirement community is a product of a multitude of forces, including the growth in the size of the elderly population, rising incomes in later life, and more importantly, the recognition of the mature population as a substantial yet underexploited market segment. To borrow from Sharon

Zukin, retirement communities are 'spaces of consumption' (1988: 825), or purposely constructed landscapes or spaces that promote consumption (Featherstone and Hepworth 1995; Zukin 1988). Retirement communities are part of a package of 'consumer accoutrements' (Featherstone and Hepworth 1995: 374) that define the active retiree lifestyle and by extension successful aging and the ageless self.

Images are intimately tied to ideologies. Ideologies determine what images are used to depict and represent people, places, and lifestyles. Images, then, make it possible to visualize people, places, and ways of living, e.g., the advertisement for a retirement community on a Florida I-75 billboard that read, 'the community too busy to retire'. (Bourdieu 1990; Schwartz and Ryan 2003). The (re)conceptualization and (re)visioning of retirement, aging, and family have led not only to the use of positive images to represent later life but the emplacement of those images in retirement communities. Retirement communities are therefore the spaces and places where the attainment of these ideals is possible.

Through a content analysis of the photographs used to market retirement communities we examine how 'family' in later life is being (re) imagined and how these (re)imaginings shape our perceptions of possibilities. Central to what we do here is to suggest that retirement communities are not unlike other 'cultural' products targeted to specific audiences—exclusionary, not neutral, and invested with meaning and intentionality. Next we consider how these communities sell the good life in the 'third age'. Like everything made available through the culture industry, location and space are critical. Retirement communities, not unlike suburban communities, have been used to shape and market ideas about aging and family that ultimately had to be made both popular and possible. Through skilful marketing, these spaces have quickly become the quintessential visual representations of particular ideas regarding aging and every aspect of later life including the family. Following this discussion, we offer a discussion of the built environment of retirement communities and how it participates in the process of in situ reconstruction of ideas around family. In the final sections we provide a description of the data, methodology, analysis, results, and conclusions.

Retirement Communities as Sites for (Re)constructing the Meaning of Old Age

Spatial concentrations of the elderly population can be divided into two broad types: naturally occurring retirement communities (NORCs) (Hunt and Gunter-Hunt 1985; Hunt and Ross 1990) and retirement destination localities (Golant 1992; McCarthy 1983). As their name suggests NORCs involve the aging-in-place of an elderly population that has either remained behind as younger cohorts moved (aged-left-behind localities), or has simply aged in their current residences (aging-in-place localities) (Golant 1992; McCarthy 1983). Retirement destination localities in contrast, exist because of the in-migration of an elderly population that moved purposefully in order to take advantage of the 'active adult' lifestyle and recreational amenities offered and the desire for health, recreation, safety, security, and a sense of community in the post-retirement population (Litwak and Longino 1987; Longino 1981; Streib et al. 1984). Retirement destination localities range in size from states, counties, and at a smaller scale, campus-style developments and individual buildings or retirement communities. Defined by Lucas (2002: 3) as 'planned development[s] consisting of a group of housing units that has at least one shared service or facility, target-marketed to individuals over a specified age'. Such retirement communities are the focus of this research.

Retirement communities are not neutral spaces. The space and built environment of retirement communities, like other age-segregated spaces, are implicated in the construction, reproduction, and emplacement of specific related anti-ageist ideologies (successful aging and the ageless self) and aged identities (Fitzgerald 1986; Kastenbaum 1993; Kaufman 1986; Laws 1995; Lucas 2002, 2004; McHugh 2000, McHugh and Larson-Keagy 2005). Successful aging involves the maintenance in later life of 'activity patterns and values typical of middle age' (Atchley 1980: 239), including levels of physical activity, the preservation of family and social roles, and last but by no means least, consumption levels. The related idea of the ageless self emphasizes continuity, activity, and the absence of the aged body in descriptions of old age (Kaufman 1986). Taken together these two ideologies suggest that old age (and decline) can be kept at bay by prolonging middle age and more importantly, old age can only, and most successfully, be kept at bay by residing in a retirement community.

The aged identities constructed and emplaced in retirement communities are likewise anti-ageist and very different from those constructed and reproduced by almshouses and private homes for the aged in the past and by nursing and care homes today (Laws 1993). Rather than constructing old age as a time of physical and cognitive decline, dependence, and social isolation, retirement communities construct old age as a time of opportunity to enjoy a plethora of leisure activities (Ekerdt 1986; Katz 2000; Laws 1993, 1995; McHugh 2000). Retirement community residents, **active retirees**, are depicted as comparatively wealthy and healthy, physically active, busy individuals enjoying various social and recreational pursuits. Activity, be it physical movement, the pursuit of hobbies, or participation in social activities is considered the key to individual adjustment to physical aging (Cavan et al. 1949; Ekerdt 1986; Havighurst and Albrecht 1953; Moody 1988) and is used to manage everyday life for active retirees in much the same way work regulates life for the working population (Ekerdt 1986; Katz 2000; Moody 1988; Neugarten et al. 1961). Equally importantly, active retirees are freed from other responsibilities, particularly caring for children, homemaking, and home maintenance (Katz 2000). Active retirees have no responsibilities beyond enjoying life.

Retirement communities can also be understood as exclusionary spaces where societal ideas around political beliefs, race, class, and gender are played out. Kastenbaum (1993) for example shows that residents of Sun City share specific racist views that were instrumental in defeating the adoption of the Martin Luther King, Jr. holiday in Arizona. Widows/widowers, seasonal, and frail residents are also socially marginalized and excluded in retirement communities and other age-segregated housing developments (Longino and Lipman 1982; Perkinson and Rockemann 1996; Sullivan 1985; van den Hoonaard 1984, 1994, 2002). Seasonal residents or snowbirds are likewise excluded and socially marginalized (Longino and Lipman 1982; Sullivan 1985; van den Hoonaard 1984, 1994, 2002). More recently, an emerging literature on older gay, lesbian, bisexual, and transgendered seniors shows that the socially and demographically homogeneous spaces of age-segregated housing in general and retirement communities in particular ostracize homosexuals (Hervey 2005; Johnson et al. 2005; Lee 1989; Lucco 1987).

Retirement Communities: Selling the Good Life in the Third Age

Langmeyer (1993) argues that all advertisements document reality and reflect society's attitudes about social groups and individuals, including the elderly. Others (Belk and Pollay 1985; Gerbner et al. 1986; Leiss et al. 1990; O'Guinn et al. 1989) argue that advertisements do not simply passively reflect society's attitudes but create, change, and reinforce individual and societal perceptions about social groups, the ownership/use of objects, and particular behaviours. This view of advertising suggests that the negative and positive images employed in advertisements not only reflect society's attitudes towards aging and the elderly but also aid in the production and popularization of new images of and ideas about later life.

The (re)conceptualization of later life and retirement as times of self-fulfillment and the promotion of successful aging and the ageless self has led to a veritable explosion in the use of positive images to describe later life. Cole (1992) argues that an important force behind the promotion of these new ideologies (successful aging and the ageless self) and the associated use of positive imagery to describe old age is the development of the aging 'industry'. Not unlike other cultural industries, the development of the aging industry is predicated on the recognition of the mature population as a substantial and relatively under-exploited market segment. The use of positive images in advertisements selling goods and services to the older population suggests that successful aging can be achieved and individuals may come to embody the ageless self through the purchase of the goods and services on offer (Featherstone and Hepworth 1995; Long 1998; Moschis et al. 1997; Sawchuck 1995).

Retirement communities are very much part of the aging industry and are unequivocally implicated in the promotion of successful aging and the ageless self. Like shopping malls, coffee shops, art galleries, and designer clothing boutiques, retirement communities are highly symbolic 'spaces of consumption' (Zukin 1988: 825) that reflect new patterns of leisure and **cultural production** that disguise the ultimate purpose of generating profit through the consumption of space and the creation/promotion of distinct identities and lifestyles in later life. We further suggest that the same images can tell us a great deal about family life and family roles in later life.

The Changing Nature of Family Structure and Roles Across the Life Course

An appropriate starting point for our discussion of how retirement community promotional materials (re)envision and (re)present 'family' in later life is the ongoing debate about what type of social grouping constitutes a legitimate 'family'. These debates began with the identification and definition of the 'nuclear family' by Murdock (1949). Defined as consisting of a first-married man and woman maintaining a heterosexual relationship and sharing a residence with their biological child(ren), the nuclear family Murdock (1949) argued was natural, universal, and desirable. Buoyed initially by its majority status, popularized by television shows like *Leave it to Beaver*, and supported by a plethora of laws, public policies, and institutional practices, the idea of the nuclear or traditional family rapidly gained hegemonic status during the 1950s (Coontz 1997; Dowd 1997; Erera 2002; Hill 1995; Stacey 1990). As a result, the nuclear family became (and remains) the idealized family form and the standard against which all other family structures are compared (Allen and Baber 1992; Erera 2002; Glick 1989).

During the 1960s and 1970s economic, demographic (falling mortality and fertility, increased life expectancy), and social changes, for example increased rates of divorce, single-parenthood, and cohabitation produced dramatic changes in the composition of families (DeVanzo and Rahman 1993; Fields 2003; McLanahan and Casper 1995; Teachman et al. 2000). Most notably the number of non-family and small households increased from 1970 to 2000 while the percentage of individuals living in family households, particularly married-couple households with children (the nuclear family) fell (Casper and Bianchi 2002; Fields 2003; Teachman et al. 2000). Over the same 30-year period the number of unmarried-partner or cohabiting households and one-parent households also increased in the United States (Casper and Bianchi 2002; Fields 2003; Teachman et al. 2000). Among one-parent households, the numbers of female-headed households in particular have increased dramatically since 1970; reaching 18 per cent of American family households in 1980 (Casper and Bianchi 2002; Teachman et al. 2000). Likewise, the proportion of unmarried men and women cohabiting increased from 3 to 5 per cent respectively in 1978 to 9 and 12 per cent in 1998 (Casper and Cohen 2000). Though estimates are difficult to come by, the number and diversity of families headed by lesbians or gays has also increased in recent years (Flaks et al. 1995; Pies 1989). As a result of these changes the diversity and pluralism of family structure has increased markedly since the 1960s producing in the process what Fineman (1995) and Coontz (1992, 1997) respectively refer to as the 'post-traditional' or 'non-traditional' family.

What about 'family' in later life? Have the same demographic and social changes that produced the post-traditional family also affected the structure of families made up of older individuals and produced, in the process, increased compositional diversity? Studies indicate that family structure in the second half of life has indeed undergone considerable change in recent decades and that as a result families in later life are as diverse as younger families (Allen et al. 1999; Bedford and Blieszner 1997; Fields 2003; Johnson and Barer 1997; Peek et al. 2004). Among the most significant changes is a pronounced decline in the incidence of co-residence between generations (Ruggles 1994, 2007), a steady increase in the number of older adults, particularly women, living in single-person households, and a growth in the number and variety of co-resident grandparent families (Bryson and Casper 1997, 1999; Fields 2003; Johnson and Barer 1997). Four-fifths of co-resident grandparent families are grandparent maintained, meaning that they maintain the households in which their children and grandchildren reside (Bryson and Casper 1999). Of these households, the largest proportion is headed by either grandparents or grandmothers, with a much smaller but growing percentage headed by grandfathers.

Allen et al. (1999) add colour to the usual statistical analysis of family structure by using in-depth interviews to examine compositional diversity not just among older adults but among their children as well. Allen et al. (1999) found that nearly half of the older adults they interviewed reported pluralistic or diverse marriage or parenting patterns that transcend those of the traditional or nuclear family. These older adults described family structures characterized by step-parentage, remarriage, divorce, cohabitation, and/non-marital childbearing (Allen 1989; Allen et al. 1999). Other studies report frequent **kin conversion**, adoption after teen pregnancies and parental surrogacy (Allen 1989; Johnson and Barer 1997; Stacey 1990; Troll 1996).

Significant changes have also occurred in the type of relationships that characterize families in later life. Johnson and Barer (1997) and Allen et al. (1999) found that family in later life is increasingly characterized by non-traditional kin relationships including **fictive kin** and **upgraded kin** rather than traditional blood relationships between parent and child. Fictive and upgraded kin relationships involve non-family members providing emotional and material support typically supplied by biological children. Allen et al. (1999) found that such relationships were commonplace among older families and were manifest in a variety of forms.

Changes in the nature of grandparenting have also affected the structure of familial roles in the second half of life. The nuclear family generally excluded and devalued grandparents and grandparents' roles in the family. Over time, grandparents became companions to their grandchildren with neither authority nor responsibility (Apple 1956; Kahana and Kahana 1971; Neugarten and Weinstein 1964; Szinovacz and DeVinney 1999a, b). In recent decades, however, grandparents have become (re)incorporated into the family as sources of support or as caregivers (Goodman 2003; Goodman and Silverstein 2002; Hayslip and Kaminski 2005a, b; Smith and Beltran 2003; Szinovacz and DeVinney 1999a, b). Variously referred to as 'surrogate parents', 'co-parents' or 'custodial grandparents', Cox (2000) noted a 44 per cent increase in custodial grandparenting between 1980 and 1990, while Smith and Beltran (2003) found that the incidence of custodial grandparenting increased 30 per cent between 1990 and 2000 in the United States.

Retirement Communities: (Re)constructing Family

The definition of family, the idea of what is required for family life, and the function of the family in later years (particularly with the absolution of child-rearing responsibilities) is of long-standing interest to those who study retirement communities. We suggest here that retirement communities result from the cultivation of identity, a desire for security, and (re)imaginings and (re)definitions of family life that have occurred in recent decades in North America. Unlike Packard (1972) and Laws (1993) who suggest that age-segregated retirement communities reflect the rootlessness of contemporary society and serve the interest of private developers who can capitalize on the desire of the elderly to separate themselves from younger people, here we look at these communities as cultural landscapes, spaces where the notion of family is lifted from its usual moorings and (re)conceived with greater elasticity.

Just as space, place, and the built environment have contributed to the (re)construction and emplacement of gendered (Domosh and Seager 2001; Dyck 1989, 1990; England 1993; Gilbert 1998; Hanson and Johnston 1985; Hanson and Pratt 1988, 1990, 1991) and aged identities (Laws 1993, 1994, 1995), space, place, and the built environment also (re)construct and emplace ideas about and definitions of family. In particular, the suburbs quickly became associated with a very specific definition of family (Fava 1980; Hartley 1997; Jeffery 1972; Miller 1983). Beginning in the 1800s, the suburbs were championed as the most appropriate environment for what would become known as the nuclear family (Markensen 1980; Marsh 1990; Palen 1995; Spigel 1992). The nuclear family was unequivocally defined as a cohabiting married man and woman and their children. Grandparents and other relatives were excluded and assigned to other spaces (Laws 1993).

Age-segregated environments in general and retirement communities in particular also help (re)define family. These (re)definitions, though rarely articulated as such, are fundamentally exclusionary. As Lucas (2002) shows, the textual images of family used in retirement community marketing materials (re)produce and (re) construct accepted notions of family that (re) produce the exclusion and marginalization of grandparents. Grandparents remain visitors in their grandchildren's lives who provide entertainment, companionship, and limited care (Apple 1956; Kahana and Kahana 1971; Neugarten and Weinstein 1964; Szinovacz and DeVinney 1999a, b). As Fitzgerald (1986) shows in her study of a Florida retirement community, the separation of grandparents and grandchildren is mutual. Retirement community residents who are dependent on their grandchildren (or children) are denounced and cast as weak; their dependency is described as a disease: 'gramma-itis' (243). Physical separation/ distance implies independence.

The exclusion of grandparents from the nuclear family is accepted and well documented. What is new about the (re)presentations of family produced by retirement communities is that family now excludes children. The textual descriptions of family analyzed by Lucas (2002) describe heterosexual couples unencumbered by biological kin, grandchildren, or children. Family as defined by the retirement community advertisements and emplaced in the space of retirement communities has shrunk from 'extended' to 'nuclear' to its present post-traditional form of two individuals dedicated to their mutual support, mutual growth, and self-actualization in partnership *sans* parenting. As a result, the obligations, resources, and legitimacy of the traditional family are extended in the post-traditional family to an even smaller group.

Methods and Analysis

Data

Marketing literature provided to potential retirement community residents is the focus of the study discussed here. This marketing literature is produced for the sole purpose of selling the active retiree lifestyle to older individuals (a specific market segment) and therefore meets the basic definition of an advertisement. However, unlike traditional print advertisements the marketing material analyzed in this study includes packages containing multiple inserts and booklets. These packages provide potential residents with information on dwelling specifications and floor plans, on-site leisure and recreational facilities, community social activities, building layouts, price lists, and construction plans. As with traditional print advertisements, the retirement community marketing literature contains a mixture of text and images. The focus of this study is the images contained in the marketing brochures.

Marketing literature of 13 retirement communities is used to examine the images employed to describe, define, and depict family. The sample consists of nine for-profit communities and four not-for-profit communities. For-profit communities are a market-driven housing option for older individuals. All the for-profit communities included in this study are independent living facilities; residents live in their own dwellings (single-family, townhomes or apartments) and receive no personal or nursing care. Built and managed by not-for-profit corporations and faith-based groups, the not-for-profit communities provide affordable housing and a limited range of leisure facilities.

The retirement communities studied offer a variety of housing options, including townhomes, apartments, and single-family dwellings in a variety of city and suburban locations within the urban regions of the northeastern United States and southern Ontario, Canada. Because of their location, all the retirement communities are close to the amenities offered by a number of major urban places (including Philadelphia and its suburbs, Toronto, and Kitchener-Waterloo) and the surrounding rural areas. The social and recreational amenities provided vary from indoor activities based in a clubhouse (e.g., card evenings and dances) to a full range of adult-learning classes as well as indoor and outdoor recreational activities. Accordingly, the facilities available in an individual retirement community can be restricted to a clubhouse or include a swimming pool, fully equipped gym, craft and wood-working rooms, tennis courts, golf courses, drawing studios, and classrooms.

Content Analysis

In their seminal work, *Reading National Geographic*, Lutz and Collins (1993) employ content analysis to analyze over 500 photographs published in National Geographic magazine. Lutz and Collins (1993) suggest that while the act of counting and developing codes to detect frequencies does not substitute for a critical qualitative analysis, it uncovers patterns that are 'too subtle to be visible on casual inspection' (xi). They note that a significant proportion of the ideas and text-based knowledge that we employ to make sense of the world is from visual images contained in magazines and popular media. This suggests that there is much to learn about the ways in which photographs, in conjunction with other kinds of images and texts, shape distinctive images of place, time, identity, and **positionality**.

Traditionally, in geographical research, photographs are used primarily to shed light on the past or enable viewers to visualize the world beyond their doorstep. This chapter pursues a different objective and examines the images embedded in the advertising brochures/promotional materials produced by retirement communities. While typically these images are analyzed with regard to their effectiveness in increasing sales, our goal here is to contextualize the photographs in discourses around family and aging in order to describe a deeper, perhaps more authentic world of social relations and family life in retirement communities. As such, photographs are primary documents, or moments in a particular set of ideologies, practices, and discourses surrounding aging and family which must be subjected to a process of interpretation and decoding.

Defined by Berelson (1952: 74) as 'a research technique for the objective, systematic, and quantitative description of manifest content of communications', content analysis is used to quantify the occurrence of certain words, concepts, themes, phrases, characters, or sentences within texts. Texts can include any primary documents: essays, interviews, newspaper headlines and articles, historical documents, speeches, advertisements and a variety of images be they print or video. The first step in carrying out a content analysis on text or images is to code the text, or break it down into a variety of categories or themes. The text can be coded in order to establish the existence and frequency of concepts or themes in a text (conceptual content analysis), or to identify the relationships among concepts in a text (Krippendorf 1980; Riffe et al. 1998). The second step is to make inferences about the messages contained in the text based on either the frequency of themes or relationships between themes.

In the present context, we employ conceptual or descriptive content analysis to examine how family is (re)visioned, (re)thought, and (re) presented in photographs used in retirement community advertisements. We examined 92 photographs containing people from the brochures of 13 retirement communities. The largest percentage of the photographs, 77 per cent or 72 photographs, are in brochures produced by for-profit communities. The remaining 20 images are in brochures produced by not-for-profit communities. Only photographs that contained one image were included in the analysis. Composite images, images obscured by text, artist impressions of buildings etc., and images containing no people are not included in the analysis.

The photographs included in the analysis we coded using six descriptive themes: setting, number of participants in image, race, gender, activity, and family role. Five of these themes, setting, number of participants in image, race, gender, and activity (sitting, standing, learning, reading, recreational, spending time with family members, social, and consumption) allow us to explore and understand the ideological milieu that created each image. Family role, in contrast, describes both the composition of 'family' and the nature of 'family' relationships or roles as constructed and reconstructed in retirement community marketing brochures.

In order to examine how family is conceptualized in retirement community brochures we first identified relevant dyadic units and then specific family relationships/roles. A **dyad** is usually defined as a social group comprised of two individuals who are related or connected. The definitions of both 'dyad' and the types of connections between members are not (pre)determined or fixed. Because of this flexibility, Trost (1988) and Levin and Trost (1992) suggest that dyadic units provide a practical approach to defining family that can easily reflect the impact of all the processes that affect how family is defined.

Each dyadic unit is then described with reference to the type of familial relationship/role depicted. The first type of familial relationship/role identified for each dyadic unit is 'caring'. 'Caring' is defined as any expression of concern, familiarity, or non-sexual affection between members of a dyad and/or a third party other than medical caring, for example an adult child administering medication to a parent. A father and son embracing before or after playing a game of tennis or a grandparent playing with a grandchild are examples of caring relationships/roles. The decision to exclude medical caring from the definition of this role was guided by the simple fact that there were no images depicting this activity. The second type of familial relationship/role identified is 'social' and describes each dyad with reference to other individuals or dyads. Examples include talking, engaging in a physical activity together, sharing a meal, and walking. 'Alone' is the third role identified. As the descriptor suggests, this relationship describes situations in which a dyad is shown in isolation.

In order to meet the requirements of objectivity suggested by Berelson (1952) and systematization stipulated by Kassarjian (1977) that consistent coding rules be applied and that the results be generalizable and replicable, each image or photograph was coded in two stages. In the first stage, two individuals coded each image independently. In the second stage, the results of the individual content analyses were compared and combined. Differences were resolved by (re)coding the images. It should be noted, that both coders agreed on the themes represented in the photographs.

Results and Discussion

The images embedded in the marketing brochures of retirement communities, whether for-profit or not-for-profit, reassure the audience of desirous consumers that life in these communities is busy and comfortable. Retirement is portrayed as a time of good health, physical and emotional independence, leisure, a time free from the traditional strictures of family life and the responsibilities of home ownership (the idea of maintenance-free living). Residents are socially engaged and surrounded by friends. Consequently, 82 per cent of images contained in the advertisements analyzed show residents in groups. Only in 18 per cent of images are of residents shown alone.

Rather than being organized around paid work, retirement community life is organized around activity: physical movement, the pursuit of everyday interests, and social participation. Days are spent enjoying a plethora of learning, recreational, and social activities with partners and friends. Using the schema for describing activity discussed by Katz (2000), nearly half of the images in the brochures show people engaged in some form of recreational pursuit or everyday interests and 18 per cent show residents with family members or engaged in social activities. Overwhelmingly, photographs picture residents outside (66 per cent) (Table 10.1). Of those inside, most are engaged in learning activities (e.g., computer classes, painting), reading, exercise classes, or social activities (e.g., talking). Those outside are shown biking, hiking, playing golf or tennis, swimming, sitting on the beach, or playing croquet. Photographs of residents posed in front of the camera are rare.

The homogeneity of retirement community populations with respect to race (white), sexuality

Table 10.1 Advertising Brochures: Description of Communities, Residents, and Activity

	All Communities		For-Profit		Not-for-Profit	
	Number	Per cent	Number	Per cent	Number	Per cent
Setting						
Inside	31	34	21	29	10	50
Outside	61	66	51	71	10	50
Number of individuals in image						
One individual	17	18	12	17	5	25
Two individuals	42	46	36	50	6	30
Group 3 or more	33	36	24	33	9	45
Race						
One race	84	91	67	93	17	85
Multiple	8	9	5	7	3	15
Gender						
Male	12	13	8	11	4	20
Female	18	20	15	21	3	15
Mixed	62	67	49	68	13	65
Activity						
Sitting (posed)	3	3	1	1	2	10
Standing	3	3	3	4	0	0
Reading	3	3	2	3	1	5
Learning	2	2	2	3	0	0
Recreational	45	49	37	51	8	40
Family	18	20	14	19	4	20
Social	15	16	12	17	3	15
Consumption	3	3	1	1	2	10

(heterosexual) and, indirectly, marital status noted elsewhere (Fitzgerald 1986; Kastenbaum 1993; Laws 1995; McHugh and Larson-Keagy 2005) is replicated in the images analyzed (Table 10.2). Less than 1 per cent of the photographs analyzed show multiple races. The remaining images contain white individuals. The racial homogeneity of retirement community residents clearly does not reflect the racial and ethnic heterogeneity of the US and Canadian populations. References to heterosexuality and marriage are numerous. Over half of all the images analyzed contain heterosexual couples (Table 10.2). No images contain same-sex couples. As a result, the analysis and discussion of family undertaken here should be understood as an evaluation of white older families and the specific ideologies that inform their structure and produce the images analyzed.

What about family? How is family represented? What images of family in later life are produced by and emplaced in the spaces of retirement communities? Family as depicted and described in retirement community promotional brochures is rigid and homogeneous. The 'retirement community family' in later life is defined as consisting of persons related by blood, (first) marriage, and traditional **kinship** ties. The use of these images (re)affirms an understanding of family based on cohabitation, heterosexuality, and marriage. References to heterosexuality, and marriage are palpable: wedding rings, physical proximity, and affectionate touching. In several of these images, couples are shown gazing into each other's eyes, holding hands, touching, and embracing. The clear implication from these images is that each couple is married, even if their wedding rings are not visible. Indirect references to heterosexuality and marriage include four images of single adult females who are married but alone in the photograph (Table 10.2). In each of these images, the individual's wedding ring is prominently displayed. Cohabitation is suggested by the predominance of domestic settings (garden patios, kitchens, living rooms) and images of couples engaged in domestic tasks, for example cooking or gardening together.

Thus, two dyads dominate the images examined: the heterosexual spousal dyad consisting of a cohabiting heterosexual married man and woman (Photo 10.2) and the single adult (male or female) dyad. Over half of the images used in the brochures analyzed contain the heterosexual spousal dyad consisting of a married man and woman (Table

Photo: iStockphoto/Stephan Zabel

Photo 10.1 Active retirees on the move

Table 10.2 Family Type, Family Role, and Relationship

	All communities (n=92)		For Profit (n=92)		Not for Profit (n=92)	
	Number	Per cent	Number	Per cent	Number	Per cent
Social relationship	36	39	27	29	9	10
Hetero couple with adult children	1	1	1	1	0	0
Hetero couple with other hetero couple	11	12	8	11	3	15
Hetero couple with multi-generational family	2	2	1	1	1	5
Hetero couple with single adult female	2	2	0	0	2	10
Hetero couple with single adult male	1	1	0	0	1	5
Hetero couple with group	1	1	1	1	0	0
Unrelated adults	18	20	16	22	2	10
Caring relationship	12	13	8	9	4	4
Single adult male caring for children/grandchildren	7	8	4	6	3	15
Single adult female caring for children/grandchildren	1	1	0	0	1	5
Hetero couple caring grand-children	4	4	4	6	0	0
Alone relationship	44	48	37	40	7	8
Single adult female married but alone in photo	4	4	4	6	0	0
Single adult female	6	7	4	6	2	10
Single adult male	3	3	2	3	1	5
Hetero couple alone	29	32	25	35	4	20
No adults	2	2	2	3	0	0

10.2). The heterosexual spousal dyad appears more frequently in images produced by for-profit communities (40 per cent) than in images produced by not-for-profit communities (12 per cent). Consisting of a single adult male or female (Photo 10.3), the single adult dyad appears in 42 per cent of images. Images of the single adult dyad appear with greater frequency in the marketing brochures produced by for-profit communities (33 per cent) than they do in brochures created by not-for-profit communities (10 per cent of all images).

References to heterosexuality and marriage directly suggest child-bearing within the nuclear family and the existence of children and grandchildren. However, adult children and grandchildren appear in only five images, and only two photographs show a multi-generational family (Table 10.2). The comparative lack of images containing children or grandchildren indicates that both are generally excluded from direct inclusion in the family as envisioned by the marketers of retirement communities.

The association of both the heterosexual spousal dyad with the domestic realm of the home on the one hand reproduces the conceptualization of the home and the nuclear family as the only valid sites for child-bearing and child-rearing. At the same time, rather than being concerned with childbearing and childrearing, family life in retirement communities is promoted as being focused on self-fulfillment and activity, especially organized activities based in the retirement community. Extending discussions on the importance and role of activity in later life, particularly for active retirees/retirement community residents, and considering the large number of photographs that show residents engaged in various activities, we suggest that the heterosexual spousal dyad and home are (re)envisioned as the basic organizing units around which such activities are arranged.

The prominence of the single adult dyad in the representations of family reflects the reality of old age: that many older people, particularly women, live in single-person households. Single adult males appear in only 3 per cent of images, whereas single adult females appear in 7 per cent of images (Table 10.2). The single adult dyad appears in more images describing not-for-profit retirement communities than for-profit communities. The images of the single adult dyad again (re)affirm an understanding of family based on heterosexuality, marriage, and ties of blood. However, the

Photo: iStockphoto/Sheldon Kralstein

Photo 10.2 A retired couple engage in 'active aging' on the golf course

Photo: iStockphoto/Chris Schmidt

Photo 10.3 Gardening is a common activity in retirement community brochure images of single older adults

references made are by definition indirect because the individuals are alone. For example, marriage and heterosexuality are implied by images of solitary women wearing clearly visible wedding rings and the presence of children and grandchildren. Even if the spouse is absent, the implication is that the single adult dyad was once part of a heterosexual spousal dyad.

Activity is again central to the definition of the single adult dyad. However, the type of activity is different. Individuals are shown engaging in sedentary activities including pottery classes, flower gardening, and quilting (Photo 10.3). Social participation mostly takes the form of spending time with family rather than participating in organized recreational activities that require dexterity and energy. For example, individuals are shown reading to their grandchildren, or teaching them to walk.

Based on the previous discussions, a number of somewhat contradictory conclusions can be drawn about the (re)imaginings of family (re) presented by retirement community advertisements. Overwhelmingly, the dominant image of family (re)imagined and (re)presented is that of the traditional or nuclear family. As such, the visions of family emplaced in and commodified by retirement communities consist of a co-resident, generally first-married man and woman with dependent child(ren). Grandchildren and other relatives are excluded. This is in keeping with the original conceptualizations of the nuclear family and the evolution of the family noted earlier from 'extended' to 'traditional' and the long-term decline of the multi-generational family consisting of co-resident elderly members (Casper and Bianchi 2002; Ruggles 1994). Even when a single adult dyad is shown, the clear implication is that they were once part of a nuclear family.

The advertisements analyzed also (re)present notions about family roles and intra-family interactions that conform to those associated with the traditional family. The nuclear family was and remains the only context within which legitimate sexual relationships, child-bearing, and child-rearing could take place. Direct evidence of sexual relationships abounds and includes couples locked in passionate embraces and gazing into each other's eyes. The existence of both children and grandchildren (though both are physically distant) and prevailing societal norms during the 1950s and 1960s suggest that in the past both family dyads' sexual relationships, child-bearing, and child-rearing took place in the context of the traditional family. Likewise, grandparents and the roles they fulfill are commensurate with traditional notions of grand-parenting associated with the nuclear family. Grandparents (and parents), now retirement community residents, are playmates who lack authority over and responsibility for their grandchildren (and children) (Cherlin and Furstenberg 1985, 1986; Neugarten and Weinstein 1964).

As well as (re)presenting traditional notions of the nuclear family, the (re)imaginings of family in retirement community advertisements also offer modified images of the nuclear family. Unlike the version of the nuclear family emplaced in the suburbs where children are present, the image of family constructed and marketed by retirement communities conceptualize both dyads as unencumbered by children and concomitantly by the responsibilities that accompany child-rearing and family. The emergence of this type of family is the result of a number of demographic changes (increased life expectancy, delayed and decreased fertility, and delayed mortality), economic factors, and social trends (including the (re)conceptualization of retirement as a time of self-fulfillment, independence, and leisure) that have fuelled the growth in the number of individuals/couples living in households containing no dependent children (Casper and Bianchi 2002; DaVanzo and Rahman 1993; McLanahan and Casper 1995; Ruggles 2007).

In addition to providing information about how family is defined within the specific space of retirement communities, the images examined also provide information about the nature of family roles in later life. According to Johnson (2000), and Bedford and Blieszner (1997) describing the types of roles and relationships among family members may provide a useful means of understanding the family in later life and by extension, the types of familial roles and relationships considered marketable. Table 10.2 shows that the frequency of each of the family roles or relationships previously identified. Of the three roles identified, 'alone' appears more frequently in the retirement community brochures than any other. Images showing both dyads in settings where they are alone appear most frequently in not-for-profit communities. Images showing the second type of familial relationship/role, 'social', are the next most common (Table 10.2). Again, these images are most common in for-profit communities. Caring relationships and roles appear in only 4 per cent of images of not-for-profit communities, 9 per cent of images of for-profit communities, and 13 per cent of images of all communities.

Defined as situations in which either dyad is shown in isolation, (re)presentations of the role 'alone' demonstrate both consistency and variability. An examination of the images containing the heterosexual spousal dyad alone shows most depict the expression of heterosexual love and are sensual in nature. Couples are shown embracing, holding hands, touching, and gazing into each other's eyes. These displays of affection are situated in various indoor and outdoor environments and are connected with a host of activities, specifically eating, walking, cooking, and participating in recreational activities. In for-profit communities the displays of affection are more sensual and passionate than in not-for-profit communities. In not-for-profit communities, in contrast, couples expressing heterosexual love do so with restraint. The emphasis appears to be on tenderness rather than passion.

The second most common role shown is 'social' (Photo 10.1). Table 10.2 shows that the highest percentage of images showing social roles appears in for-profit community advertisements (22 compared to 10 for not-for-profit communities). Unrelated adult dyads and two cohabiting heterosexual spousal dyads (two couples) are most often shown in social roles. The types of social roles represented emphasize togetherness, friendliness, as well as physical and emotional autonomy. Being social, spending time with other people, participating in group activities, is a fundamental part of retirement community life (Biggs et al. 2000; Ekerdt 1986; Katz 2000). The combination of spending time with others and being seen as part of a smaller family-type unit is described by Biggs et al., (2000) in their study of a British retirement community as 'sociability with autonomy' (658). The idea of sociability with autonomy is important. The idea suggests that, while there are boundaries to social interaction, couples and individuals are still part of the greater community. The communication of this message is particularly important in for-profit communities because the promotion of social interaction and participation in group activities is central to their marketing strategies. Furthermore, such images play a crucial role in positioning what might be a problematic status in a retirement community, being a single (perhaps previously married but now widowed) person, within the greater social milieu of the community. The message is clear: being a single individual is not a problem.

Engaging in social roles, talking, eating, being with friends, or engaging in recreational activities, rather than working or caregiving are identified as important family roles for both the heterosexual spousal and the single adult dyads. The types of social roles each dyad is shown engaging in is consistent across both types of communities. Moreover, in both types of communities social roles are often shown in combination with recreational or leisure activities; for example couples are shown chatting over a garden wall with gardening implements in hand. As mentioned previously, such an association suggests that both are central to the retirement community lifestyle and family life in retirement communities. However, differences exist in the type of social roles shown in combination with leisure/recreational activities. In images used by for-profit communities social roles are connected to more strenuous physical activities than in images used by not-for-profit communities. Two examples are illustrative of the difference: a group of three heterosexual spousal dyads living in a for-profit community chatting, smiling, and embracing while holding skis compared to a group of two cohabiting heterosexual spousal dyads in not-for-profit communities chatting and playing cards.

The least common role shown in images embedded in the promotional literature is 'caring'. Defined as any expression of concern, familiarity, support, or non-sexual affection between members of a dyad and/or a third party, images depicting caring roles account for only 13 per cent of all images (Table 10.2). Caring roles are more common in images of life in for-profit communities where they appear in 9 per cent of photographs (Table 10.2). In not-for-profit communities caring roles appear in only 4 per cent of photographs (Table 10.2). Biggs et al. (2000) suggest that references to caring are incongruous with autonomy and independence, conceptions central to the active retiree-aged identity and ageless self and attempts to popularize and market both identities and ideologies. Furthermore, the lack of independence or the expression of dependence on a family member and the (assumed) need for care is often described in negative terms. Fitzgerald (1986: 242–43), for example, points out that residents of the retirement community she studied use the language of disease, 'gramma-itis', to describe a situation where a woman spent time with her daughter outside the community. Consequently, retirement community advertisements generally make scant reference to such familial roles.

The type of caring roles pictured is instructive. In all the images analyzed only one type of caring role is depicted: grandparent/s with their children/grandchild. In these images, grandparents are typically shown sharing a picnic, meal, or walk. Grandparents are also shown teaching their grandchildren to read, walk, or fish. Such characterizations of the grandparent-grandchild relationship replicate the roles assigned in the original conceptualizations of the nuclear family (Cherlin and Furstenberg 1985, 1986; Neugarten and Weinstein 1964). Using the grandparent role classification scheme developed by Cherlin and Furstenberg (1985, 1986) the images of grandparenting suggest active-supportive grandparenting; a situation in which grandparents babysit, play, and entertain their grandchildren but demonstrate little parent-like behaviour.

The (re)visioning and (re)presentation of family embedded within the retirement community brochures studied reproduce ideas and understandings of family based on the nuclear or traditional family. The sheer numbers of direct and indirect, manifest and subtle references to the nuclear family indicate that within the space of retirement communities the nuclear family retains its hegemonic and ideological status. As such the 'retirement community family' is the only acceptable family style and more than that, the ideal against which all other family forms are compared and ultimately found wanting. Given the age of retirement community residents (the youngest are typically around 55 years), the nuclear family and equally importantly the portrayals of that family style in the popular media are fundamental elements of their world. Moreover, the norms and values, the economic, demographic, and social processes that produced the nuclear family moulded their experiences as children, individuals, and parents. Therefore, (re)visioning family as the nuclear or traditional family in retirement community advertisements makes sense. The provided images of family return the audience to a simpler time of clear gender roles, heterosexual sex, universally understood and followed child-bearing and rearing practices, and lifelong marriages ended by death rather than divorce.

The (re)visioning and (re)presenting of the family presented in retirement community advertisements no matter how reassuring denies the pluralism and diversity of contemporary family structure. The form of family presented is particularly homogeneous with regard to race/ethnicity and likewise rejects structural pluralism among older families. Beginning with the latter, the images presented indicate that older families are essentially modified traditional families. Few images portray multi-generational families. No images portray grandparent-maintained or parent-maintained families with co-resident grandparents and children. Grandparenting styles or roles reflect the traditional conceptualization of grandparents' roles as playmates and babysitters with no authority. Though many images contain the single adult dyad and such individuals probably live alone, it is clear that these individuals were once part of a cohabiting heterosexual dyad. If the limited but growing research on older family diversity is to be believed, then the portrayals of family in retirement community brochures are simply too homogeneous.

The recognition of increased society-wide pluralism in family structure during the 1960s led to the increased acknowledgment and study of racial differences in family structure and the roles assigned to family members. Studies of changing patterns of African American and Latino families have noted significant differences between African

American, Latino, and white women in the age at which sexual intercourse first takes place (Duany and Pittman 1990; Staples 1981), the use of contraception (Zelnick and Kantner 1977), premarital births (Aquirre, Molina, and Parra 1995; Duany and Pittman 1990; US Census Bureau 1983), age at first marriage (Haines 1996; Koball 1998) and the likelihood of divorce (McLanahan and Casper 1995; US Census Bureau 1983). Along with institutionalized racism, poverty, and high levels of unemployment these differences are theorized to have contributed to an increased number of female-headed households containing children, multi-generational families, co-resident grandparent-maintained households, and custodial or surrogate grandparenting in the African American and Latino populations (Casper and Bianchi 2002; Collier and Williams 1982; Goodman 2003; Goodman and Silverstein 2002; Hayslip and Kaminski 2005a, b; Ruggles 1994, 2007; Smith and Beltran 2003; Sottomayer 1989; Staples 1987; Staples and Mirande 1980; Szinovacz and DeVinney 1999a, b). For Latino families, place of origin and problems adjusting to life in the US may, in addition to the sources of familial change listed above, cause significant pluralism and variation in family structure compared to both the white and African American populations (Gil at al. 1994; Szapocznik and Hernandez 1988; Szapocznik and Kurtines 1988; Vega and Amaro 1994; Wallace 1986).

Among the most significant differences in family structure identified between white, Latino, and African American families are the number of extended/multi-generational families with co-resident older family members, grandparent-maintained multi-generational families, and single-person households, particularly female-headed single-person households. With reference to racial differences in the number of multi-generational families, Casper and Bianchi (2002) found that African Americans were more likely in 2000 to live in multi-generational families than both Latinos and whites. Goldscheider and Bures (2003) confirm changes and differences in the incidence of multi-generational families with co-resident elderly members between the white and African American populations identified by among others, Angel and Tienda (1982), Beck and Beck (1989) and Ruggles (1994, 2004). Goldscheider and Bures (2003) show that between 1900 and 1990, while the occurrence of multi-generational families containing co-resident elderly members has fallen for both African American and white families, the rate of decline was much faster for whites. Among Latino families, approximately 20 per cent of households were multi-generational in 2000 (Casper and Bianchi 2002).

The composition of multi-generational families also varies considerably by race. Grandparent-maintained households, households where the grandparents take in their adult children and minor grandchildren are the most common multi-generational family particularly for African Americans and Latinos (Casper and Bianchi 2002). Parent-maintained households, where an elderly individual resides in a household maintained by his/her children are more common among whites (Casper and Bianchi 2002). The incidence of single mothers living in grandparent-maintained households is greatest for Latinos and African Americans, particularly when the father is absent (Casper and Bianchi 2002; Fields 2003; Goldscheider and Bures 2003).

Both the incidence of single-parenthood/female single-parenting and racial differences in both family types has increased substantially since the 1960s (Bianchi 1995; Fields 2003; Teachman et al. 2000). In 1992, 57 per cent of African American families were single-parent families, compared to 30 per cent of Latino families and 20 per cent of white families (Norton and Miller 1992). The percentage of single-parent female-headed households has been highest among the Latino and

African American populations since the 1960s (Bianchi 1995; Ortiz 1995). Among African Americans and Latinos the percentages of single-parent female-headed households rose from 28 and 15 per cent respectively in 1970 to 44 and 23 per cent respectively in 1990 (Ortiz 1995). For the same two periods, the incidence of single-parent female-headed households among the white population remained 9 and 13 per cent (Ortiz 1995).

Conclusions

> The groupings that are called families are socially constructed rather than naturally or biologically given. (Jagger and Wright 1999: 3)

As a social construct, how 'family' is defined varies over time and across space and is a product of a particular set of circumstances and purposes (Sprey 1999). The (re)visioning of family presented by the images embedded in the retirement community advertisements analyzed in this chapter are moments in a particular set of ideologies that are continually (re)shaping our understanding of retirement, later life, and family. The images presented to readers of retirement community advertisements/brochures are not neutral. They are selected in order to sell a very specific vision of later life, a view often summarized as 'the retirement community lifestyle' or 'the active retiree lifestyle'. Our purpose in analyzing the images embedded in retirement community brochures was to examine how 'family' in later life is (re)imagined, (re)visioned and (re)presented in the specific spaces of retirement communities.

Our findings indicate that the 'retirement community family' is essentially the nuclear or traditional family, modified to largely exclude children as well as grandchildren. 'Family' as marketed by retirement communities consists of either a cohabitating heterosexual couple enjoying a sexual relationship, or a single once-married individual unencumbered by the responsibilities of children or caregiving relationships. Family and home life is organized around recreational and social activities with other couples rather than around waged or domestic labour. The (re)envisioned, (re)presented, and commodified retirement community family is a monolithic entity that reproduces a hegemonic view of what type of social unit constitutes a legitimate family and denies the diversity and pluralism of contemporary families. More specifically, this monolithic view of family denies the complexity and diversity of families in later life. The pleasant images and the skilful marketing that underpin the production and presentation of these images mean that the images of family in later life have become for many the quintessential representations of the good life in later life.

The visions of family (and indeed later life) embedded in the advertisements analyzed are far from natural or neutral. They are one (re)vision of what constitutes 'family' in later life—a vision that valorizes white heterosexual families. Differences in family composition among ethnic and racial groups, same-sex families, and even the older population are denied and marginalized. The unpacking of such variations in family composition lies beyond the scope of this research but nevertheless needs to be undertaken. Interesting and under-explored avenues of research include the analysis of images of older African American and Latino families and the spaces in which they are emplaced. Moreover, with the construction of an increasing number of retirement communities for same-sex families, efforts need to be made to analyze the images used to portray family in later life emplaced in these spaces. Comparative research on the (re)visioning and (re)presentation of family in NORCs, retirement communities, and **continuing care communities** would shed additional light on what particular ideas about family are emplaced in each of these spaces/communities.

Questions for Further Thought

1. Ideology guides content analysis. Ideologies serve as meta-narratives that give meaning to text and images. What particular ideologies and common narratives configure and reconfigure our thinking regarding aging and later life?
2. As spatial and cultural sites, retirement communities and suburbs are landscapes of interest to both social scientists and those interested in the built environment. Both retirement communities and suburbs reflect the dominance of specific ideologies and the capacity of society's hegemonic group to produce urban landscapes that emplace specific groups in well-defined spaces. Briefly discuss the cultural history of both and the place they occupy in the contemporary urban and social condition.
3. In recent years aging, old age, and retirement have been redefined by the realization that older adults represent a key and comparably under-exploited market segment. Discuss how the 'discovery' of the aged market has caused the use of specific images (including but not limited to advertisements) to describe the third age and the construction of different spaces of consumption in which to enjoy later life.
4. It has been suggested that the shift from the extended family to the nuclear to the current post-traditional family is a symptom of a much larger crisis of social obligation that has produced and is producing significant changes in the nature and extent of familial roles and relationships. Do you agree or disagree? Support your position.
5. Laws (1995) describes the physical environment and residents of retirement communities as 'plasticized'; places and populations of almost total homogeneity. Discuss the various aspects of social, demographic, and cultural homogeneity with reference to retirement communities.

References

Aguirre-Molina, M., and P.A. Parra. 1995. 'Latino youth and families as active participants in planning change'. Pp. 130–53 in R.E. Zambrana, ed. *Understanding Latino Families: Scholarship, Policy and Practice.* Thousand Oaks, CA: Sage.

Allen, K. 1989. *Single Women/Family Ties: Life Histories of Older Women.* Newbury Park, CA: Sage.

———, and K.M. Barber. 1992. 'Starting the revolution in family life education: a feminist vision'. *Family Relations* 42: 378–84.

———, R. Blieszner, K. Roberto, E. Farnsworth, and K. Wilcox. 1999. 'Older adults and their children: family patterns of structural diversity'. *Family Relations* 48: 151–57.

Andrews, M. 1999. 'The seductiveness of agelessness'. *Ageing and Society* 19: 301–18.

Angel, R., and M. Tienda. 1982. 'Determinants of extended family structure: cultural pattern or economic need?'. *American Journal of Sociology* 87: 1360–83.

Apple, D. 1956. 'The social structure of grandparenthood'. *American Anthropologist* 58: 656–63.

Atchley, R. 1980. *The Social Forces in Later Life.* Belmont, CA: Wadsworth.

Beck, R.W., and S.H. Beck. 1989. 'The incidence of extended households among middle-aged black and white women: estimates from a 5-year panel study'. *Journal of Family Issues* 10: 147–68.

Bedford, V., and R. Blieszner. 1997. 'Personal relationships in later life families'. Pp. 523–39 in S. Duck, ed. *Handbook of Personal Relationships.* 2nd edn. New York: Wiley.

Belk, R.W., and R.W. Pollay. 1985. 'Images of ourselves: the good life in twentieth century advertising'. *Journal of Consumer Research* 11: 887–67.

Berelson, B. 1952. *Content Analysis in Communications Research.* Glencoe, Ill: The Free Press.

Bianchi, S.M. 1995. 'The changing demographic and socioeconomic characteristics of single parent families'. *Marriage and Family Review* 20, 1–2: 71–97.

Biggs, S., M. Bernard, P. Kingston, and H. Nettleton. 2000. 'Lifestyles of belief: narrative and culture in a retirement community'. *Ageing & Society* 20, 6: 649–72.

Bourdieu, P. 1990. *Photography: A Middle Brow Art.* Stanford University Press.

Bryson, K., and L.M. Casper. 1997. 'Household and family characteristics: March 1997'. Current Population Reports, P20–509. Washington, DC: Government Printing Office.

———. 1999. 'Co-resident grandparents and their grandchildren'. Current Population Reports, pp.20–198. Washington, DC: Government Printing Office.

Casper, L.M., and S.M. Bianchi. 2002. *Continuity and Change in the American Family.* Thousand Oaks, CA: Sage.

———, and P.N. Cohen. 2000. 'How does POSSLQ measure up? Historical estimates of cohabitation'. *Demography* 37: 237–45.

Cavan, R.S., E.W. Burgess, R.J. Havighurst, and H. Goldhamer. 1949. *Personal Adjustment in Old Age.* Chicago: Science Research Associates.

Cherlin, A.J., and F.F. Furstenberg, Jr. 1985. 'Styles and strategies of grandparenting'. Pp. 97–116 in V.L. Bengston and J.F. Robertson, eds. *Grandparenthood.* Beverly Hills, CA: Sage.

———. 1986. *The New American Grandparent: A Place in the Family, A Life Apart.* New York: Basic Books.

Cole, T. R. 1992. *The Journey of Life: A Cultural History of Aging in America.* Cambridge: Cambridge University Press.

Collier, J., and L. Williams. 1982. 'The economic status of the black male: a myth exploded'. *Journal of Black Studies* 12, June: 487–98.

Coontz, S. 1992. *The Way we Never Were: American Families and the Nostalgia Trap.* New York: Basic Books.

———. 1997. *The Way we Really Are: Coming to Terms with American's Changing Family.* New York: Basic Books.

Cox, C.B. 2000. 'Why grandchildren are going to and staying at grandmother's house and what happens when they get there?'. Pp. 3–19 in C.B. Cox, ed. *To Grandmother's House We Go and Stay; Perspectives on Custodial Grandparents.* New York: Springer.

DaVanzo, J., and M.O. Rahman. 1993. 'American families: trends and correlates'. *Population Index* 56: 350–86.

Domosh, M., and J. Seager. 2001. *Putting Women in Place: Feminist Geographer Make Sense of the World.* New York: Guildford Press.

Dowd, N.E. 1997. *In Defense of Single-Parent Families.* New York: New York University Press.

Duany, L., and K. Pittman. 1990. *Latino Youths at a Crossroads.* Washington, DC: Children's Defense Fund.

Ekerdt, D. 1986. 'The busy ethic: moral continuity between work and retirement'. *The Gerontologist* 26: 239–44.

England, K.V.L. 1993. 'Suburban pink collar ghettos: the spatial entrapment of women?'. *Annals of the Association of American Geographers* 83: 225–42.

Erera, P. 2002. *Family Diversity: Continuity and Change in the Contemporary Family.* Thousand Oaks, CA: Sage Publications.

Fava, S.F. 1980. 'Women's place in the new suburbia'. Pp. 129–49 in G. Wekerle, R. Peterson, and D. Morley, eds. *New Space for Women.* Boulder, CO: Westview Press.

Featherstone, M., and M. Hepworth. 1995. 'Images of positive aging: a case study of *Retirement Choice* magazine'. Pp. 29–47 in M. Featherstone and A. Wernick, eds. *Images of Aging: Cultural Representations of Later Life.* London: Routledge.

Fields, J. 2003. 'America's families and living arrangements: 2003'. Current Population Reports, 20–553. Washington, DC: US Census Bureau.

Fineman, M. 1995. 'Masking dependency: the political role of family rhetoric'. *Virginia Law Review* 81, 8: 2181–215.

Fitzgerald, F. 1986. *Cities on a Hill: A Journey Through Contemporary American Culture.* New York: Simon and Schuster.

Flaks, D.K., I. Ficher, F. Masterpasqua, and G. Joseph. 1995. 'Lesbians choosing motherhood: a comparative study of lesbian and heterosexual parents and their children'. *Developmental Psychology* 31, 1: 105–25.

Friedan, B. 1963. *The Feminist Mystique.* New York: Norton.

Gerbner, G., L. Gross, M. Morgan, and N. Signorielli. 1986. 'Aging with television: the dynamics of the cultivation process'. Pp. 17–40 in J. Bryant and D. Zillman, eds. *Perspectives on Media Effects.* Hillsdale, NJ: Lawrence Erlbaum.

Gill, A.G., W.A. Vega, and J.M. Dimas. 1994. 'Acculturative stress and personal adjustment among hispanic adolescent boys'. *Journal Community Psychology* 22: 43–53.

Glass, J., and T. Huneycutt. 2002a. 'Grandparents raising grandchildren: the courts, custody, and educational implications'. *Educational Gerontology* 28, 3: 237–51.

———. 2002b. 'Grandparents parenting grandchildren: extent of the situation, issues involved, and educational implications'. *Educational Gerontology* 28, 2: 139–61.

Glick, P.C. 1997. 'Remarried families, stepfamilies and stepchildren: a brief demographic analysis'. *Family Relations* 38: 24–7.

Golant, S. 1992. *Housing America's Elderly Many Possibilities/Few Choices.* Newbury Park: Sage Publications, Inc.

Goldscheider, F.K., and R.G. Bures. 2003 'The racial crossover in family complexity in the United States'. *Demography* 40, 3: 369–587,

Goodman, C. 2003. 'Intergenerational triads in grandparent-headed families'. *Journal of Gerontology: Social Sciences* 58B: S281–S289.

———, and M. Silverstein. 2002. 'Grandparents raising grandchildren: family structure and well-being in culturally diverse families'. *The Gerontologist* 42: 676–89.

Haines, M.R. 1996. 'Long-term marriage patterns in the United States from colonial times to the present'. *History of the Family* 1: 15–39.

Hartley, J. 1997. 'The sexualization of suburbia: the diffusion of knowledge in the postmodern public sphere.' Pp. 108–31 in R. Silverstone, ed. *Visions of Suburbia*. London: Routledge.

Havighurst, R.J., and R. Albrecht. 1953. *Older People*. New York: Longmans, Green.

Hayslip, B., and P. Kaminski. 2005a. 'Grandparents raising their grandchildren: a review of the literature and suggestions for practice'. *Gerontology* 42, 2: 262–69.

———. 2005b. 'Grandparents raising their grandchildren'. *Marriage & Family Review* 37, 1/2: 147–69.

Hervey, P. 2005. 'Finding community in retirement'. *Urban Land* May: 75–9.

Hill, M. 1995. 'When is a family a family? Evidence from survey data and implications for family policy'. *Journal of Family and Economic Issues* 16, 1: 35–64.

Hunt, M.E., and G. Gunter-Hunt. 1985. 'Naturally occurring retirement communities'. *Journal of Housing for the Elderly* 3: 3–21.

———, and L.E. Ross. 1990. 'Naturally occurring retirement communities: a multivariate examination of desirability factors'. *The Gerontologist* 30, 5: 667–74.

Jagger, G., and C. Wright. 1999. 'Introduction: changing family values'. Pp. 1–16 in G. Jagger and C. Wright, eds. *Changing Family Values*, London: Routledge.

Jeffery, K. 1972. 'The family: a utopian retreat from the city: the nineteenth century contribution'. *Soundings* 55: 21–41.

Johnson, C. 2000. 'Perspectives on American kinship in the late 1990s'. *Journal of Marriage and the Family* 62: 623–39.

———, and B. Barer. 1997. *Life Beyond 85 years: The Aura of Survivorship*. New York: Springer.

Johnson, M.J., N.C. Jackson, J.K. Arnette, and S.D. Koffman. 2005. 'Gay and lesbian perceptions of discrimination in retirement care facilities'. *Journal of Homosexuality* 49, 2: 83–164.

Kahana, E., and B. Kahana. 1971. 'Theoretical and research perspectives on grandparenthood'. *International Journal of Aging and Human Development*, 2: 261–68.

Kassarjian, H.H. 1977. 'Content analysis in consumer research'. *Journal of Consumer Research* 4: 8–18.

Kastenbaum, R. 1993. 'Encrusted elders: Arizona and the political spirit of postmodem aging'. Pp. 160–83 in T. Cole, W. Achenbaum, P. Jacobi, and R. Kastenbaum, eds. *Voices and Visions of Aging: Toward a Critical Gerontology*, New York: Springer Publishing.

Katz, S. 2000. 'Busy bodies: activity, aging, and the management of everyday life'. *Journal of Aging Studies*, 14: 135–52.

Kaufman, S. 1986. *The Ageless Self: Source of Meaning in Late Life*. New York: Meridian.

Koball, H. 1998. 'Have African American men become less committed to marriage? Explaining the twentieth-century racial crossover in men's marriage timing'. *Demography* 35: 251–58.

Krippendorff, K. 1980. *Content Analysis: An Introduction to its Methodology*. London: Sage Publications, Inc.

Langmeyer, L. 1993. 'Advertising images of mature adults: an update'. *Journal of Current Issues in Advertising* 15, 2: 81–91.

Laws, G. 1993. 'The land of old age: society's changing attitudes toward the built environments for elderly people'. *Annals of the Association of American Geographers* 83, 4: 672–93.

———. 1994. 'Aging, contested meanings and the built environment'. *Environment and Planning A* 26: 1787–802.

———. 1995. 'Embodiment and emplacement: identities, representation and landscape in Sun City retirement communities'. *International Journal of Aging and Human Development* 40: 253–80.

Lee, J.A. 1989. 'Invisible men: Canada's aging homosexuals: can they be assimilated into Canada's "liberated" gay communities?'. *Canadian Journal on Aging* 8: 79–97.

Leiss, W., S. Kline, and S. Jhally. 1990. *Social Communication in Advertising*. Scarborough, Ontario: Nelson.

Levin, I., and J. Trost. 1992. 'Understanding the concept of family'. *Family Relation* 41: 348–51.

Litwak, E., and C.F. Longino. 1987. 'Migration patterns among the elderly: a developmental perspective'. *The Gerontologist* 27: 266–72.

Long, N. 1998. 'Broken down by age and sex: exploring the ways we approach the elderly consumer'. *Journal of Market Research* 40, 2: 73–91.

Longino, C.F. 1981. 'Retirement communities'. Pp. 391–418 in F.J. Berghorn and D.E. Schafer, eds. *The Dynamics of Aging Original Essays on the Processes and Experiences of Growing Old*. Boulder: Westview Press.

———, and A. Lipman. 1982. 'The married, the formerly married and the never married: support system differentials of older women in planned retirement communities'. *International Journal of Aging and Human Development* 14, 4: 285–97.

Lucas, S. 2002. 'Retirement communities and the changing suburban dream'. *Canadian Journal of Urban Research* 11, 2: 1–16.

———. 2004. 'The images used to "sell" and represent retirement communities'. *The Professional Geographer* 56, 4: 449–59.

Lucco, A.J. 1987. 'Planned retirement housing preferences of older homosexuals'. *Journal of Homosexuality* 14: 35–56.
Lutz, C., and J. Collins. 1980. *Reading National Geographic.* Chicago: University of Chicago Press.
Markasen, A.R. 1980. 'City spatial structure, women's household work and national urban policy'. *Signs* 5, 31: 21–41.
Marsh, M. 1990. *Suburban Lives.* New Brunswick, New Jersey: Rutgers University Press.
McCarthy, K.F. 1983. *The Elderly Population's Changing Spatial Distribution.* Santa Monica: Rand Corporation.
McHugh, K.E. 2000. 'The ageless self? Emplacement of identities in Sun Belt retirement communities'. *Journal of Aging Studies* 14, 1: 103–15.
———, and E.M. Larson-Keagy. 2005. 'These white walls: the dialectic of retirement communities.' *Journal of Aging Studies*, 19, 2: 241–56.
McLanahan, S., and L. Casper. 1995. 'Growing diversity and inequality in the American family'. Pp. 1–46 in R. Farley, ed. *State of the Union: American in the 1990s. Volume 2: Social Trends.* New York: Russell Sage Foundation.
Milan, A., M. Vézina, and C. Wells. 2007. 'Family portrait: continuity and change in Canadian families and households in 2006'. Catalogue no. 97–553-XIE. Ottawa: Statistics Canada.
Miller, R. 1983. 'The Hoover in the garden: middle-class women and suburbanization'. *Environment and Planning D: Society and Space* 1: 73–87.
Moody, H.R. 1988. *Abundance of Human Life: Human Development Policies for an Aging Society.* New York: Columbia University Press.
Moschis, G., L. Euehun, and A. Mather. 1997. 'Targeting the mature market: opportunities and challenges'. *Journal of Consumer Marketing* 14, 4: 282–94.
Murdock, G. 1949. *Social Structure.* New York: The Free Press.
Neugarten, B.L., R.J. Havighurst, and S.S. Tobin. 1961. 'The measurement of life satisfaction'. *Journal of Gerontology* 16: 134–43.
———, and K. Weinstein. 1964. 'The changing American grandparents'. *Journal of Marriage and the Family* 26: 199–204.
Norton, A.J., and L.F. Miller. 1989. 'Marriage, divorce and remarriage in the 1990s'. Current Population Reports, Series P-23, No. 180. Washington, DC: Government Printing Office.
O'Guinn, T., R. Faber, J. Curias, and K. Smith. 'The cultivation of consumer norms'. Pp. 779–85 in T.K. Srull, ed. *Advances in Consumer Research*, Vol. 16, Provo, UT: Association for Consumer Research.
Ortiz, V. 1995. 'The diversity of Latino families'. Pp. 18–39 in R.E. Zambrana, ed. *Understanding Latino Families: Scholarship, Policy and Practice.* Thousand Oaks, CA: Sage.
Palen, J. 1995. *The Suburbs.* New York: McGraw-Hill, Inc.
Peek, C.W., T. Koropeckyj-Cox, B.A. Zsembik, and R.T. Coward. 2004. 'Race comparisons of the household dynamics of older adults'. *Research on Aging* 26, 2: 179–201.
Perkinson, M., and D. Rockemann. 1996. 'Older women living in a continuing care retirement community: marital status and friendship'. *Journal of Women and Aging* 8, 3/4: 159–78.
Riffe, D., S. Lacy, and F.G. Fico. 1998. *Analyzing Messages: Using Quantitative Content Analysis in Research.* New York: Lawrence Erlbaum Associates Inc.
Rowe, J.W., and R.L. Kahn. 1997. 'Successful aging'. *The Gerontologist* 37: 433–40.
———. 1998. *Successful Aging.* New York: Pantheon Books.
Ruggles, S. 1994. 'The transformation of American family structure'. *American Historical Review* Feb.: 103–28.
———. 2007. 'The decline of intergenerational co-residence in the United States, 1850 to 2000'. *American Sociological Review* 27: 964–89.
Sawchuck, K.A. 1995. 'From gloom to boom: age, identity and target marketing'. Pp. 173–87 in M. Featherstone and A. Wernick, eds. *Images of Aging: Cultural Representations of Later Life.* London: Routledge.
Schlesinger, B. 1996. 'The sexless years or sex rediscovered'. *Journal of Gerontological Social Work* 26, 1/2: 117–31.
Schwartz, J., and J.R. Ryan. 2003. *Picturing Place: Photography and the Geographic Imagination.* London and New York: I.B. Tauris.
Smith, C., and A. Beltran. 2003. 'Grandparents raising grandchildren: challenges faced by these growing numbers of families'. *Journal of Aging and Social Policy* 12, 1: 7–18.
Sottomayor, M. 1989. 'The Hispanic elderly and the intergenerational family'. *Journal of Children in Contemporary Society* 20, 3–4: 55–65.
Spigel, L. 1992. 'The suburban home companion: television and the neighborhood ideal in postwar America'. Pp. 184–217 in B. Colomina, ed. *Sexuality and Space.* New York: Princeton Architectural Press.
Stacey, J. 1990. *Brave New Families.* New York: Basic.
Staples, R. 1987. 'Social structure and black family life: an analysis of current trends'. *Journal of Black Studies* 17, 3: 267–86.
———, and A. Mirande. 1980. 'Racial and cultural variations among American families: a decennial review of the literature on minority families'. *Journal of Marriage and the Family* 42, November: 887–903.
Streib, G.F., W. Folts, and M. Hilker. 1984. *Old Homes - New Families: Shared Living for the Elderly.* New York: Columbia University Press.

Sullivan, D. 1985. 'The ties that bind: differentials between seasonal and permanent migrants to retirement communities'. *Research on Aging* 7, 2: 235–50.

Szapocznik, J., and R. Hernandez. 1997 'The Cuban American family'. Pp. 160–72 in C.H. Mindel, R.W. Habenstein, and R. Wright, Jr., eds. *Ethnic Families in America*. New York: Elsevier.

———, and W. Kurtines. 1980 'Acculturation, bilingualism and adjustment among Cuban Americans'. Pp. 47–83 in A.M. Padilla, ed. *Acculturation Theory, Models and Some New Findings*. Boulder, CO: Westview Press.

Szinovacz, M., and S. DeVinney. 1999. 'The retiree identity'. *Journals of Gerontology Series B: Psychological Sciences and Social Services* 54B, 4: 207–19.

———. 1999. 'Effects of surrogate parenting on grandparents' wellbeing'. *Journals of Gerontology Series B: Psychological Sciences and Social Services* 54B, 6: 376–89.

Teachman, J.D., L.M. Tedrow, and K.D. Crowder. 2000. 'The changing demography of America's families'. *Journal of Marriage and the Family* 62, 4: 1234–46.

Troll, L.E. 1996. 'Modified-extended families over time: discontinuity in parts, continuity in wholes'. Pp. 246–70 in V. Bengston, ed. *Adulthood and Aging: Research on Continuities and Discontinuities*. New York: Springer.

Trost, J. 1988. 'Conceptualizing the family'. *International Sociology* 3: 301–08.

US Census Bureau. 1983. 'American's black population, 1970–1982: a statistical view'. Series P10/POP83. Washington, DC: Government Printing Office.

van den Hoonaard, D. 1984. 'Every day is sunday: compliance and the development of social values in a Florida retirement community'. Master's thesis, University of New Brunswick.

———. 1994. 'Paradise Lost: Widowhood in a Florida retirement community'. *Journal of Aging Studies* 8, 1: 121–32.

———. 2002. 'Life on the margins of a Florida retirement community: the experience of snowbirds, newcomers, and widowed persons'. *Research on Aging* 24: 50–66.

Vega, W.A., and H. Amaro. 1994. 'Latino outlook: good health, uncertain prognosis'. *Annual Review of Public Health* 15: 39–67.

Wallace, S.P. 1986. 'Central American and Mexican immigrant characteristics and economic incorporation in California'. *International Migration Review* 20, 3: 657–71.

Zelnik, M., and J.F. Kantner. 1977. 'Sexual and contraceptive experience of young unmarried women in the United States, 1976 and 1971'. *Family Planning Perspectives* 9, May/June: 55–9.

Zukin, S. 1988. 'Urban lifestyles: diversity and standardization in spaces of consumption'. *Urban Studies* 35, 6 (1988): 825–36.

Chapter Eleven

Becoming an Informal Caregiver for a Dying Family Member

Valorie A. Crooks and Allison Williams

Introduction

The informal provision of care by one family member for another during periods of illness or in other times of need is something that has been done for centuries, across continents and cultural groups. One period when such care has commonly been provided is while someone is dying. Caring for family members during this period (i.e., at the end of life) can last anywhere from a matter of days to years depending on the progression of illness and resulting impairments (McMillan 2005). This **informal care** can take on many forms, ranging from the hands-on care needed for pain and symptom management to spiritual support and simply physical co-presence. Thus, while some family members have only limited involvement in caring for a few hours per week by performing indirect caregiving tasks such as cooking, cleaning, banking, and shopping, others are on call twenty-four hours per day and provide complex direct care such as administering medications, bathing, feeding, and toileting (Pyper 2006; Waldrop et al. 2005).

In previous decades, caregiving in general was increasingly institutionalized in developed nations such as Canada (Armstrong and Kits 2001). After the onset of life-limiting illness and the resulting need for care, dying individuals were typically moved to hospitals, residential care facilities, or hospices to be monitored and have on-site access to health care. While this is still occurring, there is a greater recognition that dying individuals should be supported in their homes or home communities for as long as possible (Stajduhar et al. 2008). One way to make this possible has been to increase the reliance on informal caregiving provided by family members. This is part of the overall movement in Canada to deinstitutionalize caregiving which in turn is necessitating that family members take on the caregiver role and often times the responsibility for care coordination (Armstrong and Kits 2001; Skinner and Rosenberg 2006). Oppositely, not having access to an informal caregiver who can take on significant care responsibilities in the home typically leads to greater reliance on formal health care, including that which is provided in institutional settings such as hospitals (Cranswick 2003).

Family members who take on the responsibility of providing some or all of the needed care for a dying loved one encounter many challenges. The increased 'medicalization' of death during the twentieth century has meant that many informal caregivers have not been exposed to dying individuals and the types of care required before providing such care themselves. This is contrasted with the century previous when it was common that the skills required to care for ill or dying individuals were passed between family members (Armstrong and Kits 2001). Further, the provision of informal care within the family unit remains a highly **gendered** undertaking in that women take on the bulk of this responsibility (Williams and Crooks 2008). Meanwhile, in the mid-to-late twentieth century women increasingly involved themselves in the paid labour market. An outcome of this is that many women now manage the dual responsibilities of providing care for a dying family member while simultaneously maintaining paid employment (MacBride-King 1999).

The provision of all forms of informal care (e.g., childcare, eldercare, end-of-life care) between family members is clearly a 'family geography'. Such care frequently (but not exclusively) takes place in the private space of the home (Yantzi et al. 2007), involves the negotiation of care provision and coordination across settings (Milligan 2006), is responsive to particular community contexts (Cloutier-Fisher and Joseph 2000; Milligan et al. 2007), and often times requires travel and the reorganization of the socio-spatial lives of both caregiver and care recipient (Hallman and Joseph 1999; Wiles 2003). These are but a few of the more overtly spatial elements of this particular family geography identified in the work of health geographers. In the remainder of this chapter the focus is explicitly on the provision of informal family **caregiving at end of life** by drawing on interviews conducted with 25 Canadian family members who provided informal care for a dying relative. This is one of the types of caregiving that takes place within the larger family geographies of care.

In the section that follows we provide some spatio-temporal context regarding family caregiving and argue that caring for a dying individual differs from other forms of family-based care because of three particular factors. We then move to introduce our study and specifically examine the ways in which taking on the 'family caregiver' role involves negotiating a challenging socio-spatial process that centres on balancing disparate identities and geographies. We conclude this chapter by considering the importance of understanding this **socio-spatial process** and its **spatio-temporal context** for social policies and programs, and specifically Canada's **Compassionate Care Benefit** (CCB)—a social program that provides job security and limited-income support for Canadians to take a temporary leave from paid employment to care for a dying family member. Our discussion of social policies and programs, and the CCB in particular, assists in demonstrating why and how family geographies are relevant to the provision of support to family caregivers.

Unique Spatio-Temporal Context of Caring for a Dying Family Member

The ways in which experiences are lived out through time and in space is an important consideration that is central to family geographies. This is also true for the geographies of family caregiving. In this section of the chapter we consider the spatio-temporal context of providing care for a dying loved one. Specifically, we identify three factors that, together, create the unique spatio-temporality of this caregiving period, specifically (1) distinct temporality, (2) little accumulated knowledge, and (3) distance between caregiver and care recipient. To do this we draw on the relatively new (Andershed 2005; Dunbrack 2005; Harding and Higginson 2003; Hudson 2004), yet burgeoning, literature focused on the experiences of individuals who have provided care for dying family members. When considered together, these factors provide important context for the discussion in the remainder of the chapter.

Temporality

Research has picked up on multiple issues related to the passing of time throughout the experience of serving as an informal caregiver for a dying loved one. **Temporality**, for example, is a critical aspect of caring for a dying family member since well-timed approaches to care are crucial for people who only have a short time to live as their health often deteriorates rapidly (Francke and Willems 2005). Complicating this issue is the well-known challenge of obtaining accurate prognostication, thus leaving caregivers with little knowledge as to when death is likely to occur. The passing of time is also crucial to the informal caregiving experience as roles and responsibilities typically change in response to the care recipient's disease progression and symptoms (Herbert and Schultz 2006). Having caregivers come to understand the magnitude of their responsibilities and the reality that their loved one is dying also typically takes time (Hauser and Kramer 2004). Related to this, Farber et al. (2003) have found that caring for a dying loved one is a time of uncertainty and unpredictability, with feelings of powerlessness about what happens from one day to the next being common.

At a more micro-scale, family caregivers often report feeling a lack of control over their everyday lives, disturbance of personal routines, and diminished opportunities for leisure activities while trying to manage the multiple demands of caring for loved ones (Hudson 2004; Stajduhar et al. 2008). Yet, when death is imminent these concerns become less important as everything else typically takes on a new perspective and the concept of time often becomes surreal. In this peri-death period, caregivers are faced with unpredictability about the future and the impending death of their loved ones (Herbert and Schultz 2006). Herein lays the profound distinction between caring for a dying loved one and other forms of family-based care: the need for hands-on caregiving will be diminished, and will then come to an end, sometimes over just a short period of time. It is for this reason we have identified temporality to be one of the distinctive features of the spatio-temporal context of this form of family caregiving.

Informational Needs

Existing research has shown that as caregiving for dying loved ones has been increasingly transferred to family members, it has become quite important for these caregivers to understand illness progression, the range and scope of formal services available, and their own roles and competencies as providers of care (Doherty et al. 2008). Gaining such knowledge can be a challenge for multiple reasons. For example, family caregivers may not be emotionally ready to accept certain kinds of information (Dunbrack 2005). Furthermore, Ashpole (2004) has pointed out that **informational needs** have a distinct temporality because informal caregivers need to know different things at different times as their roles and responsibilities evolve. Thus, they may be challenged by the need to constantly update what they know (see also Hauser and Kramer 2004; Thielemann 2000). Another challenge is the speed at which information must be accessed. This is because, as Ashpole (2004: 32) contends, '. . . informal caregivers are faced with taking on a new job with no previous experience, no job description to guide them or employee incentive program to encourage them, and where their own health and well-being may be at risk'. The frequent lack of 'previous experience' makes informational needs particularly heightened in that they must be addressed quickly given the often temporally-limited care period.

A central theme in the literature indicates that communication between caregivers and health professionals is often negligible in terms of assessing the kinds of information caregivers need (Dunbrack 2005; McMillan 2005). Another barrier to accessing needed or wanted information is caregivers' reluctance to ask questions or inabilities to even know which questions to ask (Andershed 2005; Herbert and Schultz 2006; Hudson 2004). This is complicated by the fact that the type of information needed by caregivers is often immediate and wide-ranging. Such information includes, for example, details regarding how to operate medical equipment (e.g., an oxygen machine) and home-based aides (e.g., poles and lifts) through to symptom and pain management, personal care and psycho-social issues brought about by the impending death. Such communication barriers must, however, be overcome quickly because unlike some other forms of informal family caregiving that extend over long periods of time (e.g., raising children, caring for individuals with congenital impairments), caring for a dying loved one may start abruptly and not last for an extended temporal period. For example, there is not typically a long period over which relationships of information-sharing between informal family caregivers and formal care providers can be established. Because of this we have identified informational needs to be one of the distinctive features of the spatio-temporal context of this form of family caregiving. Though less explicitly spatial or temporal than the other defining features reviewed in this section of the chapter, information transfer is in fact a highly geographic process that occurs over time and through space (Crooks et al. 2007).

Proximity

Proximity or the literal distance between the caregiver and care recipient, is a central feature of providing informal care (Cranswick 2003; Pyper 2006). Both parties may share at times a single house or other dwelling, while in other instances the two parties may be separated by hundreds of kilometres, with on-site care being provided sporadically. There is, however, little discussion of proximity in the caregiving literature, particularly that specific to caring for a dying loved one. With regard to eldercare in general, the Canadian General Social Survey (Health Canada 2002) indicates that those living with others have the greatest probability of using only informal sources of care, and that almost half of seniors receive all of their care from family and friends. This was noted to be due to the shift of care away from institutions into the home, where family and friends perform the bulk of caregiving duties.

As noted above, the literature about care provision for dying individuals does not adequately examine issues of proximity between informal caregivers and care recipients, with the exception of a few studies that discuss the location of care (Gomes and Higginson 2006; Waldrop et al. 2005) and its relevance to care provision (Stajduhar et al. 2008). However, with a large majority of individuals reporting that they want to spend their final days at home (Doherty at al. 2008; Stajduhar et al. 2008), for many, being able to receive informal care from a loved one in this specific space is a central part of the dying experience. The distance family members have to travel to provide care for dying loved ones is an important factor in providing care at home (Gomes and Higginson 2006). When proximity is considered in relation to the above discussion regarding temporality, the specific context related to providing care for a dying individual becomes clearer. Because such care is typically not long term, an established caring relationship over time will likely not have been set up. An outcome of this is that caregivers and care recipients may be quite distant from one another, thus necessitating temporally-limited relocation in order to facilitate care provisioning. It is for this reason we have identified proximity to be one of the distinctive features of the spatio-temporal context of this form of family caregiving.

Study Overview

In the remainder of this chapter we draw on the findings of a pilot study that involved 25 interviews conducted with Canadians who had provided care for a dying family member. The purpose of the pilot study was to examine the usefulness of the Compassionate Care Benefit for family caregivers providing care for a dying loved one in different kinds of care contexts (we elaborate on the CCB later on in this chapter). Details about the study design and other findings can be found in other publications and so are not elaborated upon here (see Crooks et al. 2007; Williams et al. 2005, 2006). Briefly, the study methodology was Patton's (1997) utilization-focused approach to evaluation. This approach dictates that an evaluation needs to focus on 'intended use by intended users' (Patton 1997: 20). Family caregivers from across Canada participated in semi-structured phone interviews. Recruitment of these individuals was undertaken using multiple methods. The interviews were taped and transcribed verbatim. The transcripts were managed using NVivo, a qualitative data management software program, and were analyzed using the constant comparative technique. In general, the pilot evaluation was highly successful and has since led to the undertaking of a full evaluation of the CCB from the perspective of family caregivers (see Crooks and Williams 2008).

Becoming a Family Caregiver

While the focus of the study outlined briefly above was not on the general lived experiences of caregivers providing care for a dying family member, conversations with the interviewees about such care lent naturally to discussion of numerous contextual issues. It was revealed that becoming a family caregiver is a process (i.e., something that happens over space and time) and that engaging in this process involves negotiating multiple identities and competing demands and expectations. In the remainder of this section of the chapter we share what the interviewees had to say about such negotiation. Importantly, it must be recognized that this negotiation takes place within the context of the distinct spatio-temporal characteristics which are hallmarks of providing family caregiving for a dying loved one, as were outlined previously.

Negotiating Multiple Identities

Many interviewees were challenged by needing to negotiate multiple identities simultaneously, particularly (1) that of established family member (e.g., parent, sibling); (2) that of informal caregiver (e.g., emotional support, physical care); and (3) those affiliated with other endeavours (e.g., volunteer, worker). Some of these identities were well established but played out in new ways during the caregiving period, while others were new. While it is extremely common for people to have multiple identities and roles that they live out at the same time, it was the newness of the added 'informal caregiver' identity and role that was most commonly a source of struggle and thus a hallmark of the process of becoming a family caregiver for a dying family member. Further, it was found that various identities and roles suddenly had competing expectations and demands that had to be negotiated as successfully as possible.

It was noted previously that many family caregivers maintain involvement in paid labour while providing informal care for dying family members. Managing both this reality and also one's own desire to fulfill the responsibilities associated with both identities (i.e., paid worker *and* informal caregiver) was discussed as being stressful by some interviewees. As one daughter who cared for her mother explained,

> 'Just because, you know, someone's dying in your home, doesn't mean that you can stop

> paying your mortgage or your bills . . . an income had to be somehow sustained, so I continued to work and I actually made myself very, very ill and had to go off work regardless, due to you know like a . . . they're calling it a medical stress leave kinda thing'.

This quote alludes to one of the many sacrifices made by informal caregivers facing this situation: the erosion of their own health and well-being. In other instances, however, removing oneself from paid work during the temporally-limited caregiving period was thought to be most desirable.

> 'I said to my employer "my sister is divorced and she has no other family and she is very close to me and I feel a real need to be there for her and if it is necessary can I have time off work?" And my employer was very understanding and generous and said to please ask when I needed time and they would consider it.'

For this caregiver, the absence of other family members who could share the caregiving responsibilities meant that the sibling identity and role had to be prioritized over that of paid worker during the caregiving period. It was often leisure activities and paid labour and their resulting identities (e.g., teammate, co-worker, economically productive family member) that were put aside in order to successfully negotiate caregiving responsibilities.

Through taking on a new identity and managing it along with other more established ones some respondents gained a new sense of self.

> 'Certainly I've learned to be more open and not to be selfish and you know I really appreciated more what women do at home. I always knew, but until you actually have to do it yourself. And I figure I'm a pretty good juggler, but this is, you know. And then you just have to decide what your priorities are.'

In the case of this caregiver, her revised identity as a woman 'juggling' traditional women's work led to a greater openness to the choices people make in taking on certain roles. However, while many respondents reported experiencing personal growth and gaining a great deal of satisfaction from including 'family caregiver' as one of their multiple identities, one son caring for his mother was quick to point out,

> 'Well I see myself as a full time family caregiver. Other people would see me as unemployed; I don't really know how to answer. It all depends what side of the fence you are on'.

From this we can understand that family caregivers must contend with a dual assignment of identity: those they take on, by choice or otherwise, and those others assign them. Each has implications for any alteration to the sense of self that is a result of becoming a family caregiver.

Negotiating Competing Demands and Expectations

Perhaps not surprisingly, interviewees frequently explained how they had to negotiate multiple demands on their spatio-temporality and expectations of the responsibilities they would take on, some of which were competing. Those most frequently discussed by the respondent group were (1) social constructions, (2) familial expectations, (3) others' expectations, and (4) fluidity between identities. Interestingly, the three elements of the unique context of family caregiving for a dying loved one discussed previously were frequently cited in interviewees' discussions of demands and expectations.

When multiple relatives were potentially available to provide care, discussions between family members often ensued regarding who should and would take on primary responsibility for caregiving. In many instances the interviewees felt that the expectations placed on them by other family members had led to them becoming informal caregivers. As one respondent noted,

> 'Well, I have a brother who's in this area and sister-in-law but it just seems that . . . my parents' health care has been solely on my shoulders and I don't know how that happened but anyways that's the way it is. There's always one family member that kind of takes over and takes control and makes decisions and sort of like that'.

Here this caregiver considers her close proximity to the care recipient to be one of the factors that led to others' expectations of her to provide care. Finances also sometimes played a factor,

> 'There's only me and my brother, so basically I guess from the medical end I was the one chosen to. He was self-employed type-thing. I guess I'm the one that could support her the most because it was hard on him [financially]'.

In this instance, this woman caring for her mother describes becoming a caregiver as both a default, in that nobody else was available, and a choice. Regarding familial expectations, being close in proximity to the care recipient and being able to pay for needed supplies and possibly one's ability to afford to take time off work were key factors in determining who should provide care, aside from an individual's own desire to take on such responsibility.

For some respondents, negotiating social constructions and others' expectations were also part of the process of becoming a family caregiver. Caregiving experiences that did not typify the provision of care for a family member were sometimes challenged. Although only 5 of the 25 interviewees were men, there was some discussion among this group that the family caregiver role is socially constructed as feminine and women's work. This sometimes led to feelings of stigma. In other instances, however, it was pointed out that some family relations of caregiving were more acceptable than others. A sister who cared for her brother expressed feeling a lack of understanding from others.

> 'And I also found with me, because it was a brother, your employer wasn't very sympathetic. It had to be your husband or your own son or daughter, it's like it's accepted. It's fine, you can go and everybody's got sympathy; but because it was a brother, as a matter of fact, I was even reprimanded for it once [by employer], which I found very difficult.'

In this instance it was her employer's impression of her caregiving relationship that she had to carefully negotiate in order to be able to maintain her position as a paid worker.

In the previous subsection, it was explained that the interviewees all had to negotiate multiple identities during the caregiving period. Not surprisingly, each identity has its own responsibilities and expectations. A sub-group of interviewees felt particularly challenged by others' expectations that they would take on the role of becoming an informal caregiver because of their specific paid worker roles: those employed in the health care field. These interviewees felt that it was understandable why their family members expected them to take on the bulk of the responsibility for informal caregiving, but at the same time sometimes found that doing so conflicted with their training and experiences as formal care providers (e.g., they were looked to for medical advice from family members regarding treatments and medications when they were not serving as a formal, paid caregiver of the dying individual). The contextual factors of temporality and informational needs were typically used to justify why others wanted them to take on this role. Specifically, their training would likely lessen their informational needs and, because of the need for on-the-job learning due to the limited temporality of the caregiving period, they were in the best position to successfully negotiate providing care. As one respondent explained,

> 'I'm the only family member who has a health background so, um . . . I know that people would look at me and if I you know and if I didn't flinch then "oh, then everything's ok". But if I had, you know, burst into tears or something, which I never did, people would think "oh my god, this is really, really bad". So I knew that people were looking at me sort of to try and understand what was going on and what would be going on in the next while and throughout her whole illness actually'.

In this instance, this interviewee's training and job meant that her family expected her to serve as a caregiver. However, it also placed her in the awkward position of being looked to by family as a formal provider at a time when she wanted to connect with such individuals in her identity as a family member.

Why Family Geographies Matter: Considering Canada's Compassionate Care Benefit

In the previous section it was shown that taking on the 'family caregiver' role involves negotiating a challenging process that centres on balancing disparate identities and geographies of demand and expectation and is shaped by the spatio-temporal context of providing care for a dying loved one. While these findings help us to gain an understanding of the lived experience of family caregiving, the question remains: what is the relevance of knowing about these family geographies? In this section of the chapter we address this question through focusing on why and how understanding family geographies is important to providing appropriate supports for family members caring for a dying relative. Our specific focus is on the CCB as an exemplar of a particular type of support.

The CCB is a Canadian social program that came into effect in 2004 (Osborne and Margo 2005). The goal of the program is to enable family members and close others who are employed to take a temporary secured leave from work to care for a dying individual (Crooks and Williams 2008). It is administered through the federal Employment Insurance program as a 'special benefit'. Successful applicants to the program can receive up to 55 per cent of their average insured earnings over a six-week period to provide care to a loved one at risk of dying within six months. To qualify for the Benefit applicants must have worked a minimum of 600 insurable employment hours over the previous 52 weeks and have access to a medical certificate from the dying individual's doctor indicating that death is imminent (HRSDC 2008).

Unique Spatio-Temporal Context

Above we drew on existing literature to identify factors that together form the unique spatio-temporal context of providing family caregiving specifically at end of life. We identified three specific factors (1) temporality, (2) informational needs, and (3) proximity. Given the importance of these factors in shaping the experience of providing family caregiving, they must clearly be considered in creating programs such as the CCB to support individual family caregivers.

Temporality dictates the very nature of the CCB program. We noted above that the benefit program is available for a temporally-limited period of time. This falls in line with the temporally-limited nature of much informal caregiving for dying loved ones. However, in instances where caregiving extends beyond the specific temporal limits of the CCB program (i.e., lasting longer than six weeks), individuals can no longer be supported through the benefit program (see Williams et al. 2006). This was viewed as problematic by the family caregivers we interviewed, particularly given the challenges of prognostication. These prognostication challenges were also implicated in caregiver's decisions around when to start the leave from work. While considering the temporality of family caregiving specifically for a dying family member is important for the CCB, dictating an exact length of support period was thus viewed as a limitation by the caregivers interviewed for this study.

From the literature reviewed above it is clear that family caregivers have a great number of informational needs. Clearly, supportive programs such as the CCB must do their best to not add to caregivers' informational burdens. The family caregivers we interviewed clearly expected information about the benefit program to be easily accessible and understandable. What they found, however, is that this was not always the case (see Crooks et al. 2007). Needed information was sometimes difficult to obtain and some of the specific program regulations were difficult to understand. In response, sometimes extra time was spent verifying information that had been accessed. In other instances, lack of information negatively affected caregivers in their attempts to access support through the CCB. The process of learning about and applying for the CCB, thus, seemed to add to the informational burdens of the interviewees.

While proximity does not dictate the nature of the CCB program as temporality does, it is something that is literally written into the program. Specifically, the benefit program enables caregivers to take a secured employment leave in order to provide care for dying family members who are non-local. It thus does not require caregivers and care recipients to be proximal. Caregivers can travel to the dying loved one in order to provide care during the secured leave afforded by the CCB. Travelling in order to provide care was something that 9 of the 25 people we interviewed reported having done. One provided care for someone out-of-country, two provided care out-of-province, and six provided care outside of the towns and cities in which they lived. The CCB also does not dictate the site at which care must be provided; the dying loved one can be cared for at home, at the care provider's residence, in an institutional setting, or in any other place. This is another way in which the CCB is quite responsive to issues of proximity shaping the unique experience of providing informal care for a dying family member.

The Process of Becoming a Caregiver

In the above subsection we have identified some of the ways in which the CCB program is and is not responsive to the spatio-temporal context of providing care for a dying family member. The interview findings also pointed to the fact that becoming a caregiver is a process that involves negotiating multiple factors that are shaped by this spatio-temporal context, specifically (1) multiple identities, and (2) competing demands and expectations. Because of the centrality of such negotiations to the process of becoming a caregiver, it is important for them to be considered in creating supportive programs, such as the CCB.

It is understandably difficult for supportive programs to do away with some of the identity and role issues that challenge family caregivers, or even to take these factors into consideration. What the CCB does enable, though, is the maintenance of the valued 'paid worker' role and identity in caregivers' lives through the provision of a secured leave. Specifically, the CCB allows caregivers to take time away from work without having to leave their jobs permanently in order to provide care. The very nature of the program may also assist in lessening the stress of needing to simultaneously maintain paid work and one's identity as a 'productive' individual (i.e., via earning money) during the six-week caregiving period allowed by the leave; this was reported by some of the interviewees and experienced by many family members providing care for dying relatives.

Interviewees identified negotiating demands and expectations, including those that were sometimes competing, to be part of the process of becoming a family caregiver over time and space. In general, programs designed to support family caregivers are likely created with the intent of lessening some of the multiple demands placed on these individuals. The CCB specifically reduces the simultaneous dual-demands of needing to be involved in paid labour and serving as a caregiver. This program is more challenged in its ability to assist caregivers in overcoming some of the other types of demands and expectations shared in the interviews. Some interviewees, for example, felt as though their employers had expectations that they were not able to meet during the caregiving period (e.g., to not discuss caregiving while at work, to maintain paid work while caregiving). While the CCB legislates that employers allow employees to take a secured leave to provide care for a dying loved one, it cannot dictate the response that such caregivers will get from their employers as a result of taking up the benefit program. Finally, the six weeks of secured leave and income assistance provided through the CCB can be shared across eligible family members or other loved ones of the dying individual. While this regulation may add some complexity to determining who will use the program, it may also assist in lessening the expectations and/or demands placed on a single caregiver by others.

Conclusions

As the place of care for dying individuals has shifted from institutionalized settings into the community in recent years in Canada, the need for informal caregivers such as family members to assist with care provisioning has increased. Our review of the literature revealed three factors which together create the unique spatio-temporal context of providing informal care for a dying family member: a limited temporality, in that caring for a dying individual is often not as long of a care commitment as some other forms of care (e.g., eldercare); high informational needs that must be addressed during a short period of 'on the job' learning; and, issues of proximity, due simply to the fact that relationship of care has not necessarily been established, which means that caregivers and care recipients may be quite distant from one another in advance of the caregiving period. These factors assist with contextualizing the family geographies of this form of caregiving.

Twenty-five interviews with Canadians who have provided care for a dying family member pointed to the fact that becoming a family caregiver is a process that is shaped by multiple factors, including managing multiple identities, and negotiating demands and expectations which are sometimes competing. Such negotiation happens within the unique context of providing informal care, as identified in the literature. In other words, it was shown that becoming a family caregiver involves negotiating a challenging process that centres on balancing disparate identities and geographies of demand and expectation and is shaped by the spatio-temporal context of providing care for a dying loved one. Such negotiation is, then, a hallmark of the family geographies of caregiving at end of life.

Certainly all families, however configured, increasingly are expected to find ways to negotiate caregiving for dependents across the life trajectory, from care in infancy through to eldercare and finally, end-of-life care. Negotiating this role is one of the many challenges families face and one that can be repeatedly experienced numerous times. Successful negotiation of this role is often the result of consistent communication and equitable sharing of a wide range of responsibilities amongst family members. Unfortunately, not all families experience success in negotiating this role, often leaving a single individual solely responsible to manage. This is often unsustainable as this person 'burns out' and ultimately can no longer provide the care required. As mentioned, this role is shaped by spatio-temporal contexts and centres on balancing disparate identities and geographies of demand and expectation, all of which serve to challenge family geographies as an analytical category.

In an attempt to demonstrate the relevance of identifying and articulating family geographies central to the caregiving experience, we turned our attention specifically to Canada's CCB. Specifically, the ways in which the program does, or does not, respond to caregivers' needs and geographies were considered. Regarding this, it was found that

1. By its very nature, the CCB is shaped around the typically temporally-limited period of providing care for a dying individual;
2. By stating a specific temporal period (i.e., six weeks of support) the CCB is not responsive to the needs of those providing care for longer periods;
3. There were issues encountered regarding accessing information about the CCB program which, in turn, added to the informational burdens experienced by family caregivers;
4. The CCB is quite responsive to issues of proximity through enabling distant family members to provide care and allowing care to be provided in a number of sites;
5. The CCB allows successful applicants to maintain or put on hold their 'paid worker' identities and roles when providing care and focusing on familial roles and responsibilities;
6. The CCB reduces the simultaneous demands of working while providing care for a dying loved one; and
7. There is the potential to lessen the expectations and demands placed on a single caregiver in that the CCB program allows the benefit program to be shared between family members.

Our intent in discussing the CCB was not to provide a thorough review or critique of the program; rather, it was to assist with demonstrating why family geographies are relevant through tracing the program's responsiveness to the socio-temporal context of caregiving and some of the factors such informal family caregivers must negotiate. From this summary it can be seen that some of the program's features or requirements respond quite well to hallmarks of what we have identified in this caregiving experience while others pose challenges.

Acknowledgements

The authors are grateful for the participation of the family caregivers in this study. A New Emerging Team Grant in Family Caregiving in Palliative/End-of-Life Care by the Canadian Institutes of Health Research (CIHR) provided funding. Valorie A. Crooks was supported by a Canadian Health Service Research Foundation/CIHR Postdoctoral Fellowship in Health Services Research and a CIHR Strategic Training Postdoctoral Fellowship in Health Care, Technology, and Place during part of the study period. Allison Williams was supported by a personnel award from CIHR during the entire study period.

Questions for Further Thought

1. What other familial roles, besides a family caregiver at end of life, can be described as a process (i.e., something that happens over space and time), where an individual is required to negotiate multiple identities and competing demands and expectations?
2. In your own familial history have you experienced or observed the negotiation of multiple identities in the caregiver role? How can each of these identities be described and defined? Within the caregiver role, what competing demands and expectations were at play and how were they negotiated?
3. What are the relevant program implications of this research for the Compassionate Care Benefit? If you were a policy-maker, what suggestions would you make to revise the CCB? Based on what you know about family geographies, what other types of caregiver policies/programs would you support the development of, if any?
4. Apply a similar approach, with respect to understanding the socio-spatial process and its spatio-temporal context, to a social policy or program other than Canada's Compassionate Care Benefit that is of relevance to families.

Further Reading

Crooks, V.A., A. Williams, K.I. Stajduhar, D.E. Allan, and S.R. Cohen. 2007. 'The information transfer and knowledge acquisition geographies of family caregivers: an analysis of Canada's Compassionate Care Benefit'. *Canadian Journal of Nursing Research* 39, 3: 36–54.

Health Canada. 2002. *National Profile on Family Caregivers in Canada*. Ottawa: Executive Summary-Final Report.

MacBride-King, J. 1999. *Caring about caregiving: The eldercare responsibilities of Canadian workers and the impact on employers.* Ottawa: Conference Board of Canada.

Stajduhar, K., W.L. Martin, D. Barwick, G. Fyles. 2008. 'Factors influencing family caregiver's ability to cope with providing end-of-life cancer care at home'. *Cancer Nursing* 31, 1: 77–85.

Wiles, J. 2003. 'Daily geographies of caregivers: mobility, routine, scale'. *Social Science & Medicine*, 57, 7: 1307–25.

Williams, A., V.A. Crooks. 2008. 'Introduction: space, place and the geographies of women's caregiving work'. *Gender, Place & Culture,* 15, 3: 243–47.

References

Andershed, B. 2005. 'Relatives in end-of-life care—Part 1: a systematic review of the literature the last five years, January 1999–February 2004'. *Journal of Clinical Nursing,* 15: 1158–69.

Armstrong, P., and O. Kits. 2001. *One hundred years of caregiving.* Prepared for the Law Commission of Canada.

Ashpole, B.R. 2004. *The Informational Needs of Informal Caregivers.* Ottawa, ON: Secretariat on Palliative & End-of-Life Care (Health Canada).

Cloutier-Fisher, D., and A.E. Joseph. 2000. 'Long-term care restructuring in rural Ontario: retrieving community service user and provider narratives'. *Social Science & Medicine,* 50, 7–8 : 1037–45.

Cranswick, K. 2003. *Caring for an aging society.* Housing, Family and Social Statistics Division. Statistics Canada. General Social Survey Cycle 16: Catalogue no. 89–582.

Crooks, V.A., A. Williams, K.I. Stajduhar, D.E. Allan, and S.R. Cohen. 2007. 'The information transfer and knowledge acquisition geographies of family caregivers: an analysis of Canada's Compassionate Care Benefit'. *Canadian Journal of Nursing Research* 39, 3:36–54.

Crooks, V.A., and A. Williams. 2008. 'An evaluation of Canada's Compassionate Care Benefit from a family caregiver's perspective at end of life'. *BMC Palliative Care,* 7, 14.

Doherty, A., A. Owens, M. Asadi-Lari, R. Petchey, J. Williams, and Y.H. Carter. 2008. 'Knowledge and information needs of informal caregivers in palliative care: a qualitative systematic review'. *Palliative Medicine,* 22: 153–71.

Dunbrack, J. 2005. *The information needs of informal caregivers involved in providing support to a critically ill loved one. (A synthesis report prepared for Health Canada.)*. Retrieved online August 15, 2008 from: www.hc-sc.gc.ca/hcs-sss/pubs/home-domicile/2005-info-caregiver-aidant/index-eng.php

Farber, S.J., T.R. Egnew, J. Herman-Bertsch, T.R. Taylor, and G.E. Guldin. 2003. 'Issues in end-of-life care: patient, caregiver, and clinician perspectives'. *Journal of Palliative Medicine,* 6, 1: 19–32.

Francke, A.L., and D.L. Willems. 2005. 'Terminal patients' awareness of impending death: the impact upon requesting adequate care'. *Cancer Nursing,* 28 3: 241–47.

Gomes, B., and I.J. Higginson. 2006. 'Factors influencing death at home in terminally ill patients with cancer: systematic review'. *BMJ,* 332: 515–21.

Hallman, B.C., and A.E. Joseph. 1999. 'Getting there: mapping the gendered geography of caregiving to elderly relatives'. *Canadian Journal on Aging,* 18, 4: 397–414.

Harding, R., and I.J. Higginson. 2003. 'What is the best way to help caregivers in cancer and palliative care? A systematic literature review of interventions and their effectiveness'. *Palliative Medicine,* 17: 63–74.

Hauser, J.M., and B.J. Kramer. 2004. 'Family caregivers in palliative care clinics'. *Geriatric Medicine,* 20: 671–88.

Health Canada. 2002. *National Profile on Family Caregivers in Canada.* Ottawa: Executive Summary-Final Report 2002.

Herbert, R.S., and R. Schultz. 2006. 'Caregiving at the end of life'. *Journal of Palliative Medicine,* 9, 5: 1174–87.

Hudson, P. 2004. 'A critical review of supportive interventions for family caregivers of patients with palliative-stage cancer'. *Journal of Psychosocial Oncology,* 22, 4: 77–92

Human Resources Social Development Canada (HRSDC). 2008. *Compassionate Care Benefit* (2008). Retrieved online March 31, 2008 from www.hrsdc.gc.ca/en/ei/types/compassionate_care.shtml

MacBride-King. J. 1999. *Caring about caregiving: The eldercare responsibilities of Canadian workers and the impact on employers.* Ottawa: Conference Board of Canada.

McMillan, S.C. 2005. 'Interventions to facilitate family caregiving at the end of life'. *Journal of Palliative Medicine,* 8, 1: S-132-140.

Milligan, C. 2006. 'Caring for older people in the 21st century: notes from a small island'. *Health and Place,* 12, 3: 320–31.

Milligan, C., S. Atkinson, M.W. Skinner, and J. Wiles. 2007. 'Geographies of care: a commentary'. *New Zealand Geographer,* 63: 135–45.

Osborne, K., and N. Margo. 2005. *Compassionate Care Benefit: Analysis and Evaluation.* Toronto, Canada: Health Council of Canada.

Patton, M.Q. 1997. *Utilization-Focused Evaluation.* 3rd ed. Thousand Oaks, California, Sage.

Pyper, W. 2006. 'Balancing career and care'. *Perspectives.* Ottawa: Statistics Canada-Catalogue no. 75-001-XIE. 2006: 5–15.

Skinner, M.W., and M.W. Rosenberg. 2006. 'Informal and voluntary care in Canada: caught in the Act?'. Pp. 91–114 in C. Milligan and D. Conradson, eds. *Landscapes of Voluntarism: New Spaces of Health, Welfare and Governance.* Bristol, Policy Press.

Stajduhar, K., W.L. Martin, D. Barwick, and G. Fyles. 2008. 'Factors influencing family caregiver's ability to cope with providing end-of-life cancer care at home'. *Cancer Nursing* 31, 1: 77–85.

Thielemann, P. 2000. 'Educational needs of home caregivers of terminally ill patients: Literature review'. *American Journal of Hospice & Palliative Care,* 17, 4: 253–57.

Waldrop, D., B.J. Kramer, J.A. Skretny, R.A. Milch, and W. Finn. 2005. 'Final transitions: family caregiving at the end of life'. *Journal of Palliative Medicine, 8,* 3: 623–38.

Wiles, J. 2003. 'Daily geographies of caregivers: mobility, routine, scale'. *Social Science & Medicine,* 57, 7: 1307–25.

Williams, A., and V.A. Crooks. 2008. 'Introduction: space, place and the geographies of women's caregiving work'. *Gender, Place & Culture,* 15, 3: 243–47.

Williams, A., V.A. Crooks, K. Stajduhar, R. Cohen, and D. Allan. 2005. 'A pilot evaluation of the Compassionate Care Benefit - Research Report / Évaluation pilote des prestations de compassion subventionnée - Rapport de recherché'. Available at www.coag.uvic.ca/eolcare/compassionate_care.htm [Date of access: January 6, 2006].

———. 2006. 'Canada's Compassionate Care Benefit: views of family caregivers in chronic illness'. International Journal of *Palliative Nursing,* 12, 8: 438–49.

Yantzi, N.M., M.W. Rosenberg, and P. McKeever. 2007. 'Getting out of the house: the challenges mothers face when children have long-term care needs'. *Health and Social Care in the Community,* 15, 1: 45–55.

Conclusion

Not Really a Conclusion But a Beginning

Bonnie C. Hallman

Goals and Accomplishments

From the first days of development of this volume, a primary goal of the project was to provide strong 'waypoints' that give some indication of, and substance to, the (permeable) boundaries that define this intellectual terrain called 'family geographies'. Though not the first scholars to use this term (e.g., see Aitken 1998), a guiding purpose in this work has been to give more form, more depth, and a clearer outline of this focus that would also denote the analytical promise of the concept of family geographies. In the process, we hope to inspire other researchers as well as students to see where their work can contribute to, challenge, and (re) develop this focal area for the study of the everyday, yet complex, geographies of families and family life.

As presented in the introduction, a central focus of several of the chapters in this volume, and a key theme in family geographies/geographies of family life is the furthering of our understanding of how, and to what outcomes and effects, our day-to-day activities, responsibilities, and opinions regarding family life are grounded in the spaces and places that constitute family geographies. Families actively create the spaces and places of family life, the locales of their 'being in the world', as resources for the activities and responsibilities of their life together (Aitken 1998). Family members do this within the structures and strictures of their cultures and societies particularly as they pertain to class, age, and gender norms, and situated within particular economic and political contexts. Thus, a geography that focuses on families and family life must be sensitive to the contexts within which family life is mediated or negotiated through, and recognize that this is a dynamic and fluid reality, constantly being reproduced and reconstructed by every new event and challenge—from migration decision-making and settlement (e.g., see Huot and Dodson, and Samuel, this volume), to the negotiating and managing conducted by parents in organizing the activities of youth athletes (e.g., Williams and Crumplin, this volume).

A focus on the everyday and supposedly mundane is not the sole theme developed here however, because we must and can also make the linkages between everyday activity and larger regional, national, and international/global processes. This is not only because it is at the local scale of the neighbourhood, of the family, and at the intimate geography of the body that global processes are made 'real', but, as Massey (1995) argued, the increasingly 'stretched' social relations of families across space illustrate how the seemingly everyday activities of family life are not simply local or private. Rather they are both the product and the producer of social, cultural, and economic relations over space. Families, of all ages, ethnic backgrounds, socio-economic statuses, and sexual orientations, react to and are agents of economic changes and their simultaneous alterations of social and cultural norms and values.

Given the range of scales at which family geographies can be examined—from the intimate relations between family members to the influences of, for example, global trends in economic development and political conditions, and associated population movements—the diversity of possible topics to be fruitfully explored can seem somewhat overwhelming. Add to this the possibility of research themes that range across the family life course, and the breadth of what work may yet be done to develop our understanding of the geographies of family life is daunting to say the least. This volume starts a process of formulating areas of foci, presenting two initial themes to guide the development of family geographies as an analytical category.

First, we have developed the theme of *places of family interaction and identity construction* (e.g., the sports fields, the teleworking home, or the images of the later-life family in retirement community advertising). The second theme developed here, in several chapters, is *mobility and migration, both regional and international, and its effects on family life*, through the examples of, for example, women's lives on 1930s New Brunswick family farms, the life experiences of South Asian women migrating to Canada, and migrant families from Newfoundland struggling to adjust after economic-based migration. However, we have but indicated paths here, with numerous possible themes and topics yet to be explored. There will also need to be new theories applied, new concepts tested, and applications of new methodologies, as well as productive collaboration between geographers and researchers in cognate fields such as nursing, sociology, gerontology, native studies, and anthropology, to name but a few, for a more mature family-geographies scholarship to develop. Throughout, it will be a challenge, but it will be a valuable contribution to ensure that reflexive examinations of our own family geographies and the forces at work as we negotiate the spaces and places of family life and its intersections with other aspects of our social, economic, and political contexts are incorporated in future research theme developments.

While the possibilities of future directions may seem at first somewhat boundless, the acknowledged gaps in this collection of family geography research give some indication of where new research, and new researchers, may be inspired to take this field of inquiry. Specifically, three possible themes are indicated in the following section, each with brief discussions of journal papers, most very recent, as examples. These are presented to illustrate some of the ways that the study of family geographies may be broadened and at the same time made more inclusive.

Addressing Absences

While our work here gives some 'signposts' indicating areas of focus within the study of family geographies, they are by no means meant to be restrictive or exclusionary. There are clear and recognized gaps and absences in the topics gathered here, especially in the range of possible family forms and social groups represented, as well as the geographic focus of the research presented in the preceding chapters. In some cases, there were researchers that simply could not contribute at this time. However, a significant contribution to geographies of the family and family life are made by the studies highlighted here, and they represent a set of potentially important directions for moving the study of family geographies forward.

Geographies of Children and Adolescents

In dominant western discourse children belong within families, homes and private

> space. Since the 19th century, children have progressively been confined to special spaces of childhood. (Ansell 2009: 193).

There would seem to be an inherent connection between a focus on the geography of families and the research that has been produced over the last ten years in a resurgence of studies in children's geographies (e.g., Ansell 2009; Crewe and Collins 2006; Philo 2000; Tucker, Gilliland, and Irwin 2007). As noted by Ansell (2009) children's geographies analyses have been dominated by studies of the micro-geographies of the everyday environments of children's lives such as neighbourhoods, schools, homes, and playgrounds. Recent research such as that by Tucker, Gilliland, and Irwin (2007) and Tucker, Irwin, Gilliland, and He (2008), for example, has examined the qualities (e.g., seats in the shade, proximity to school or home) and amenities (e.g., splash pads, pools, availability of basketball nets and bicycles) that can be inherent to park and home-built environments and which encourage physical activity amongst children and adolescents. These researchers have done this from the perspectives of the parents of both young children and of youth aged 12 to 14 years. This area of work has generally focused on children and adolescents as social actors who perceive and engage with the world in interesting ways different from those of adults, and across diverse geographic settings, which they encounter in their lives (see Tucker, Irwin, Gilliland, and He 2008). This has meant a 'focus on the local processes, practices, and events that shape children's lives' (Ansell 2009: 191). Such work potentially contributes to family geographies by providing a wider representation of the everyday geographies of children and adolescents, studied specifically from the perspectives and experiences of these groups, within the contexts of their families and communities.

As noted for children's geographies, within the terrain of family geographies there also exists a focus on the places of everyday life, and the processes and practices that give meaning and value to families interacting together and engaging in activities in designated 'family' places (see in this volume Hallman and Benbow; Williams and Crumplin), or family relations between each other and to family spaces (including children's spaces as a subcategory within that) that must be renegotiated when work space must be incorporated into the family home (see Andrey and Johnson, this volume).

Family geographies should not merely absorb children's geographies however, but rather by recognizing that the larger family unit (however it is defined) is shaped by the collected processes, practices, and events that influence the geographies of all of its members, the analytical concept of 'family geography' may aide in making the connections between the everyday lives and lived spaces of children and adolescents with wider processes, discourses, and institutions (starting with the family itself). Understanding children's geographies as a dynamic component in the geography of the larger, multi-person family unit and associated family spaces can assist in the project of considering these geographies as the 'context for understanding social and material transformations' (Aitken 2001: 123). These transformations occur on the local scale, and in the lived experiences of all members of a family, inclusive of the social and material lives of children and adolescents within families.

Ansell (2009), following calls by Philo (2000), urges the development of an equal focus on the macro-scale, structural forces and institutions (social, cultural, economic, and political) that influence and possibly transform the cultural and material conditions of societies, and in turn construct notions of family, of childhood, and of appropriate family relationships, interactions, and child behaviour in, and movement through, spaces and in places. These are the same forces which change family structures and relations, for example, as political shifts or globally influenced economic cycles shape employment opportunities, creating the conditions for family (im)migration (see Huot and Dodson, and Samuel, this volume). Focusing on families encourages the examination of outcomes and changes on all family members, as well as on their socio-spatial interrelationships across generations and age groups (i.e., differences in perception and experience between parents and children, or between that of younger and older children and adolescents in the same/similar family contexts). It is in these interconnections and interdependencies between generations and families, grounded in specific geographical contexts, in which there seems to be a rich area for future work in the geographies of families and family life. The more we understand of the micro and macro-scale geographies of children, the more detailed and nuanced will be our conceptions and understandings of family geographies/the geographies of family life.

Queer Spaces and Family Places

As noted in the introduction, one of the first recognized and unfortunately persistent absences in this project has been research that focuses on the family geographies/geographies of family life of diverse family forms and types. Central to addressing this gap is the development and publishing of work that focuses on the spatiality of families headed by/including homosexual members (i.e., children, parent(s), or extended/fictive kin). Examples of such work that might indicate some paths forward have been difficult to find, however the work of sociologist Jacqui Gabb, particularly her 2005 article '*Locating lesbian parent families: everyday negotiations of lesbian motherhood in Britain*', and the paper '*Queering the family home: narratives from gay, lesbian and bisexual youth coming out in supportive family homes in Australia*', by geographer A. Gorman-Murray (2008) offer some very interesting indications of how family geographies and queer geographies can beneficially come together and mutually be informed and enriched.

Specifically, Gabb (2005: 1) 'interrogates how lesbian parent families negotiate everyday places, such as the street and schools and how they inhabit and produce space'. Such work not only contributes to the broadening of the scope of family geographies, but augments our understanding of how family spaces are constructed, valued, and inscribed with meaning. In her work, Gabb (2005) found that the home in particular was critical to lesbian parents' sense of self, as it is one of the few places where both their sexual and parental identities are reconciled. She argues effectively that the 'multiple identifications and subject positions of lesbian mothers and their families need to be acknowledged . . . in the queer cartography of lesbian and gay space'. (Gabb 2005: 1). In the interest of developing a richer, better informed family geography, we might turn this around to say that they also must be acknowledged in order to capture the true importance, diversity, and complexity of the cartography of family spaces and places.

Gorman-Murray (2008) critiques the relevant geographical literature as, in near normative fashion, presenting the heterosexual nuclear family home as an environment that is inevitably an oppressive and abusive one for gay, lesbian, and bisexual (GLB) youth. Certainly, violence and eventual expulsion from the family home and estrangement from the nuclear family are horrific, yet not uncommon outcomes of coming out in too many situations. However, as Gorman-Murray argues, consideration must also be given to the experiences and meaning of home for GLB youth whose disclosure is affirmed and supported by their immediate family members, particularly by their parents and siblings. This is because, 'through the support of parents and siblings, family homes can become sites of resistance to wider practices of heterosexism, and support for the GLB youth' (Gorman-Murray 2008: 1). Thus, the family homes of supported GLB youth are inscribed with positive and supportive meanings and become powerful sites or 'fissures' for sexual difference to exist peaceably within overarching, potentially monolithic, structures of heteronormativity. Through these processes the nuclear family and the family home may even act to 'queer' this family place and family geographies more broadly.

Minorities and Family Geographies

Another of the absences that researchers doing work on family geographies might address, involves broadening a focus that is represented in this volume on the spaces and places of family interaction within and amongst minority families, and in so doing examine more directly how racial and ethnic background/identity shape family geographies. Additionally, this work needs to go further in examining the intersections of family life and associated environments with other activities and locations, beyond an initial migration event. More attention needs to be paid to the spatialities of social life and adjustment to Canada not only in the short term and amongst first generation immigrants, but also encompassing the experiences of second and third generation minority group members. Specifically, here inspiration for suggesting possible avenues for growth is being drawn from work by Walton-Roberts and Hiebert (1997) and Walton-Roberts and Pratt (2005); two research papers which are exemplars of the kinds of themes that could fruitfully inform a more racially and ethnically inclusive geography of families and family life.

In the earlier paper, Walton-Roberts and Hiebert (1997), in their examination of ethnic entrepreneurship in the Indo-Canadian community in Vancouver, British Columbia, call for a research focus on the role of the nuclear and extended family in this and other ethnic communities, in order to understand its influence in the development and growth of ethnic enterprise, dominated as it is by smaller, family-owned and operated businesses. Walton-Roberts and Hiebert (1997) argue that the family plays a crucial role in sustaining ethnic small business. Their case study of Indo-Canadian construction businesses demonstrated that the 'extended family is a central motif in the creation and maintenance of businesses, and various forms of spousal support are fundamental to the operation of these firms' (Walton-Roberts and Hiebert 1997: 21). For example, family context was cited by a number of their research participants as the very reason their businesses exist. Many indicated a family history of entrepreneurship in India and other countries in which family members had lived. Other research participants indicated that relatives in business, including those outside of Canada, were a primary source for raising capital, especially if business loans were proving difficult to come by through Canadian banks. Thus, Walton-Roberts and Hiebert (1997) effectively argue that families and businesses become intertwined, and that this situation can create business opportunities as well as prospects for strengthening family ties and relationships. However, it can also create significant limitations for ethnic entrepreneurs and tension within their families, particularly between older and younger generations if unacceptably patriarchal and/or potentially exploitive or marginalizing relations are reproduced through the ways that family enterprises operate.

A needed area of further investigation that will inform a more racially and ethnically diverse geography of the family is that of family immigration, settlement, and possible transnational activities, with particular emphasis on the post-settlement experience well beyond the initial few years as well as beyond the first generation. Research conducted by Walton-Roberts and Pratt (2005), with its focus on how class, gender, and sexuality have influenced the settlement experience of Sikh immigrants to Canada, gives some interesting indications of where such research might proceed within the framework of family geographies. One area that could beneficially inform family geographies is Walton-Roberts and Pratt's discussion of some of the challenges faced in maintaining transnational business networks and a transnational household, particularly as extended family members become increasingly important business partners running enterprises in the home or destination country when the migrant family members are not present. Another key area involves the exploration of what the (trans)migration experience means to different members of the family, as well as the (inter) dependencies and (inter)relationships between family members in both the home and destination countries that support or challenge the transnational migrant family.

Through interviews with 'Meena', an upper-class Punjabi woman splitting her time and business interests between the Punjab and British Columbia, Walton-Roberts and Pratt (2005) show that her successful beauty and fashion businesses in the Punjab could not be maintained without the assistance of her mother and sister. In Canada, new businesses are developed from within the family home, employ a daughter-in-law full-time, and at the same time keep 'Meena' in Canada where she is close to her two sons. While always intending to maintain a transnational lifestyle, 'Meena' reports being more comfortable in Canada than is her husband; in fact she clearly states that 'home' is where her children are and so most of the time they stay in Vancouver, going to India largely for business purposes. In this case then, family and business are clearly intertwined; however it is family ties of affection and motherhood, which most strongly influence the commitment to, and sense of comfort in the destination country. Research in a similar vein can inform family geographies by further examining post-migration event adjustments within various minority groups and can work to articulate the ways gender, class, and ethnicity impact or effect the differential experiences of migration and settlement amongst family members.

Additional Ways Forward

The previous discussions of suggested topic areas that would address some of the recognized gaps and absences in family geographies scholarship, particularly as represented in this volume, are just that—suggestions. Many other topics, inspired by ongoing research in other disciplines, using other methodologies and guided by additional theoretical and conceptual frameworks, certainly can, and we hope will, contribute to, and challenge, the parameters of family geographies we have started to demarcate here in this collection. One particular 'coming together' that seems to bear particular promise is the connection between the fields of emotional and family geographies. The next section briefly outlines some possibilities for this 'commingling'. While any suggestions made here are necessarily partial and can only be tentative, they are made to illustrate a possibly productive avenue for furthering the development of family geographies.

Connecting Family and Emotional Geographies

> Clearly, our emotions matter. They affect the way we sense the substance of our past, present and future; all can seem bright, dull or darkened by our emotional outlook . . . the emotional geographies of our lives are dynamic, transformed by our procession through childhood, adolescence, middle and old age, and by more immediately

> destabilizing events such as birth or bereavement, or the start or end of a relationship. Whether joyful, heartbreaking or numbing, emotion has the power to transform the shape of our lives, expanding or contracting our horizons, creating new fissures or fixtures, we never expected to find'. (Bondi, Davidson, and Smith 2005).

The above quote, while rather lengthy, gives some important indications of the key interconnections between emotions and families that can enrich both areas of geographical inquiry. Most obviously perhaps, emotions influence our relationships, responsibilities, and interactions in space and place at the various points in the family life course in potentially different and spatially variable ways. We as individuals are embedded in family contexts that shift and change over time and also therefore as we age. In meeting the changing needs of our family members or perhaps in response to societal, political, and economic changes, our emotional responses and reactions colour our perceptions, influence our decision-making, and lie at the heart of our attachments, or lack thereof, to specific places of meaning in our family lives. Significant events, such as births and bereavements as noted in the quote, but also events such as school completion, retirement, and (im)migration trigger not only spatial change, but are also experienced as momentous emotional events. The positive or negative emotions that are evoked (e.g., anxiety, fear, sense of isolation, or disappointment, as well as possible positive feelings of fulfilment, joy, and relief) must influence people's everyday experiences and relationships within their families and of their individual and family geographies.

Exploring the geographies of families and family life may well prove to enable researchers to better investigate and communicate the 'affective elements at play beneath the topographies of everyday life' (Bondi, Davidson, and Smith 2005: 1). Evidence of this may be found in the research presented here, despite the fact that the exploration of the emotional geographies of families and family life was not an explicit focus in any of the research projects reported on in this collection. A few illustrative examples will clarify this point.

First, consider Cloutier's chapter (Chapter 3), which explores how family geographies are influenced by parental perceptions of child-pedestrian risk. While risk of injury can be quantitatively measured in terms of numbers of vehicular accidents involving child pedestrians or the numbers of child-pedestrian injuries recorded in a specific location or jurisdiction, perception of that risk and how this perception relates to the 'expanding or contracting [of] our [children's geographical] horizons' (Bondi, Davidson, and Smith 2005) clearly must be influenced by the emotional responses of parents. Specifically, their fear and anxiety associated with possible threats of harm to their child(ren) may be correlated with the conditions in their urban neighbourhoods, such as high traffic flows and busy intersections. Exploring the emotions of parents as they influence their decisions regarding the freedom with which they allow their children to move unaccompanied in their neighbourhoods will not only better inform studies of pedestrian safety, but analyses of children's and family geographies more broadly by unearthing the underlying ideas and beliefs shaping the occurrence (or absence) of children's spatial activity.

Another illustrative example can be drawn from Samuel's research (Chapter 7) focusing on the effects of transnationalism on family life amongst a group of South Asian women resident in Canada since the 1970s. At several points, strong emotions come into play as the research subjects relate their experiences of adjustment to settlement in Canada, particularly as reflected in their relationships with spouses and with adolescent and young adult children. For example, one woman expresses experiencing 'terror' that impacted her psychologically when her children became 'Canadian teenagers' and did not follow the social norms and behaviours she expected, particularly around dating and the selection of potential marriage partners. Other research participants reported disappointment and disillusionment with spouses who could not find work, or otherwise could not adjust to life in Canada, in at least one case to the point of returning to India and ending the marriage. Disillusionment also was apparent amongst some of the South Asian women studied as they entered their retirement years and assessed their lives as immigrants to Canada. For several of these women, while they were generally pleased with the material successes and educational achievements they and their children had achieved, emotionally they expressed disappointment and fear that, when it came to a sense of family solidarity and support as they aged, they might have been better off staying in India. Here again we can see the significance of a focus on the emotional geographies of these families as they struggle to make new lives for themselves in Canada. Material success and advancement may indicate, in a quantitative way, a positive immigration outcome for a family, however, there are potentially significant emotional costs that influence and give shape to family geographies that may point to quite different and troubling more negative immigration outcomes.

Thus, while a recent focus on emotions has emerged across the social science academic disciplines, despite a continuing reticence to discuss in print the impact of emotions on our work and ourselves (Bondi 2005) (see Walton-Roberts, this volume for a refreshing exception), it would seem this must be an integral part of the further development of scholarship in family geographies. Analysis of the geographies of family life will be impoverished without the insights and increased sensitivity to the nuances of interconnections, relationships, and behaviours between family members, across distances, and within places of meaning and import in the lives of families collectively and of their individual members. A welcomed and increased capacity to reflect on emotional experiences can bring new energy as well as new insights into research into the geographies of family life, as it recognizes the import of family relations in the emotional lives, and not just the spatial existence, of those we study.

Additionally, an inclusive, non-prescriptive reflection on the emotional responses of the researcher, including feelings evoked in and by the research process, the research subjects encountered, and responses to their own family experiences and context as it influences their work, can also only serve to enrich family geography research practices and outcomes. As we move forward in the development of family geographies as a focal area in human geography and in interdisciplinary research with cognate disciplines, may we also embrace not only sensitivity to emotions in family life and in our research practice, but develop the emotional geographies of families as a central component of the field.

Final Words

The production of this book has been a multi-year odyssey for the contributors and editor alike; one that has been a profitable and fulfilling endeavour for us all. We encourage our readers, researchers and students alike, to gain inspiration from what we have accomplished, to delve into the additional readings we have suggested, to pursue answers to the questions we have posed, and to consider the directions we have indicated in order to fill some of the absences we have noted in what we have been able to produce. As a group we look forward to seeing where others take the study of the geographies of families and family life in the near future, and to continuing with our own contributions to this growing area within human geography and its cognate fields of scholarly inquiry.

References

Aitken, S. 1998. *Family Fantasies and Community Space.* New Brunswick, NJ: Rutgers University Press.

Aitken, S. 2001. 'Global crises of childhood: rights, justice and the unchildlike child'. *Area.* 33, 2, 119–27.

Ansell, N. 2009. 'Childhood and the politics of scale: descaling children's geographies?' *Progress in Human Geography.* 33, 190–209.

Bondi, L. 2005. 'The place of emotions in research: from partitioning emotion and reason to the emotional dynamics of research relationships.' Pp. 231–46 in J. Davidson, L. Bondi, and M. Smith, eds. *Emotional Geographies.* Aldershot UK: Ashgate.

Bondi, L., J. Davidson, and M. Smith. 2005. 'Introduction: geography's emotional turn.' Pp. 1–16 in J. Davidson, L. Bondi, and M. Smith, eds. *Emotional Geographies.* Aldershot UK: Ashgate.

Crewe, L., and P. Collins. 2006. 'Commodifying children: fashion, space, and the production of the profitable child'. *Environment and Planning A.* 38, 7–24.

Gabb, J. 2005. 'Locating lesbian parent families: everyday negotiations of lesbian motherhood in Britain'. *Gender, Place and Culture.* 12, 4, 419–32.

Gorman-Murray, A. 2008. 'Queering the family home: narratives from gay, lesbian and bisexual youth coming out in supportive family homes in Australia'. *Gender, Place and Culture.* 15, 1, 31–44.

Massey, D. 1995. 'The conceptualization of place'. Pp. 45–85 in D. Massey and P. Jess, eds. *A Place in the World? Places, Culture and Globalization.* Oxford: The Open University Press.

Philo, C. 2000. 'The corner-stone of my world: editorial introduction to special issue on spaces of childhood'. *Childhood.* 7, 243–56.

Tucker, P., J. Gilliland, and J. Irwin. 2007. 'Splashpads, swings, and shade: parent's preferences for neighbourhood parks'. *Canadian Journal of Public Health.* 98, 3, 198–202.

Tucker, P., J. Irwin, J. Gilliland, and M. He. 2008. 'Adolescents' perspectives of home, school and neighbourhood environmental influences on physical activity and dietary behaviours'. *Children, Youth and Environments.* 18, 2, 12–35.

Walton-Roberts, M., and D. Hiebert. 1997. 'Immigration, entrepreneurship, and the family: Indo-Canadian enterprise in the construction industry of Greater Vancouver'. *Canadian Journal of Regional Science.* 20, 1–2, 19–39.

Walton-Roberts, M., and G. Pratt. 2005. 'Mobile modernities: a South Asian family negotiates immigration, gender and class in Canada'. *Gender, Place and Culture.* 12, 2, 173–95.

Glossary

active retiree
an individual who maintains the activity levels of middle age into old age (see Ch.10)

active transport
any form of human-powered transportation, including modes such as walking and cycling. (see Ch. 3)

ageless self
the expression of a sense of self that exhibits an ability to maintain in later life the values and consumption patterns of middle age (see Ch. 10)

apartheid
a political ideology and era in South Africa, beginning in 1948 and ending in 1984, during which time people were classified and separated according to race (see Ch. 8)

automobility
having physical and social independence of movement. It can be associated with the increased use of private transportation such as the car from the 1920s onwards, however a range of socio-economic circumstances influence automobility to the present day (see Ch. 5).

boundary work
approaches used by individuals to maintain or ease the separation between the public world of work and the private sphere of home, such as using a BlackBerry© to keep track of work obligations, and an iPhone© for family activities (see Ch. 4)

brain drain
the loss of skilled people from a country or region due to emigration which can result in a skills shortage within the source area (see Ch. 8)

caregiving at end of life
the provision of care to a terminally ill individual by a family member and/or health professional (see Ch. 11)

Compassionate Care Benefit (CCB)
a benefit program enacted in 2004 by the Canadian Federal Government to provide family caregivers a job-secured time away from work with six weeks of employment insurance benefits while they provide care to an ill or dying family member (see Ch. 11)

continuing care communities
retirement communities that provide multiple levels of care, as dictated by medical necessity (see Ch. 10)

control theory
the theory that someone, knowing they are being observed in a task or in a social situation, will perform as closely as possible to the relevant social expectations (see 'Court of Motherhood/Fatherhood', Ch. 2)

countertopography
the concept of drawing analytic contours between places typically encountered as discrete to offer a means of building a vigorous and geographically imaginative practical response to the contemporary processes of globalization, which keep apart places with common problems and shared interests (see Ch. 9)

Court of Motherhood (Fatherhood)
the phenomenon where, in a given social context, a parent will feel judged while attempting to be a good parent/good sport parent. Many parents will modify or feel they should modify their behaviour in certain places because they know others are observing and judging them to a specific standard (see 'control theory', Ch. 2).

cultural landscape
the material and non-material aspects of a surrounding area, including how it is represented, structured, and/or symbolized. Cultural landscapes are rooted in the everyday geographies of their inhabitants through the expression of local attachments, activities, and rituals (see Ch. 1).

cultural production
the relatively recent idea that the practices, behaviours, institutions, and ideas which are developed out of in-group collaboration provide the context for meaning in the lives of individuals and groups. Also, the act of making the products of culture industries such as publishing and filmmaking (see Ch. 10).

deterritorialization
a term, emerging from the literature on transnational migrants, which refers to the attachments and connections people have to different communities despite not sharing a common physical space. Thus, sharing territory is not a necessary element in the construction of transnational identities (see Ch. 7).

diaspora
the dispersion of communities of people through migration, and their continued linkages with each other and their originating homeland (see Ch. 7)

dyad
any two partnered individuals, such as a spousal couple. These social units reflect both the social and demographic processes that have produced and changed notions of family. The term is often used in research specific to later life (see Ch. 10).

emotional geography
the concept that people respond to geographic spaces and places emotionally, and experience different emotions in and about specific places and spaces (see Ch. 2)

emotional work
actions performed with the intention of improving the psychological well-being of others, including efforts to be understanding and empathetic (see Ch. 4)

e-work
an abbreviation of 'electronic work', this inclusive term refers to the use of information and telecommunication technologies to enable work production virtually anywhere (see Ch. 4)

exposure to risk
the amount of 'opportunity' for accidents (or other health consequences) which a protagonist (e.g., a driver; a pedestrian) or a location (e.g., an intersection) experiences. This concept is essential in understanding to what extent people are at risk of situations which may result in an accident or injury. For example, national statistics can be compared on the number of car accidents against the estimated amount of road travel (hundreds of vehicle-kilometres per year) (see Ch. 3).

familial ideology

the idea that the nuclear family unit, comprised of husband, wife, and their dependent children, is the 'normal' and most favoured form of family (see Ch. 7)

family

a group of people who may, or may not, live under one roof or be related by blood, but who voluntarily consent to sharing rights and responsibilities to each other and for the socialization and enculturation of children (see Ch. 10)

a social ideal, referring to a unit of economic co-operation, typically thought to include only those related by blood. This has been revised by feminists to include those forming an economically co-operative residential unit bound by feelings of common ties and strong emotions (see Ch. 7).

family-friendly

a range of characteristics and associated meanings that imply a particular location or activity is meant for families and is thought to offer a positive benefit of some kind to them. It may also infer a certain level of safety, as well as conformity to dominant social norms (see Ch. 1).

family gaze

the 'gaze' is a form of looking that is informed by perceptions and preferences that are often simplified and stereotypical. The *family* gaze is a form of looking that frames and captures images and impressions of family relations and family activities that reinforce enduring views of preferred family life. Recording family activities (such as in family photography) that reflect the family gaze produces a preferred narrative that creates and reinforces positive memories and perceptions of family functioning (see Ch. 1).

family of orientation

sometimes called the 'family of origin', this is the family unit within which a person grows up (see Ch. 7)

family photography

equated with amateur photograph production and the display of the images of domestic scenes and activities that are produced. Such 'family snapshots' are generally only of value to the photographer and his or her (extended) kin (see Ch. 1).

fictive kin

the relationship between individuals who do not necessarily share biological, cultural, or historical origin/descent, but who assume roles and responsibilities otherwise ascribed to biological kin, such as caregiving (see Ch. 1 and 10)

gender dichotomies

a 'breadwinner-homemaker' model of strict gendered division of labour whereby one partner, typically masculine, has primary responsibility for paid employment and another, typically feminine, is responsible for household and family (see Ch. 4)

gender roles

socially constructed attitudes and behaviours, usually a dichotomy of masculinity/femininity and based on the cultural expectations associated with understandings of gender and gender differences (see Ch. 7)

gendered
having or making distinctions between femininity and masculinity for a given role, activity, or behaviour (see Ch. 11)

gendered mobility
the notion that the movement of individuals is conditioned by their gender and that this is subject to contestation and negotiation (see Ch. 5)

gendered spaces
the observation that in a given social or cultural context some spaces are most often populated by members of one gender. The expectations of the people that populate a given place frequently are a reflection of the gender roles of that society. The behaviours exhibited in the gendered space can be considered to reinforce or 'construct' gendered identities and social roles (see Ch. 2).

Global North/South
the divide between wealthy developed nations and poorer developing nations; not necessarily a geographical reference, as some countries in the south are wealthy (Australia), and some areas in the north are less wealthy (parts of China and Russia) (see Ch. 9)

good parent
the idea that certain behaviours, expectations, and norms exist in a society which mothers and fathers are to follow. These must be seen, by others and the self, to be met in order to be identified as 'good mothers' or 'good fathers', or to self-identity as such (see Ch. 2).

good sport parent
an extension of the 'good parent' idea to include the behaviours and expectations of parents as they provide for the needs of competitive athlete children and their teams (see Ch. 2)

haven
a refuge offering protection from external threats; typically the domestic environment is envisaged as a shelter from the outside world (see Ch. 4)

homemaking
the set of responsibilities and activities associated with domestic work, including child and eldercare, cleaning, cooking, entertaining, and decorating (see Ch. 4)

informal care
care and assistance provided by a family member that is typically non-paid and not administered by a government agency (see Ch. 11)

informational needs
the need to acquire appropriate sets of skills or knowledge to navigate the complexity of a(n) (health) issue (see Ch. 11)

information society
a label given to contemporary Western civilization because of the importance given to the creation, distribution, and use of knowledge (see Ch. 4)

intensive parenting
a relatively recent model of child-rearing that views parenting as a deliberate undertaking and involves investing enormous amounts of time in the day-to-day care of children (see Ch. 4)

kinship
the relationship between individuals who share biological, cultural, and/or historical origin and/or descent (see Ch. 10)

kin conversion
the process whereby an individual assumes, or is given, the rights and responsibilities of a blood relative. This 'conversion' may occur as a result of adoption, or in recognition of long association (see Ch. 10).

knowledge-intensive sectors
industries that have relatively high levels of research and development and/or high concentrations of educated workers employed in sectors such as science, engineering, education, medicine, and law; also referred to as high-knowledge sectors by Statistics Canada (see Ch. 4)

lifeworlds
a culturally defined spatio-temporal setting of everyday life, that involves the totality of people's experiences of place and environment. It is in the lifeworld that meaning is given to external phenomena through our intuitive experiences and relationships with them. (For a full discussion see *Dictionary of Human Geography*, ed. R.J. Johnston, D. Gregory, and D. Smith. 4th ed. 1997.) (see Ch. 6).

Likert scale
a psychometric scale commonly used in questionnaires and survey research. Respondents specify their level of agreement, usually along 5 or 7 points, with a statement (e.g., from 'always dangerous' to 'never dangerous'). The scale is named after its inventor, psychologist Rensis Likert (see Ch. 3).

masculinist normativity
the suggestion that there is a male-dominated norm in analysis. Most feminist scholars argue that this conceptual norm has provided a theoretical justification for ignoring women's voices; feminist geographers extend this point to include erasure of women's presence from the socio-cultural and political landscapes (see Ch. 5).

migration
spatial movement across a boundary of an areal unit. A precise definition is difficult because it is possible for no distinct areal units to be identified despite a long-distance move. Distinctions are often made between internal migration (as between provinces in Canada) and international migration (between countries). There is also a need to distinguish between temporary, often described as seasonal and permanent migration (see Ch. 6 and 8).

mobile work
the performance of work in multiple locations such as customer sites, the corporate office, and the home (see Ch. 4)

mobility
in geography, this is a general term that refers to both migration patterns and to people changing residential location. Thus, the Canadian Census, for example, has a category of 'movers' in different time periods of less than one year, and one to five years. It excludes commuters and those moving to summer homes. The use of mobility in Chapter 6 is explicitly spatial, not social (see Ch. 6).

nearest neighbour index
measures the degree of spatial dispersion of points in an area based on the minimum of the inter-feature (i.e., points) distance (see Ch. 3)

network distance
the distance (or cost, time, etc.) involved in travelling from point A to point B along any network , such as streets, pipes, or a river basin (see Ch. 3)

patriarchy
a social structure in which authority is vested in males/the masculine. It may also refer to the male control of a society or group (see Ch. 7).

perception of risk (or risk perception)
the subjective judgment that people make about the characteristics and severity of risk of a possible location, hazard, or event (e.g., nuclear power plant, road traffic accidents, etc.) (see Ch. 3)

place identity
the understood character and associated valuation of a locale or area as ascribed to it by an individual or group; often shared amongst the members of an identified group. This identity can be fluid and changeable over time, however it is generally reflective of local social practices and cultural norms (see Ch. 1).

positionality
can be both a literal and metaphorical referencing of one's place vis-à-vis race, class, gender, and age, or other distinguishing socio-demographic characteristics (see Ch. 10)

post-colonial identity
identity that emerges from multiple, and conflicting, positioning of the migrant who traces his/her origin to formerly colonial regions. Identity thus is one that is intimately connected to, and fused with, the history of colonization and colonialism (see Ch. 7).

production of knowledge
the researcher acknowledges that politics and ethics ground struggles over knowledge. Knowledge is therefore not 'neutral', or 'objective' but *produced*, and it is in the production of that knowledge that our situatedness matters. An important component of this perspective has been to critically analyze the methodological approaches used in the actual production of knowledge (see Ch. 9).

racialization
the process by which individuals, usually visible minority groups, are marginalized from a dominant society through physical features such as skin colour (see Ch. 7)

retirement community
an age-segregated residential site designated for exclusive use by retirees or seniors—people who no longer work or are over a certain age, usually 55+ or 65+ (see Ch. 10)

risk factor
an aspect of personal behaviour, an environmental exposure, or a genetic characteristic which is known to be associated with potentially negative health-related conditions (see Ch. 3)

skilled migration
the international migration of highly educated, qualified, or skilled individuals (see Ch. 8)

social deprivation
deprivation corresponding to physical, environmental, and social states of observable disadvantage relative to the local community (or the wider society) to which an individual (or a family) belongs (see Ch. 3)

social reproduction
domestic processes of labour force maintenance and reproduction such as the care of children and the elderly, care of the self, and of the domestic space. Increasingly social reproduction has become 'outsourced' through the use of domestic and childcare migrant labour, blurring the division between social reproduction and the productive economy. It also refers to the conventions and norms, considered outside of the 'productive' or market sphere, used in order for society to reproduce itself over time (see Ch .9).

socio-spatial process
the ways in which experiences are lived out through the context of the immediate social environment (place) and space (see Ch. 11)

spaces of production and reproduction
concepts drawing upon Marxist separations of productive (particularly industrial) spaces from the spaces of the home, which serve to reproduce the ability of labour to be applied to capital. While an over-simplification, many commentators note a male/female divide in these roles and geographies of the industrial nuclear family (see Ch. 5).

spatial entrapment thesis
the argument that gender constrains women's movement, trapping them in particular spaces and locations (see Ch. 5)

spatio-temporal context
the ways in which experiences are lived out through time, across space, and in place (see Ch. 11)

sporting spaces
including arenas, courts, and fields; essentially anywhere the sport is played or practiced. The term may also include the family home and vehicle (see Ch. 2).

sports culture
the set of formal and informal expected attitudes and behaviours demanded by a sport from participating athletes, parents, and coaches. These expectations frequently shape the on and off field behaviour of athletes and parents alike and are sometimes administered via 'team politics' or simply through subjective statements from parent to parent (see Ch. 2).

teleworking
use of telecommunication technologies to perform paid work for an employer from home or a telework centre (see Ch. 4)

telecommuting
the substitution of telecommunication technology for the commute to and from work location away from home (see Ch. 4)

temporality
the passing of time in the context of a specific (life) event (see Ch. 11)

transnational
moving beyond national boundaries and/or interests; an entity made up of persons of various national identities. (see Ch. 7 and 8)

transnationalism
maintenance, by immigrants, of social, economic, political, and cultural contacts with both their homeland and their host country. It also refers to occupations, activities, and personal relationships that require regular and sustained contacts over time and across national borders (see Ch. 7 and 8).

upgraded kin
typically refers to non-family members providing emotional and material support traditionally supplied by biological children or family (see Ch. 10)

Utopia
an imagined ideal model of a community or existence without flaws; a perfect world (see Ch. 4)

work–life conflict
a concept, based on role theory, which assumes that work roles are distinct from family and other non-work roles. This approach to research examines the conflicting demands of these two life spheres (see Ch. 4).

Contributors

Dr. Jean Andrey is a professor in the Department of Geography and Environmental Management at the University of Waterloo, in Ontario, Canada. Her interests in environment and sustainability go back to her farm roots in Bruce County, Ontario. While much of her research focuses on planning for safe and sustainable transportation systems, broadly speaking she is interested in societal change and with the ways in which individuals and households respond and adapt to risks and opportunities. Jean is also a mom, a grandma, a gardener, and someone who tries to live in a way that minimizes her environmental footprint.

Dr. S. Mary P. Benbow is an associate professor in the Department of Environment and Geography, and Associate Dean, Academic in the Clayton H. Riddell Faculty of Environment, Earth, and Resources at the University of Manitoba, Canada. Her research focuses on the social and cultural implications of zoos and aquariums and examines the fields of human-animal relations, environmental impacts, and conservation education.

Dr. Marie-Soleil Cloutier is an assistant professor in Urban Studies at the Centre Urbanisation Culture et Société, Institut National de Recherche Scientifique, Montreal, Quebec. Her research interests are in health geography, with an emphasis on the use of geographic information systems. Her current work focuses on the geography of risk for vulnerable road users (pedestrians, cyclists, children, elderly) and on family daily mobility in cities.

Dr. Valorie Crooks is an assistant professor in the Department of Geography at Simon Fraser University, British Columbia, Canada. Her research and teaching is focused on health geography. Much of her research work to-date has examined issues pertaining to access to health services, socio-spatial negotiations of disability and chronic illness, and health-related social policies.

Dr. William Crumplin is an assistant professor in the Geography Department at Laurentian University in Sudbury, Ontario. His research interests include critical assessments of GIS and determining geographic links between suburban lifestyles and existing rates of obesity. Research on travel, youth sports, and the family evolved from investigating links between suburban lifestyles and the largely unquestioned behaviour of parents ferrying children between home and a variety of activities.

Dr. Belinda Dodson's personal and professional migrations have led her from South Africa to Canada, where she is an associate professor in the Department of Geography at the University of Western Ontario, Canada. Her areas of specialization are international development, international migration, and gender, with a regional focus on Southern Africa. She has had a long involvement with the Southern African Migration Project (SAMP) and is currently working on a project on the Southern African diaspora in Canada.

Dr. Bonnie C. Hallman is a human geographer with principal research and teaching interests in the broad areas of social/cultural, gender, and health geographies. Early work focused on the geographies of family caregiving, particularly care to elderly relatives. Current research continues long-standing interest in the everyday geographies of families and the critical analysis of the spaces and places that structure and influence the resources available to, and behaviours of, families.

She is an associate professor in the Department of Environment and Geography in the Clayton H. Riddell Faculty of Environment, Earth, and Resources, University of Manitoba, Canada. She is also a research affiliate at the Centre on Aging, and a senior fellow in St John's College, University of Manitoba.

Suzanne Huot is a Ph.D. candidate in the Occupational Science field, Health and Rehabilitation Sciences Program at the University of Western Ontario, Canada. Her current research explores the integration experiences of French-speaking immigrants living in the Francophone Minority Community of London, Ontario. This doctoral work explores the intersections of language, gender, and 'race' with a critical ethnography focused on the participants' changing places and occupations.

Dr. Laura Johnson is a professor in the School of Planning, University of Waterloo, in Waterloo, Ontario, where her research and teaching focus on social planning issues and participatory research. Her research addresses ways that the built environment and community-based services can address the needs of women and their families. Johnson earned her master's degree and doctorate in sociology from Cornell University; her undergraduate degree is from Antioch College in Yellow Springs, Ohio. She is currently conducting a longitudinal qualitative study of social impacts of public housing redevelopment on dislocated and relocated tenants in Toronto's Regent Park.

Dr. Susan Lucas earned her Ph.D. from Wilfrid Laurier University in Ontario, Canada and her MA from Indiana University, US. She is a faculty member in the Department of Geography and Urban Studies at Temple University in Philadelphia, Pennsylvania. Her research interests include retirement communities, gentrification, and waterfront redevelopment. Her research on retirement communities has been published in the *Professional Geographer* and the *Canadian Journal on Urban Research.*

Dr. Joan Marshall is faculty lecturer in the McGill University School of Environment, and the author of *Tides of Change on Grand Manan Island: Culture and Belonging in a Fishing Community*, and *A Solitary Pillar: Montreal's Anglican Church and the Quiet Revolution*. Her research interests include rural social change, gender and youth issues, sustainable aquaculture, fishery regulatory regimes, food markets and consumption in rural areas, the integration of immigrant youth in urban areas, and qualitative research methods

Dr. Lina Samuel is a recent sociology graduate from York University, Toronto, Ontario. She has published in the area of women and work in the fishing industries of India. Her current research interests are in the area of migration, focusing on intergenerational relations and the construction of diasporic identities.

Dr. Rickie Sanders is a professor of Geography/Urban Studies at Temple University, Philadelphia, Pennsylvania. Her interests revolve around issues of gender, race, and class in geography classrooms and the question of individual difference. Sanders' publications from this work focus on obstacles to (and opportunities for) effective transmission of critical thinking in geography and how critical pedagogy can provide opportunities to more fully engage students. Her most recent book *Growing Up in America* is a comprehensive atlas examining the lives of America's children.

Dr. Robert Summerby-Murray is a cultural-historical geographer at Mount Allison University, New Brunswick, Canada. He has published articles and book chapters on urban and industrial heritage, oral history, community language, sense of place, and the historical geographies of settlement in the Tantramar region at the head of the Bay of Fundy, as well as on teaching methods involving geographic information systems and constructivist learning. His present research focuses on cultural landscapes of industrial heritage in Maritime Canada, historical cartography, and the historical geographies of marshland environments. He is Dean of Social Sciences at Mount Allison University and in 2006 was named a 3M teaching fellow.

Dr. Margaret Walton-Roberts is a human geographer with a focus on international migration. Her research has operated on two broad tracks: Indian emigration and transnational migrant networks; and immigration to second and third tier cities in Canada. Her recent research has been on nurse migration from India, with a focus on Kerala and Punjab, and was funded by Wilfrid Laurier University and the Shastri Indo-Canadian Institute. She is associate professor at the Department of Geography and Environmental Studies, Wilfrid Laurier University, Ontario, Canada, and the director of the International Migration Research Centre at the same university.

Dr. Allison Williams is an associate professor in the School of Geography and Earth Sciences at McMaster University in Hamilton, Ontario. She completed her Ph.D. work at York University in Toronto, and has held positions at both Brock University and the University of Saskatchewan. With a background in social and health geography, she is currently involved in a number of interdisciplinary teams. One such team is examining rural/remote palliative/end-of-life health service delivery in Canada.

Donna Williams has been a practicing professional geographer in the public sector for 22 years. Her private research interests in travel, sports, and the family evolved from years of supporting her children in both competitive and non-competitive sports and linking these with questions regarding social norms and structure. She has long been interested in social structure, gender, and geography since investigating gender differences in the perception of urban and suburban neighbourhoods.

Index